Plant Engineers and Managers Guide to Energy Conservation

Ninth Edition

Plant Engineers and Managers Guide to Energy Conservation

Ninth Edition

Albert Thumann, P.E., C.E.M.
Scott Dunning, Ph.D., P.E., C.E.M.

Library of Congress Cataloging-in-Publication Data

Thumann, Albert.
Plant engineers and managers guide to energy conservation / Albert Thumann. -- 9th ed.
p. cm.
Includes index.
ISBN 0-88173-555-8 (alk. paper) -- ISBN 0-88173-556-6 (electronic) -- ISBN 1-4200-5246-2 (distributor (taylor & francis) : alk. paper)
1. Factories--Energy conservation--Handbooks, manuals, etc. I. Dunning, Scott. II. Title.

TJ163.5.F3T48 2008
658.2'6--dc22

2007049784

Plant engineers and managers guide to energy conservation / Albert Thumann--9th ed.

Published by The Fairmont Press, Inc.
700 Indian Trail
Lilburn, GA 30047
tel: 770-925-9388; fax: 770-381-9865
http://www.fairmontpress.com

Distributed by Taylor & Francis Ltd.
6000 Broken Sound Parkway NW, Suite 300
Boca Raton, FL 33487, USA
E-mail: orders@crcpress.com

Distributed by Taylor & Francis Ltd.
23-25 Blades Court
Deodar Road
London SW15 2NU, UK
E-mail: uk.tandf@thomsonpublishingservices.co.uk

Printed in the United States of America
10 9 8 7 6 5 4 3 2 1

0-88173-555-8 (The Fairmont Press, Inc.)
1-4200-5246-2 (Taylor & Francis Ltd.)

Printed on recycled paper.

THIS BOOK IS DEDICATED TO THE ENGINEERS, ARCHITECTS, AND DESIGNERS WHO ARE IMPROVING ENERGY EFFICIENCY OF OPERATIONS IN A COST-EFFECTIVE MANNER.

Contents

Chapter 1 **THE ROLE OF THE PLANT ENGINEER IN ENERGY MANAGEMENT**
Survey of What Industry Is Doing, Results of Industrial Energy Utilization Programs, Organization For Energy Utilization, What Is An Industrial Energy Audit?, The Energy Utilization Program, Energy Accounting, The Language of the Energy Manager 1

Chapter 2 **ENERGY ECONOMIC DECISION MAKING**
Life Cycle Costing, Using the Payback Period Method, Using Life Cycle Costing, The Time Value of Money, Investment Decision-Making, The Job Simulation Experience, Making Decisions For Alternate Investments, Depreciation, Tax Reform Act, Computer Analysis 25

Chapter 3 **THE FACILITY SURVEY**
Comparing Catalogue Data With Actual Performance, Infrared Equipment, Measuring Electrical System Performance, Temperature Measurements, Measuring Combustion Systems, Measuring Heating, Ventilation and Air-Conditioning (HVAC) System Performance 57

Chapter 4 **ELECTRICAL SYSTEM OPTIMIZATION**
Applying Proven Techniques to Reduce the Electrical Bill, Why the Plant Manager Should Understand the Electric Rate Structure, Electrical Rate Tariff, Power Basics—The Key to Electrical Energy Reduction, Relationships Between Power, Voltage, and Current, What Are the Advantages of Power Factor Correction?, Efficient Motors, Synchronous Motors and Power Factor Correction, What Method Should Be Used to Improve the Plant Power Factor?, What Is Load Management?, What Have Been Some of the Results of Load Management?, Application of Automatic Load Shedding, How Does Load Demand Control Work?, The Confusion Over Energy Management Systems, Lighting Basics—The Key to Reducing Lighting Wastes, Lighting Illumination Requirements, The Efficient Use of Lamps, Control Equipment, Solid State Ballasts 73

Chapter 5 UTILITY AND PROCESS SYSTEM OPTIMIZATION
Basis of Thermodynamics, The Carnot Cycle, Use of the Specific Heat Concept, Practical Applications For Energy Conservation, Furnace Efficiency, Steam Tracing, Heat Recovery, The Mollier Diagram, Steam Generation Using Waste Heat Recovery, Pumps and Piping Systems, Distillation Columns, Incorporation of Energy Utilization In Procurement Specifications......................... 105

Chapter 6 HEAT TRANSFER
The Importance of Understanding the Principles of Heat Transfer, Three Ways Heat Is Transferred, How to Estimate the Heat Loss of A Vessel or Tank, How to Estimate the Heat Loss of Piping and Flat Surfaces... 147

Chapter 7 REDUCING BUILDING ENERGY LOSSES
Energy Losses Due to Heat Loss and Heat Gain, Conductivity Through Building Materials, The Effect of Sunlight, Window Treatments, Building Design Considerations............................ 169

Chapter 8 HEATING, VENTILATION AND AIR-CONDITIONING SYSTEM OPTIMIZATION
Efficient Use of Heating and Cooling Equipment Saves Dollars, Applying the Heat Pump to Save Energy, Efficient Applications of Refrigeration Equipment, Basics of Air Conditioning System Design For Energy Conservation, Applying Variable Air Volume Systems, Applying the Economizer Cycle, Applying Heat Recovery, Cool Storage System Performance, Thermal Storage Control Systems, The Ventilation Audit, Energy Analysis Utilizing Simulation Programs, Test and Balance Considerations.................... 221

Chapter 9 COGENERATION: THEORY AND PRACTICE
Definition of "Cogeneration," Components of a Cogeneration System, An Overview of Cogeneration Theory, Application of the Cogeneration Constant, Applicable Systems, Basic Thermodynamic Cycles, Detailed Feasibility Evaluation 255

Chapter 10 ESTABLISHING A MAINTENANCE PROGRAM FOR PLANT EFFICIENCY AND ENERGY SAVINGS
Good Maintenance Saves $, What Is the Effectiveness of Most

Maintenance Programs?, How to Turn Around the Maintenance Program, Stop Leaks and Save, Properly Operating Steam Traps Save Energy, Excess Air Considerations, Dirt and Lamp Lumen Depreciation Can Reduce Lighting Levels by 50%, Summary 287

Chapter 11 MANAGING AN EFFECTIVE ENERGY CONSERVATION PROGRAM
Organizing For Energy Conservation, Top Management Commitment, What to Consider When Establishing Energy Conservation Objectives, Using the Critical Path Schedule of Energy Conservation Activities, Electrical Scheduling of Plant Activities, An Effective Maintenance Program, Continuous Conservation Monitoring, Are Outside Consultants and Contractors Encouraged to Save Energy by Design?, Encouraging the Creative Process, Energy Emergency and Contingency Planning 301

Chapter 12 ELECTRIC MOTORS
Motor Types, Motor Efficiency and Power Factor, Motor Voltage, Motor Rewinds..313

Chapter 13 RELIABLE AND ECONOMIC NATURAL GAS DISTRIBUTED GENERATION TECHNOLOGIES
Elements of DG, Technologies, Market Potential 321

Chapter 14 FINANCING ENERGY EFFICIENCY PROJECTS
Financing Alternatives, General Obligation Bond, Municipal Lease, Commercial Loan, Taxable Lease 337

Chapter 15 STEAM SYSTEM OPTIMIZATION: A CASE STUDY
Savings Opportunities 355

Chapter 16 COST CONTAINMENT DESIGN FOR COMMERCIAL GEOTHERMAL HEAT PUMPS
Why GHPs? Why Now?, Design Methods to Realize Advantages, Software, Challenges in the US Market...................................375

Chapter 17 ENERGY AUDIT CASE STUDY
General Background, Energy Accounting, Energy Reduction

Recommendations, Existing Lighting, Proposed Lighting, ...389

Chapter 18 ECONOMIC EVALUATIONS FOR POWER QUALITY SOLUTIONS
The Principle Investigation, Determining the Phenomenon, Choosing the Right Equipment, Economic Analysis, Graphical Analysis, A More Direct Approach 407

Chapter 19 PURCHASING STRATEGIES FOR ELECTRICITY
AT&T vs. MCI: A Paradigm, Factors Impacting Power Prices, Three General Relationships, Who Offers These Options?, The College of Power Knowledge 421

Chapter 20 POWER QUALITY CASE STUDIES
Case Study 1, Case Study 2 433

Index 443

Preface

In the year 2007 energy again made the headlines. Energy management programs that became dormant were revitalized. Companies again became aware that the energy problems of the 1970s, 1980s and 1990s did not go away. As global demand for fossil fuels has driven prices up, the economics of energy conservation have become attractive again.

The first edition of *Plant Engineers and Managers Guide* was written in 1977 and it was the first book to address the need for industrial energy management.

The new edition of this book includes new technologies not available to the facility manager 30 years ago. Distributed generation, geoexchange and gas cooling technologies have emerged as new options available. Deregulation of the utility industry and purchasing power directly emerged only a few years ago as a new energy strategy.

The role of the energy manager is ever changing. If one lesson can be learned from the past it is that a comprehensive energy conservation program is crucial for every company.

Today the stakes are higher than ever and the plant engineer's and manager's roles in energy have never been greater.

Albert Thumann

Introduction

Plant engineers and managers of the 21st century are expected to apply new technologies, purchase energy at the best price and keep their plants running despite power outages. It is clear that energy conservation is part of every plant engineer's and manager's job.

It is also clear that applying this technology has significant rewards.

In a recent survey conducted by the Association of Energy Engineers, 22.2% of members surveyed have reduced accumulated costs by $5 million or more. The potential for additional savings is still great. Thirty-six percent of those surveyed indicated further savings amounting to over 10% were possible.

As we embark on the new century it has become clear that global competitiveness and energy conservation go hand in hand. Energy conservation means good business. Energy conservation means eliminating waste and insuring operations are more productive. Energy conservation means improving the quality of industrial facility management and preventing pollution. Energy conservation means improving the environment through pollution prevention, and minimizing global warming trends.

The role of the energy manager is ever changing. Today's energy manager must understand how to negotiate the best electric and gas contract as well as understand how to incorporate new energy-efficient technologies into plant operations. The energy manager must have a keen understanding of all aspects of plant operations from purchasing practices to organizational structure. The energy manager must seek out new financing opportunities to fund energy-efficient projects.

The challenge has always been great. The stakes, however, are higher than ever.

1

The Role of the Plant Engineer In Energy Management

Energy management is now considered part of every plant engineer's job. Today the plant engineer needs to keep abreast of changing energy factors which must be incorporated into the overall energy management program. The accomplishments of energy management have indeed been outstanding.

Safety, maintenance and now energy management are some of the areas in which a plant engineer is expected to be knowledgeable. The cook book and low cost-no cost energy conservation measures which were emphasized in the 1970s have been replaced with a more sophisticated approach.

The plant engineer of today must have a keen understanding of both the technical and managerial aspects of energy management in order to insure its success. When oil prices dropped in 1986 it was an opportunity in many plants to switch back to oil. As electric prices escalated it was an opportunity for many plants to install cogeneration facilities. In the late 1990s deregulation took hold, opening up new opportunities in energy purchasing. Since 2005, prices have risen again creating new opportunities. Thus the energy management area is ever changing.

Energy management or energy utilization has replaced the simplistic house keeping measures approach.

The intent of this book is not to make you an expert in each subject, but to illustrate how the overall pieces fit together. Each chapter illustrates the various pieces that comprise an industrial energy utilization program. The energy manager is analogous to a system engineer. Only when the total picture is viewed will the solution become obvious. Of course, it should be noted that the energy manager must seek the advice of experts or specialists when required and use their expertise accordingly.

ORGANIZATION FOR ENERGY UTILIZATION

A multi-divisional corporation usually organizes energy activities on a corporate and plant basis. On the plant basis, energy activities are in many instances added on to the duties of the plant manager.

An energy utilization program does not just happen. It needs a guiding force to "get the ball rolling." Production, energy costs, and raw material supplies are of great concern to plant managers; thus, they are usually the ones to initiate the program.

For a continual, ongoing program to develop, energy managers need to establish "the industrial assessment program" for their facilities. The term "industrial assessment" was introduced in most energy utilization programs in the late 1970s, yet it was rarely defined.

WHAT IS AN INDUSTRIAL ENERGY ASSESSMENT?

The simplest definition for an energy assessment is as follows: An energy assessment serves the purpose of identifying where a building or plant facility uses energy and identifies energy conservation opportunities.

There is a direct relationship to the cost of the assessment (amount of data collected and analyzed) and the number of energy conservation opportunities to be found. Thus, a first decision is made on the cost of the assessment, which determines the type of assessment to be performed.

The second decision is made on the type of facility. For example, a building assessment may emphasize the building envelope, lighting, heating, and ventilation requirements. On the other hand, an assessment of an industrial plant emphasizes the process requirements.

Most energy assessments fall into three categories or types: namely, *walk-through, mini-assessment,* or *detailed assessment.*

Walk-through. This type of audit is the least costly and identifies preliminary energy savings. A visual inspection of the facility is made to determine maintenance and operation energy saving opportunities plus collection of information to determine the need for a more detailed analysis. This type of assessment often employs checklists and usually yields a 1- to 2-page summary listing potential opportunities and typical savings found at other facilities.

Mini-assessment. This type of assessment requires tests and measurements to quantify energy uses and losses and determine the economics for changes. Data collection may consist of one-day snapshots of plant operaions.

Detailed assessment. This type of assessment goes one step further than the mini-assessment. It contains an evaluation of how much energy is used for each function, such as lighting or process. It also requires a model analysis, such as a computer simulation, to determine energy use patterns and predictions on a year-round basis, taking into account such variables as weather data.

The chief distinction between the mini-assessment and the walk-through assessment is that the mini-assessment requires a quantification of energy uses and losses and determining the economics for change.

The chief distinction between the maxi-assessment and the mini-audit is that the maxi-assessment requires an accounting system for energy to be established and a computer simulation.

THE ENERGY UTILIZATION PROGRAM

The energy utilization program usually contains the following steps:

1. Determine energy uses and losses; refer to checklist, Table 1-1.
2. Implement actions for energy conservation, refer to checklist, Table 1-2.
3. Continue to monitor energy conservation efforts; refer to checklist, Table 1-3.

Determine Energy Uses and Losses

Probably the most important aspect of an ongoing energy utilization program is to make individuals "accountable" for energy use.

Unfortunately, many energy managers find it difficult to economically justify "sub metering." The savings as a result of increased accountability are difficult to measure.

Table 1-1. Checklist to determine energy uses and losses.

SURVEY ENERGY USES AND LOSSES

A. Conduct first survey aimed at identifying energy wastes that can be corrected by maintenance or operations actions, for example:
   1. Leaks of steam and other utilities
   2. Furnace burners out of adjustment
   3. Repair or addition of insulation required
   4. Equipment running when not needed

B. Survey to determine where additional instruments for measurement of energy flow are needed and whether there is economic justification for the cost of their installation

C. Develop an energy balance on each process to define in detail:
   1. Energy input as raw materials and utilities
   2. Energy consumed in waste disposal
   3. Energy credit for by-products
   4. Net energy charged to the main product
   5. Energy dissipated or wasted

Note: Energy equivalents will need to be developed for all raw materials, fuels, and utilities, such as electric power, steam, etc., in order that all energy can be expressed on the common basis of Btus.

D. Analyze all process energy balances in depth:
   1. Can waste heat be recovered to generate steam or to heat water or a raw material?
   2. Can a process step be eliminated or modified in some way to reduce energy use?
   3. Can an alternate raw material with lower energy content be used?
   4. Is there a way to improve yield?
   5. Is there justification for:
      a. Replacing old equipment with new equipment requiring less energy?
      b. Replacing an obsolete, inefficient process plant with a whole new and different process using less energy?

E. Conduct weekend and night surveys periodically

F. Plan surveys on specific systems and equipment, such as:
   1. Steam system
   2. Compressed air system
   3. Electric motors
   4. Natural gas lines
   5. Heating and air conditioning system

Source: NBS Handbook 115.

Table 1-2. Checklist for energy conservation implementation.

IMPLEMENT ENERGY CONSERVATION ACTIONS

A. Correct energy wastes identified in the first survey by taking the necessary maintenance or operation actions

B. List all energy conservation projects evolving from energy balance analyses, surveys, etc.
Evaluate and select projects for implementation:
1. Calculate annual energy savings for each project
2. Project future energy costs and calculate annual dollar savings
3. Estimate project capital or expense cost
4. Evaluate investment merit of projects using measures, such as return on investment, internal rate of return, etc.
5. Assign priorities to projects based on investment merit
6. Select conservation projects for implementation and request capital authorization
7. Implement authorized projects

C. Review design of all capital projects, such as new plants, expansions, buildings, etc., to assure that efficient utilization of energy is incorporated in the design. Consider value of Energy Star or LEED certification.

Note: Include consideration of energy availability in new equipment and plant decisions.

Source: NBS Handbook 115.

Table 1-3. Checklist to develop continuous energy conservation efforts.

DEVELOP CONTINUING ENERGY CONSERVATION EFFORTS

A. Measure results:
1. Chart energy use per unit of production by department
2. Chart energy use per unit of production for the whole plant
3. Monitor and analyze charts of Btu per unit of product, taking into consideration effects of complicating variables, such as outdoor ambient air temperature, level of production rate, product mix, etc.
   a. Compare Btu/product unit with past performance and theoretical Btu/product unit
   b. Observe the impact of energy saving actions and project implementation on decreasing the Btu/unit of product

(Continued)

Table 1-3. (*Continued*)

c. Investigate, identify, and correct the cause for increases that may occur in Btu unit of product, if feasible

B. Continue energy conservation committee activities
1. Hold periodic meetings
2. Each committee member is the communication link between the committee and the department supervisors represented
3. Periodically update energy saving project lists
4. Plan and participate in energy saving surveys
5. Communicate energy conservation techniques
6. Plan and conduct a continuing program of activities and communication to keep up interest in energy conservation
7. Develop cooperation with community organizations in promoting energy conservation

C. Involve employees
1. Service on energy conservation committee
2. Energy conservation training course
3. Handbook on energy conservation
4. Suggestion awards plan
5. Recognition for energy saving achievements
6. Technical talks on lighting, insulation, steam traps, and other subjects
7. "savEnergy" posters, decals, stickers
8. Publicity in plant news, bulletins
9. Publicity in public news media
10. Letters on conservation to homes
11. Talks to local organizations

D. Evaluate program
1. Review progress in energy saving
2. Evaluate original goals
3. Consider program modifications
4. Revise goals, as necessary

Source: NBS Handbook 115.

Table 1-1 (B) indicates, as part of the initial survey, that a determination should be made as to who is responsible for which area or process and where "sub metering" would have the biggest impact.

Implement Actions for Energy Conservation

Once energy usage is known potential energy conservation projects can be identified. Each project will be recommended on the basis of the annual energy savings projected and the initial investment required.

Continue to Monitor Energy Conservation Efforts

Energy usage needs to be tracked by using a common energy consumption base per unit of production. This tracking will allow quick identification of changes in energy consumption.

The remaining portion of this chapter will illustrate the language of energy conservation and its applications.

ENERGY ACCOUNTING

An important part of the overall energy auditing program is to be able to measure where you are, and determine where you are going. It is vital to establish an energy accounting system at the beginning of the program. Figures 1-1 through 1-3 illustrate how energy is used for a typical industrial plant. It is important to account for total consumption, cost, and how energy is used for each commodity such as steam, water, air, and natural gas. This procedure is required to develop the appropriate energy conservation strategy.

The top portion of Figure 1-1 illustrates how much energy is used by fuel type and its relative percentage. The pie chart below shows how much is spent for each fuel type. Using a pie chart representation or nodal flow diagram can be very helpful in visualizing how energy is being used.

Figure 1-2, on the other hand, shows how much of the energy is used for each function such as lighting, process, and building heating and ventilation. Pie charts similar to the right-hand side of the figure should be made for each category such as air, steam, electricity, water, and natural gas.

Figure 1-3 illustrates an alternate representation for the steam distribution profile.

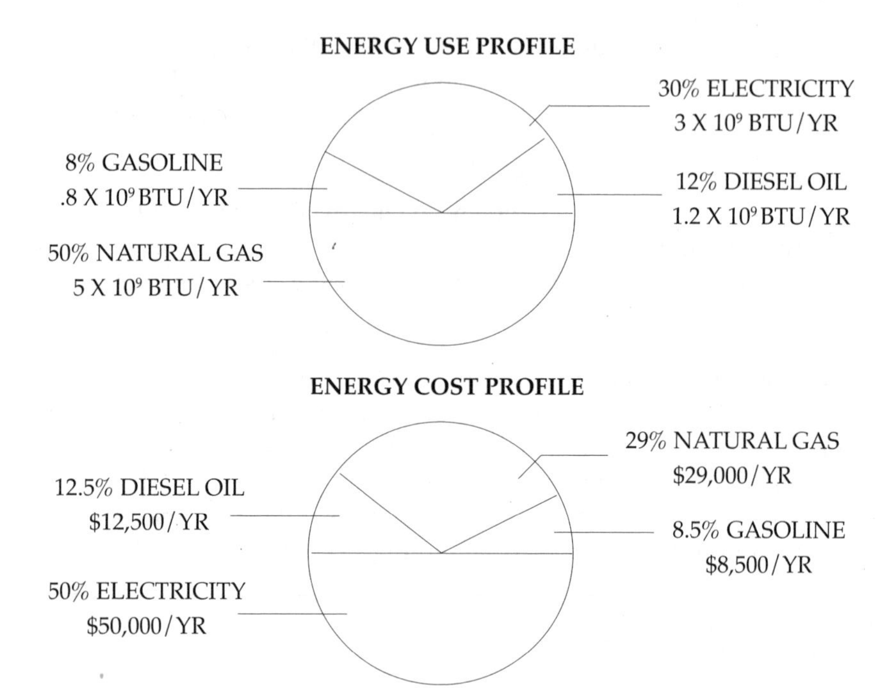

Figure 1-1. Energy use and cost profile.

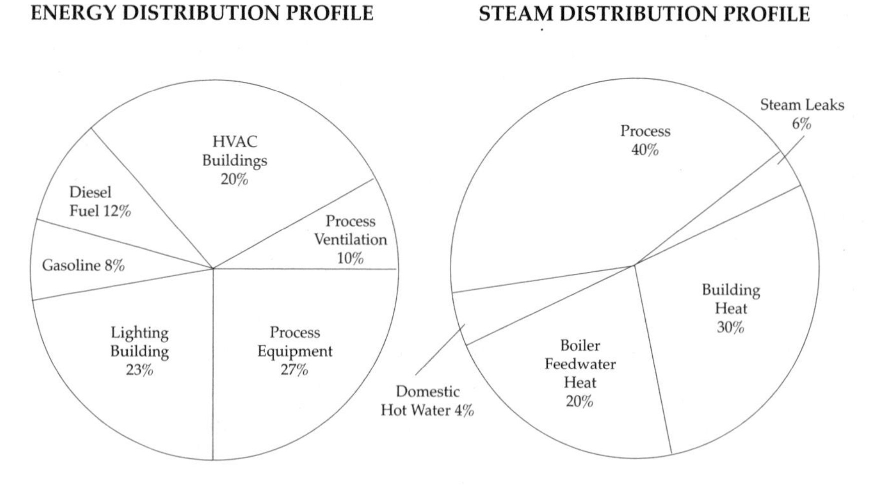

Figure 1-2. Energy profile by function.

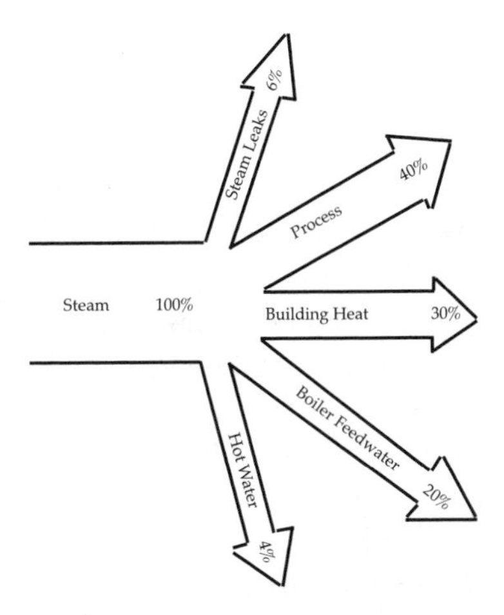

Figure 1-3. Steam distribution nodal diagram.

One of the more important aspects of energy management and conservation is measuring and accounting for energy consumption. At Carborundum an energy accounting and analysis system was developed which was unique in industry, a simple but powerful analytical, management decision-making tool. The Office of Energy Programs of the U.S. Department of Commerce asked Carborundum to work with them in developing this system into a national system, hopefully to be used in the voluntary industrial conservation program. A number of major U.S. corporations are using the system. The system is offered to those who want to use it.

Most energy accounting systems have been devised and are administered by engineers for engineers. The engineers' principal interest in developing these systems has been the display of energy consumed per unit of production. That ratio has been called "energy efficiency," and changes in energy efficiency are clearly energy conserved or wasted. The engineer focuses all of his attention on reducing energy consumed per unit of production.

An energy efficiency ratio alone, however, cannot answer the kinds of questions asked by business managers and/or government authorities:

- If we are conserving energy, why is our total energy consumption increasing?
- If we are wasting energy, why is our total energy consumption decreasing?
- If we have made no change in energy efficiency, why is our energy consumption changing?

Thus there is a need to evaluate several impacts, such as weather, volume/mix, and pollution control, which affect energy use.

Weather Impact

The effect of weather changes (colder winter or hotter summer) on energy consumption is defined as the change in degree-days in the periods under discussion times the heating or cooling efficiency in the period used as the basis for analysis. In the Carborundum system, this translates into the difference in degree-days between this year-to-date and last year-to-date, times the energy used per degree-day last year-to-date. The monetary impact of weather is the impact calculated as above times the cost per unit of energy last year-to-date. That is, the impact of weather changes on energy use or cost is the difference between this period's weather and last, times the heating/cooling energy efficiency in the last or base period. The result ignores improvements in efficiency (identified later as energy conservation effects) and inflation (identified later as price effects), and isolates the effect of weather.

Volume/Mix Impact

The impact of volume and/or product mix changes is the amount of more (or less) energy that is used currently, as opposed to previously, solely as the result of producing more (or less) product or proportionately more (or less) energy-intense products.

Pollution Control Impact

The impact of the energy increase or decrease to control pollution in the current period versus any other time period is simply the difference in the energy used in the two periods. The financial impact is the impact calculated above multiplied by the cost per unit of energy in the last period. The result ignores conservation and price effects as before, and isolates the effect of pollution control.

"Other" Impacts

The impact of other energy uses, previously defined as experimental, start-up of product lines without history, of base loads, etc., is simply the difference in energy used in the two periods being compared. The economic impact is the impact calculated above multiplied by the cost per unit of energy in the prior period. Again, the result ignores conservation and price effects and isolates the effect of these "other" uses of energy.

Figure 1-4 illustrates the data input form used in the Carborundum system.

Carborundum Energy Accounting and
Analysis System Data Input Form

Energy Management and Conservation Program

Plant ______________________
Division ______________________
Group ______________________

Plant Input Data

Today's Date ______________________
Period Covered ______________________

Description	Elec. kWh (000)*	Gas mcf	Oil gal. (000)*	Coal lbs. (000)*	Propane Gal. (000)*	Other (000)*
Total Fuel Used						
Quantity						
Cost ($)						
**Conversion Factor						
Production						
Product 1 NAME						
Production Unit						
Quant. Prod. (000)						
Fuel Used						
Product 2 NAME						
Production Unit						
Quant. Prod. (000)						
Fuel Used						
Product 3 NAME						
Production Unit						
Quant. Prod. (000)						
Fuel Used						
Product 4 NAME						
Production Unit						
Quant. Prod. (000)						
Fuel Used						
Product 5 NAME						
Production Unit						
Quant. Prod. (000)						
Fuel Used						
Heating						
Degree Days						
Fuel Used						
Cooling						
Degree Days						
Fuel Used						
Pollution Control						
Fuel Used						
Other						
Fuel Used						
**Alternate Fuel						

*All Fuel reported in thousands to two decimal places

Figure 1-4. Carborundum energy accounting and analysis system data input form.

THE LANGUAGE OF THE ENERGY MANAGER

In order to communicate energy conservation goals and to analyze the literature in the field, it is important to understand the language of the energy manager and how it is applied.

Each fuel has a heating value, expressed in terms of the British thermal unit, Btu. The Btu is the heat required to raise the temperature of one pound of water 1°F. Table 1-4 illustrates the heating values of various fuels. To compare efficiencies of various fuels, it is best to convert fuel usage in terms of Btus. Table 1-5 illustrates conversions used in energy conservation calculations. It should be noted that combustion of hydrogen creates water vapor. This additional vapor has the effect of raising the dew point of stack gas. To prevent stack damage due to corrosion, it is desirable to keep the stack temperature above the dew point. In Europe, calculations involving combustion typically use the lower heating value (LHV) for a fuel. This value ignores the heat contained in the water vapor. In the United States, the higher heating value (HHV) is used in most calculations. This value counts all the heat created during combustion. For the combustion of natural gas, the HHV is 1000 Btu/ft^3 of natural gas. The LHV is only 900 Btu/ft^3.

Table 1-4. Heating values for various fuels.

Fuel	Average Heating Value
Fuel Oil	
Kerosene	134,000 Btu/gal.
No. 2 Burner Fuel Oil	140,000 Btu/gal.
No. 4 Heavy Fuel Oil	144,000 Btu/gal.
No. 5 Heavy Fuel Oil	150,000 Btu/gal.
No. 6 Heavy Fuel Oil 2.7% sulfur	152,000 Btu/gal.
No. 6 Heavy Fuel Oil 0.3% sulfur	143,800 Btu/gal.
Coal	
Anthracite	13,900 Btu/lb.
Bituminous	14,000 Btu/lb.
Sub-bituminous	12,600 Btu/lb.
Lignite	11,000 Btu/lb.
Gas	
Natural	1,000 Btu/cu. ft.
Liquefied butane	103,300 Btu/gal.
Liquefied propane	91,600 Btu/gal.

Source: Brick & Clay Record, October 1972.

Table 1-5. List of conversion factors.

1 U.S. barrel	= 42 U.S. gallons
1 atmosphere	= 14.7 pounds per square inch absolute (psia)
1 atmosphere	= 760 mm (29.92 in) mercury with density of 13.6 grams per cubic centimeter
1 pound per square inch	= 2.04 inches head of mercury
	= 2.31 feet head of water
1 inch head of water	= 5.20 pounds per square foot
1 foot head of water	= 0.433 pound per square inch
1 British thermal unit (Btu)	= heat required to raise the temperature of 1 pound of water by 1°F
1 therm	= 100,000 Btu
1 kilowatt (kW)	= 1.341 horsepower (hp)
1 kilowatt-hour (kWh)	= 1.34 horsepower-hour
1 horsepower (hp)	= 0.746 kilowatt (kW)
1 horsepower-hour	= 0.746 kilowatt hour (kWh)
1 horsepower-hour	= 2545 Btu
1 kilowatt-hour (kWh)	= 3412 Btu
To generate 1 kilowatt-hour (kWh) requires 10,000 Btu of fuel burned by average utility	
1 ton of refrigeration	= 12,000 Btu per hr
1 ton of refrigeration requires about 1 kW (or 1.341 hp) in commercial air conditioning	
1 standard cubic foot is at standard conditions of 60°F and 14.7 psia.	
1 degree day	= 65°F minus mean temperature of the day, °F
1 year	= 8760 hours
1 year	= 365 days
1 MBtu	= 1 million Btu
1 kW	= 1000 watts
1 trillion barrels	= 1×10^{12} barrels
1 KSCF	= 1000 standard cubic feet

Note: In these conversions, inches and feet of water are measured at 62°F (16.7°C), and inches and millimeters of mercury at 32°F (0°C).

When comparing the cost of fuels, the term "cents per therm" (100,000 Btu) is commonly used.

Knowing the energy content of the plant's process is an important step in understanding how to reduce its cost. Using energy more efficiently reduces the product's cost, thus increasing profits. In order to account for the process energy content, all energy that enters and leaves a plant during a given period must be measured.

The energy content of various raw materials can be estimated by using the heating values indicated in Table 1-6.

CODES, STANDARDS & LEGISLATION

This section presents a historical perspective on key codes, standards and regulations which have impacted energy policy and are still playing a major role in shaping energy usage. The Energy Policy Act of 1992 was far-reaching and its implementation impacted electric power deregulation, building codes and energy-efficient products. Sometimes policy makers do not see the far-reaching impact of their legislation. The Energy Policy Act, for example, created an environment for retail competition. Electric utilities drastically changed the way they operated in order to provide power and lowest cost. This in turn reduced utility-sponsored incentive and rebate programs which previously influenced energy conservation adoption.

THE ENERGY INDEPENDENCE AND SECURITY ACT OF 2007 (H.R.6)

Energy Independence and Security Act of 2007 (H.R.6) was enacted into law December 19, 2007. Key provisions of the law are summarized below.

Title I Energy Security through Improved Vehicle Fuel Economy

- Corporate Average Fuel Economy (CAFE). The law sets a target of 35 miles per gallon for the combined fleet of cars and light trucks by 2020.
- The law establishes a loan guarantee program for advanced

Table 1-6. Heat of combustion for raw materials.

	Formula	Gross Heat of Combustion Btu/lb
Raw Material		
Carbon	C	14,093
Hydrogen	H_2	61,095
Carbon monoxide	CO	4,347
Paraffin Series		
Methane	CH_4	23,875
Ethane	C_2H_4	22,323
Propane	C_3H_8	21,669
n-Butane	C_4H_{10}	21,321
Isobutane	C_4H_{10}	21,271
n-Pentane	C_5H_{12}	21,095
Isopentane	C_5H_{12}	21,047
Neopentane	C_5H_{12}	20,978
n-Hexane	C_6H_{14}	20,966
Olefin Series		
Ethylene	C_2H_4	21,636
Propylene	C_3H_6	21,048
n-Butene	C_4H_8	20,854
Isobutene	C_4H_8	20,737
n-Pentene	C_5H_{10}	20,720
Aromatic Series		
Benzene	C_6H_6	18,184
Toluene	C_7H_8	18,501
Xylene	C_8H_{10}	18,651
Miscellaneous Gases		
Acetylene	C_2H_2	21,502
Naphthalene	$C_{10}H_8$	17,303
Methyl alcohol	CH_3OH	10,258
Ethyl alcohol	C_2H_5OH	13,161
Ammonia	NH_3	9,667

Source: NBS Handbook 115.

battery development, grant program for plug-in hybrid vehicles, incentives for purchasing heavy-duty hybrid vehicles for fleets and credits for various electric vehicles.

Title II Energy Security through Increased Production of Biofuels

- The law increases the Renewable Fuels Standard (RFS), which sets annual requirements for the quantity of renewable fuels produced and used in motor vehicles. RFS requires 9 billion gallons of renewable fuels in 2008, increasing to 36 billion gallons in 2022.

Title III Energy Savings Through Improved Standards for Appliances and Lighting

- The law establishes new efficiency standards for motors, external power supplies, residential clothes washers, dishwashers, dehumidifiers, refrigerators, refrigerator freezers and residential boilers.
- The law contains a set of national standards for light bulbs. The first part of the standard would increase energy efficiency of light bulbs 30% and phase out most common types of incandescent light bulb by 2012-2014.
- Requires the Federal Government to substitute energy efficient lighting for incandescent bulbs.

Title IV Energy Savings in Buildings and Industry

- The law increases funding for the Department of Energy's Weatherization Program, providing 3.75 billion dollars over five years.
- The law encourages the development of more energy efficient "green" commercial buildings. The law creates an Office of Commercial High Performance Green Buildings at the Department of Energy.
- A national goal is set to achieve zero-net energy use for new commercial buildings built after 2025. A further goal is to retrofit all pre-construction 2025 buildings to zero-net energy by 2050.
- Requires that total energy use in federal buildings relative to the

2005 level be reduced 30% by 2015.

- Requires federal facilities to conduct a comprehensive energy and water evaluation for each facility at least once every four years.
- Requires new federal buildings and major renovations to reduce fossil fuel energy use 55% relative to 2003 level by 2010 and be eliminated (100 percent reduction) by 2030.
- Requires that each federal agency ensure that major replacements of installed equipment (such as heating and cooling systems) or renovation or expansion of existing space employ the most energy efficient designs, systems, equipment and controls that are life cycle cost effective. For the purposes of calculating life cycle cost calculations, the time period will increase from 25 years in the prior law to 40 years.
- Directs the Department of Energy to conduct research to develop and demonstrate new process technologies and operating practices to significantly improve the energy efficiency of equipment and processes used by energy-intensive industries.
- Directs the Environmental Protection Agency to establish a recoverable waste energy inventory program. The program must include an ongoing survey of all major industry and large commercial combustion services in the United States.
- Includes new incentives to promote new industrial energy efficiency through the conversion of waste heat into electricity.
- Creates a grant program for Healthy High Performance Schools that aims to encourage states, local governments and school systems to build green schools.
- Creates a program of grants and loans to support energy efficiency and energy sustainability projects at public institutions.

Title V Energy Savings in Government and Public Institutions

- Promotes energy savings performance contracting in the federal government and provides flexible financing and training of federal contract officers.
- Promotes the purchase of energy efficient products and procurement of alternative fuels with lower carbon emissions for the federal government.
- Reauthorizes state energy grants for renewable energy and en-

ergy efficiency technologies through 2012.
- Establishes an energy and environmental block grant program to be used for seed money for innovative local best practices.

Title VI Alternative Research and Development

- Authorizes research and development to expand the use of geothermal energy.
- Improves the cost and effectiveness of thermal energy storage technologies that could improve the operation of concentrating solar power electric generation plants.
- Promotes research and development of technologies that produce electricity from waves, tides, currents and ocean thermal differences.
- Authorizes a development program on energy storage systems for electric drive vehicles, stationary applications, and electricity transmission and distribution.

Title VII Carbon Capture and Sequestration

- Provides grants to demonstrate technologies to capture carbon dioxide from industrial sources.
- Authorizes a nationwide assessment of geological formations capable of sequestering carbon dioxide underground.

Title VIII Improved Management of Energy Policy

- Creates a 50% matching grants program for constructing small renewable energy projects that will have an electrical generation capacity less than 15 megawatts.
- Prohibits crude oil and petroleum product wholesalers from using any technique to manipulate the market or provide false information.

Title IX International Energy Programs

- Promotes U.S. exports in clean, efficient technologies to India, China and other developing countries.
- Authorizes U.S. Agency for International Development (USAID)

to increase funding to promote clean energy technologies in developing countries.

Title X Green Jobs

- Creates an energy efficiency and renewable energy worker training program for "green collar" jobs.
- Provides training opportunities for individuals in the energy field who need to update their skills.

Title XI Energy Transportation and Infrastructure

- Establishes an office of climate change and environment to coordinate and implement strategies to reduce transportation related energy use.

Title XII Small Business Energy Programs

- Loans, grants and debentures are established to help small businesses develop, invest in, and purchase energy efficient equipment and technologies.

Title XIII Smart Grid

- Promotes a "smart electric grid" to modernize and strengthen the reliability and energy efficiency of the electricity supply. The term "Smart Grid" refers to a distribution system that allows for flow of information from a customer's meter in two directions: both inside the house to thermostats, appliances, and other devices, and from the house back to the utility.

THE ENERGY POLICY ACT OF 2005

The first major piece of national energy legislation since the Energy Policy Act of 1992; EPACT 2005 was signed by President George W. Bush on August 8, 2005 and became effective January 1, 2006. The major thrust of EPACT 2005 is energy production. However, there are many important sections of EPACT 2005 that do help promote energy efficiency and energy conservation. There are also some significant impacts on Federal Energy Management. Highlights are described below:

Federal Energy Management

- The United States is the single largest energy user with about a $10 billion energy budget. Forty-four percent of this budget was used for non mobile buildings and facilities. The United States is also the single largest product purchaser with $6 billion spent for energy using products, vehicles, and equipment.

Energy Management Goals

- An annual energy reduction goal of 2% is in place from fiscal year 2006 to fiscal year 2015 for a total energy reduction of 20%
- Electric metering is required in all federal building by the year 2012
- Energy efficient specifications are required in procurement bids and evaluations
- Energy efficient products to be listed in Federal catalogs include Energy Star and FEMP recommended products by GSA and Defense Logistics Agency
- Energy Service Performance Contracts (ESPC) are reauthorized through September 30, 2016
- New Federal buildings are required to be designed 30% below ASHRAE standard or the International Energy Code (if life-cycle cost effective.) Agencies must identify those that meet or exceed the standard
- Renewable electricity consumption by the Federal government cannot be less than: 3% from fiscal year 2007-2009, 5% from fiscal year 2010-2012, and 7.5% from fiscal year 2013-present. Double credits are earned for renewables produced on the site or on Federal lands and used at a federal facility or renewables produced on Native American lands
- The goal for photovoltaic energy is to have 20,000 solar energy systems installed in Federal buildings by the year 2012.

Tax Provisions

- Tax credits will be issued for residential solar photovoltaic and hot water heating systems. Tax deductions will be offered for highly efficient commercial buildings and highly efficient new homes. There will also be tax credits for improvements made to existing homes, including high efficiency HVAC systems, and residential fuel cell systems. Tax credits are also available for fuel cells and microturbines used in businesses.

(EPACT 1992)

THE ENERGY POLICY ACT OF 1992

This comprehensive legislation impacteds energy conservation, power generation and alternative-fuel vehicles as well as energy production. The federal as well as private sectors were impacted by this comprehensive energy act. Highlights are described below:

Energy Efficiency Provisions

Buildings

- Required states to establish minimum commercial building energy codes and to consider minimum residential codes based on current voluntary codes.

Utilities

- Required states to consider new regulatory standards that would require utilities to undertake integrated resource planning, allow efficiency programs to be at least as profitable as new supply options and encourage improvements in supply system efficiency.

Equipment Standards

- Established efficiency standards for commercial heating and air-conditioning equipment, electric motors, and lamps.

- Gives private sector an opportunity to establish voluntary efficiency information/labeling programs for windows, office equipment and luminaires, or the Department of Energy will establish such programs.

Renewable Energy

- Established a program for providing federal support on a competitive basis for renewable energy technologies. Expanded the program to promote export of these renewable energy technologies to emerging markets in developing countries.

Alternative Fuels

- Gave the Department of Energy the authority to require a private and municipal alternative fuel fleet program. Provided a federal alternative fuel fleet program with phased-in acquisition schedule; also provided a state fleet program for large fleets in large cities.

Electric Vehicles

- Established a comprehensive program for the research and development, infrastructure promotion and vehicle demonstration for electric motor vehicles.

Electricity

- Removed obstacles to wholesale power competition in the Public Utilities Holding Company Act by allowing both utilities and non-utilities to form exempt wholesale generators without triggering the PUHCA restrictions.

Global Climate Change

- Directed the Energy Information Administration to establish a baseline inventory of greenhouse gas emissions and established a program for the voluntary reporting of those emissions. Directed the Department of Energy to prepare a report analyzing the strategies for mitigating global climate change and to develop a least-cost energy strategy for reducing the generation of greenhouse gases.

Research and Development

- Directed the Department of Energy to undertake research and development on a wide range of energy technologies, including energy efficiency technologies, natural gas end-use products, renewable energy resources, heating and cooling products, and electric vehicles.

STATE CODES

More than three quarters of the states have adopted ASHRAE Standard 90-80 as a basis for their energy efficiency standard for new building design. The ASHRAE Standard 90-80 is essentially "prescriptive" in nature. For example, the energy engineer using this standard would compute the average conductive value for the building walls and compare it against the value in the standard. If the computed value is above the recommendation, the amount of glass or building construction materials would need to be changed to meet the standard.

Most states have initiated "Model Energy Codes" for efficiency standards in lighting and HVAC. Probably one of the most comprehensive building efficiency standards is California Title 24. Title 24 established lighting and HVAC efficiency standards for new construction, alterations and additions of commercial and noncommercial buildings.

ASHRAE Standard 90-80 has been updated into two new standards:

ASHRAE 90.1-1989 Energy-Efficient Design of New Buildings Except New, Low-Rise Residential Buildings

ASHRAE 90.2 Energy-Efficient Design of New, Low-Rise Residential Buildings

The purposes of ASHRAE Standard 90.1-1989 are:

(a) set minimum requirements for the energy-efficient design of new buildings so that they may be constructed, operated and maintained in a manner that minimizes the use of energy without constraining the building function or the comfort or productivity of the occupants;
(b) provide criteria for energy-efficient design and methods for determining compliance with these criteria;
(c) provide sound guidance for energy-efficient design.

In addition to recognizing advances in the performance of various components and equipment, the Standard encourages innovative energy-conserving designs. This has been accomplished by allowing the building designer to take into consideration the dynamics that exist between the many components of a building through use of the System Performance Method or the Building Energy Cost Budget Method compliance paths. The standard, which is cosponsored by the Illuminating Engineering Society of North America, includes an extensive section on lighting efficiency, utilizing the Unit Power Allowance Method.

REGULATORY & LEGISLATIVE ISSUES IMPACTING COGENERATION & INDEPENDENT POWER PRODUCTION[2]

Federal, state and local regulations must be addressed when considering any cogeneration project. This section provides an overview of the federal regulations that most significantly impact cogeneration facilities.

Federal Power Act

The Federal Power Act asserts the federal government's policy toward competition and anti-competitive activities in the electric power industry. It identifies the Federal Energy Regulatory Commission (FERC) as the agency with primary jurisdiction to prevent undesirable anti-competitive behavior with respect to electric power generation. Also, it provides cogenerators and small power producers with a judicial means to overcome obstacles put in place by electric utilities.

Public Utility Regulatory Policies Act (PURPA)

This legislation was part of the 1978 National Energy Act and has had perhaps the most significant effect on the development of cogeneration and other forms of alternative energy production in the past decade. Certain provisions of PURPA also apply to the exchange of electric power between utilities and cogenerators.

PURPA provides a number of benefits to those cogenerators who can become Qualifying Facilities (QFs) under the act. Specifically, PURPA:

- Requires utilities to purchase the power made available by cogenerators at reasonable buy-back rates (rates typically based on the utilities' cost).

- Guarantees the cogenerator or small power producer interconnection with the electric grid and backup service from the utility.

- Dictates that supplemental power requirements of the cogenerator must be provided at a reasonable cost.

- Exempts cogenerators and small power producers from federal and state utility regulations and their associated reporting requirements.

In order to assure a facility the benefits of PURPA, a cogenerator must become a Qualifying Facility. To achieve Qualifying Status, a cogenerator must generate electricity and useful thermal energy from a single fuel source. In addition, a cogeneration facility must be less than 50% owned by an electric utility or an electric utility holding company. Finally, the plant must meet the minimum annual operating efficiency standard established by FERC when using oil or natural gas as the principal fuel source. The standard is that the useful electric power output plus one half of the useful thermal output of the facility must be no less than 42.5% of the total oil or natural gas energy input. The minimum efficiency standard increases to 45% if the useful thermal energy is less than 15% of the total energy output of the plant.

Natural Gas Policy Act (NGPA)

The major objective of this legislation was to create a deregulated national market for natural gas. It provides for incremental pricing of higher-cost natural gas supplies to industrial customers who use gas, and it allows the cost of natural gas to fluctuate with the cost of fuel oil. Cogenerators classified as Qualifying Facilities under PURPA are exempt from the incremental pricing schedule established for industrial customers.

Resource Conservation and Recovery Act of 1976 (RCRA)

This act requires that disposal of non-hazardous solid waste be handled in a sanitary landfill instead of an open dump. It affects only cogenerators with biomass and coal-fired plants. This legislation has had little, if any, impact on oil and natural gas cogeneration projects.

Public Utility Holding Company Act of 1935

The Public Utility Holding Company Act of 1935 (the 35 Act) authorizes the Securities and Exchange Commission (SEC) to regulate certain utility "holding companies" and their subsidiaries in a wide range of corporate transactions.

The Energy Policy Act of 1992 created a new class of wholesale-only electric generators—"exempt wholesale generators" (EWGs)—which are exempt from the Public Utility Holding Company Act (PUHCA). The Act dramatically enhanced competition in U.S. wholesale electric generation markets, including broader participation by subsidiaries of electric utilities and holding companies. It also opened up foreign markets by exempting companies from PUHCA with respect to retail as well as wholesale sales.

Moving Towards a Deregulated Electric Power Marketplace

California was one of the first states to deregulate. Deregulation was supposed to lower prices and encourage new generation. Instead it lead to a power crisis. The power industry in California experienced a shortage of generation capacity and a tripling of electric costs. Customers in California had to deal with rolling blackouts.

How did power companies run short of power? Under deregulation, vertically integrated utilities such as SDG&E were allowed to sell their generation business and become middlemen that buy electricity on the open market from new generator operators, and distribute to their customers. With capacity shortages driving up wholesale prices, these costs are passed on to customers. Capacity shortages are the result of strong demand for electricity and utilities not building traditional power generating stations. The growing demand is due in part to the strong economy of the 1990s and that computers and hi-tech equipment account for nearly 10% of all consumption.

Utilities which used to be guaranteed 5-7% profit before deregulation are reluctant to invest in billion-dollar plants with the uncertainties deregulation has created.

The US electricity demand in 2000 jumped 23% since 1992 while capacity has risen only 6%. The Department of Energy estimates that 1,000 new power generating stations are needed in the next two decades.

The Energy Policy Act set into motion a widespread movement for utilities to become more competitive. Retail wheeling proposals were set into motion in states such as California, Wisconsin, Michigan, New Mexico, Illinois and New Jersey. Many issues are involved in a deregulated power marketplace and public service commission rulings and litigation will continue to play a major role in the power marketplace of the future.

The Energy Policy Act of 2005 further supported deregulation by repealing the Public Utility Holding Company Act of 1935 and creating a new PUHCA of 2005. The new law included changes such as the removal of Section 32 which established Exempt Wholesale Generators. It also gave review authority to FERC to access books and records of holding companies.

The swift early moves by many states towards retail deregulation have slowed. The downsizing of utilities to increase competitiveness has leveled off. The years of tight budgets created a lack of investment in transmission and distribution facilities and so utility spending is increasing once again.

The global energy demand is affecting all markets as the cost of oil and natural gas climbs to historic highs. Plant managers will need to increase energy conservation efforts to cope with the cost increases generated by this tight energy market.

- Utilities will need to become more competitive. Downsizing and minimization of costs including elimination of rebates are the current trend.

- Utilities will merge to gain a bigger market share.

- Utilities are forming new companies to broaden their services. Energy service companies, financial loan programs, mechanical contracting firms and purchasing of related companies are all part of the new utility strategy.

The California power crisis is sure to spread to other states. Plant managers need to develop a plan to cope with higher electric and natural gas prices and power outages. Energy conservation is playing a major role in companies' energy plans.

2

Energy Economic Decision Making

LIFE CYCLE COSTING

When a plant manager is assigned the role of energy manager, the first question to be asked is: "What is the economic basis for equipment purchases?"

Some companies use a simple payback method of two years or less to justify equipment purchases. Others require a life cycle cost analysis with no fuel price inflation considered. Still other companies allow for a complete life cycle cost analysis, including the impact for the fuel price inflation and the energy tax credit.

The energy manager's success is directly related to how he or she must justify energy utilization methods.

USING THE PAYBACK PERIOD METHOD

The payback period is the time required to recover the capital investment out of the earnings or savings. This method ignores all savings beyond the payback years, thus penalizing projects that have long life potentials for those that offer high savings for a relatively short period.

The payback period criterion is used when funds are limited and it is important to know how fast dollars will come back. The payback period is simply computed as:

$$\text{Payback period} = \frac{\text{initial investment}}{\text{after tax savings}} \tag{2-1}$$

The energy manager who must justify energy equipment expenditures based on a payback period of one year or less has little chance for long-range success. Some companies have set higher payback periods for energy utilization methods. These longer payback periods are justified on the basis that:

- Fuel pricing will increase at a higher rate than the general inflation rate.

- The "risk analysis" for not implementing energy utilization measures may mean loss of production and losing a competitive edge.

USING LIFE CYCLE COSTING

Life cycle costing is an analysis of the total cost of a system, device, building, machine, etc., over its anticipated useful life. The name is new but the subject has, in the past, gone by such names as "engineering economic analysis" or "total owning and operating cost summaries."

Life cycle costing has brought about a new emphasis on the comprehensive identification of all costs associated with a system. The most commonly included costs are initial in place cost, operating costs, maintenance costs, and interest on the investment. Two factors enter into appraising the life of the system: namely, the expected physical life and the period of obsolescence. The lesser factor is governing time period. The effect of interest can then be calculated by using one of several formulas which take into account the time value of money.

When comparing alternative solutions to a particular problem, the system showing the lowest life cycle cost will usually be the first choice (performance requirements are assessed as equal in value).

Life cycle costing is a tool in value engineering. Other items, such as installation time, pollution effects, aesthetic considerations, delivery time, and owner preferences will temper the rule of always choosing the system with the lowest life cycle cost. Good overall judgment is still required.

The life cycle cost analysis still contains judgment factors pertaining to interest rates, useful life, and inflation rates. Even with the judgment element, life cycle costing is the most important tool in value engineering, since the results are quantified in terms of dollars.

As the price for energy changes, and as governmental incentives are initiated, processes or alternatives which were not economically feasible will be considered. This chapter will concentrate on the principles of the life cycle cost analysis as they apply to energy conservation decision making.

THE TIME VALUE OF MONEY

Most energy saving proposals require the investment of capital to accomplish them. By investing today in energy conservation, yearly operating dollars over the life of the investment will be saved. A dollar in hand today is more valuable than one to be received at some time in the future. For this reason, a *time value* must be placed on all cash flows into and out of the company.

Money transactions are thought of as a cash flow to or from a company. Investment decisions also take into account alternate investment opportunities and the minimum return on the investment. In order to compute the rate of return on an investment, it is necessary to find the interest rate which equates payments outgoing and incoming, present and future. The method used to find the rate of return is referred to as *discounted cash flow.*

INVESTMENT DECISION-MAKING

To make investment decisions, the energy manager must follow one simple principle: Relate annual cash flows and lump sum deposits to the same time base. The six factors used for investment decision making simply convert cash from one time base to another; since each company has various financial objectives, these factors can be used to solve *any* investment problem.

Single Payment Compound Amount—F/P

The F/P factor is used to determine the future amount F that a present sum P will accumulate at i percent interest, in n years. If P (present worth) is known, and F (future worth) is to be determined, then Equation 2-2 is used.

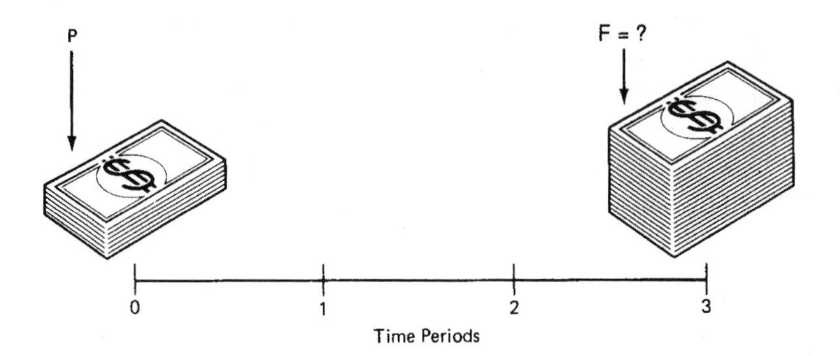

Figure 2-1. Single payment compound amount (F/P).

$$F = P \times (1 + i)^n \tag{2-2}$$

$$F/P = (1 + i)^n \tag{2-3}$$

The *F/P* can be computed by an interest formula, but usually its value is found by using the interest tables. Interest tables for interest rates of 10 to 50 percent are found at the conclusion of this chapter (Tables 2-1 through 2-8). In predicting future costs, there are many unknowns. For the accuracy of most calculations, interest rates are assumed to be compounded annually unless otherwise specified. Linear interpolation is commonly used to find values not listed in the interest tables.

Tables 2-9 through 2-12 can be used to determine the effect of fuel escalation on the life cycle cost analysis.

Single Payment Present Worth—P/F

The P/F factor is used to determine the present worth, *P*, that a future amount, *F*, will be at interest of *i*-percent, in *n* years. If *F* is known, and *P* is to be determined, then Equation 2-4 is used.

$$P = F \times 1/(1 + i)^n \tag{2-4}$$

$$P/F = \frac{1}{(1 + i)^n} \tag{2-5}$$

Table 2-1. 10% Interest factors.

Period n	Single-payment compound-amount F/P	Single-payment present-worth P/F	Uniform series compound-amount F/A	Sinking-fund payment A/F	Capital recovery A/P	Uniform-series present-worth P/A
	Future value of $1 $(1+i)^n$	Present value of $1 $\frac{1}{(1+i)^n}$	Future value of uniform series of $1 $\frac{(1+i)^n-1}{i}$	Uniform series whose future value is $1 $\frac{i}{(1+i)^n-1}$	Uniform series with present value of $1 $\frac{i(1+i)^n}{(1+i)^n-1}$	Present value of uniform series of $1 $\frac{(1+i)^n-1}{i(1+i)^n}$
1	1.100	0.9091	1.000	1.00000	1.10000	0.909
2	1.210	0.8264	2.100	0.47619	0.57619	1.736
3	1.331	0.7513	3.310	0.30211	0.40211	2.487
4	1.464	0.6830	4.641	0.21547	0.31147	3.170
5	1.611	0.6209	6.105	0.16380	0.26380	3.791
6	1.772	0.5645	7.716	0.12961	0.22961	4.355
7	1.949	0.5132	9.487	0.10541	0.20541	4.868
8	2.144	0.4665	11.436	0.08744	0.18744	5.335
9	2.358	0.4241	13.579	0.07364	0.17364	5.759
10	2.594	0.3855	15.937	0.06275	0.16275	6.144
11	2.853	0.3505	18.531	0.05396	0.15396	6.495
12	3.138	0.3186	21.384	0.04676	0.14676	6.814
13	3.452	0.2897	24.523	0.04078	0.14078	7.103
14	3.797	0.2633	27.975	0.03575	0.13575	7.367
15	4.177	0.2394	31.772	0.03147	0.13147	7.606
16	4.595	0.2176	35.950	0.02782	0.12782	7.824
17	5.054	0.1978	40.545	0.02466	0.12466	8.022
18	5.560	0.1799	45.599	0.02193	0.12193	8.201
19	6.116	0.1635	51.159	0.01955	0.11955	8.365
20	6.727	0.1486	57.275	0.01746	0.11746	8.514
21	7.400	0.1351	64.002	0.01562	0.11562	8.649
22	8.140	0.1228	71.403	0.01401	0.11401	8.772
23	8.954	0.1117	79.543	0.01257	0.11257	8.883
24	9.850	0.1015	88.497	0.01130	0.11130	8.985
25	10.835	0.0923	98.347	0.01017	0.11017	9.077
26	11.918	0.0839	109.182	0.00916	0.10916	9.161
27	13.110	0.0763	121.100	0.00826	0.10826	9.237
28	14.421	0.0693	134.210	0.00745	0.10745	9.307
29	15.863	0.0630	148.631	0.00673	0.10673	9.370
30	17.449	0.0673	164.494	0.00608	0.10608	9.427
35	28.102	0.0356	271.024	0.00369	0.10369	9.644
40	45.259	0.0221	442.593	0.00226	0.10226	9.779
45	72.890	0.0137	718.905	0.00139	0.10139	9.863
50	117.391	0.0085	1163.909	0.00086	0.10086	9.915
55	189.059	0.0053	1880.591	0.00053	0.10053	9.947
60	304.482	0.0033	3034.816	0.00033	0.10033	9.967
65	490.371	0.0020	4893.707	0.00020	0.10020	9.980
70	789.747	0.0013	7887.470	0.00013	0.10013	9.987
75	1271.895	0.0008	12708.954	0.00008	0.10008	9.992
80	2048.400	0.0005	20474.002	0.00005	0.10005	9.995
85	3298.969	0.0003	32979.690	0.00003	0.10003	9.997
90	5313.023	0.0002	53120.226	0.00002	0.10002	9.998
95	8556.676	0.0001	85556.760	0.00001	0.10001	9.999

Table 2-2. 12% Interest factors.

Period n	Single-payment compound-amount F/P	Single-payment present-worth P/F	Uniform series compound-amount F/A	Sinking-fund payment A/F	Capital recovery A/P	Uniform-series present-worth P/A
	Future value of $1	Present value of $1	Future value of uniform series of $1	Uniform series whose future value is $1	Uniform series with present value of $1	Present value of uniform series of $1
	$(1+i)^n$	$\frac{1}{(1+i)^n}$	$\frac{(1+i)^n-1}{i}$	$\frac{i}{(1+i)^n-1}$	$\frac{i(1+i)^n}{(1+i)^n-1}$	$\frac{(1+i)^n-1}{i(1+i)^n}$
1	1.120	0.8929	1.000	1.00000	1.12000	0.893
2	1.254	0.7972	2.120	0.47170	0.59170	1.690
3	1.405	0.7118	3.374	0.29635	0.41635	2.402
4	1.574	0.6355	4.779	0.20923	0.32923	3.037
5	1.762	0.5674	6.353	0.15741	0.27741	3.605
6	1.974	0.5066	8.115	0.12323	0.24323	4.111
7	2.211	0.4523	10.089	0.09912	0.21912	4.564
8	2.476	0.4039	12.300	0.08130	0.20130	4.968
9	2.773	0.3606	14.776	0.06768	0.18768	5.328
10	3.106	0.3220	17.549	0.05698	0.17698	5.650
11	3.479	0.2875	20.655	0.04842	0.16842	5.938
12	3.896	0.2567	24.133	0.04144	0.16144	6.194
13	4.363	0.2292	28.029	0.03568	0.15568	6.424
14	4.887	0.2046	32.393	0.03087	0.15087	6.628
15	5.474	0.1827	37.280	0.02682	0.14682	6.811
16	6.130	0.1631	42.753	0.02339	0.14339	6.974
17	6.866	0.1456	48.884	0.02046	0.14046	7.120
18	7.690	0.1300	55.750	0.01794	0.13794	7.250
19	8.613	0.1161	63.440	0.01576	0.13576	7.366
20	9.646	0.1037	72.052	0.01388	0.13388	7.469
21	10.804	0.0926	81.699	0.01224	0.13224	7.562
22	12.100	0.0826	92.503	0.01081	0.13081	7.645
23	13.552	0.0738	104.603	0.00956	0.12956	7.718
24	15.179	0.0659	118.155	0.00846	0.12846	7.784
25	17.000	0.0588	133.334	0.00750	0.12750	7.843
26	19.040	0.0525	150.334	0.00665	0.12665	7.896
27	21.325	0.0469	169.374	0.00590	0.12590	7.943
28	23.884	0.0419	190.699	0.00524	0.12524	7.984
29	26.750	0.0374	214.583	0.00466	0.12466	8.022
30	29.960	0.0334	241.333	0.00414	0.12414	8.055
35	52.800	0.0189	431.663	0.00232	0.12232	8.176
40	93.051	0.0107	767.091	0.00130	0.12130	8.244
45	163.988	0.0061	1358.230	0.00074	0.12074	8.283
50	289.002	0.0035	2400.018	0.00042	0.12042	8.304
55	509.321	0.0020	4236.005	0.00024	0.12024	8.317
60	897.597	0.0011	7471.641	0.00013	0.12013	8.324
65	1581.872	0.0006	13173.937	0.00008	0.12008	8.328
70	2787.800	0.0004	23223.332	0.00004	0.12004	8.330
75	4913.056	0.0002	40933.799	0.00002	0.12002	8.332
80	8658.483	0.0001	72145.692	0.00001	0.12001	8.332

Table 2-3. 15% Interest factors.

Period n	Single-payment compound-amount F/P	Single-payment present-worth P/F	Uniform series compound-amount F/A	Sinking-fund payment A/F	Capital recovery A/P	Uniform-series present-worth P/A
	Future value of $1	Present value of $1	Future value of uniform series of $1	Uniform series whose future value is $1	Uniform series with present value of $1	Present value of uniform series of $1
	$(1+i)^n$	$\frac{1}{(1+i)^n}$	$\frac{(1+i)^n-1}{i}$	$\frac{i}{(1+i)^n-1}$	$\frac{i(1+i)^n}{(1+i)^n-1}$	$\frac{(1+i)^n-1}{i(1+i)^n}$
1	1.150	0.8696	1.000	1.00000	1.15000	0.870
2	1.322	0.7561	2.150	0.46512	0.61512	1.626
3	1.521	0.6575	3.472	0.28798	0.43798	2.283
4	1.749	0.5718	4.993	0.20027	0.35027	2.855
5	2.011	0.4972	6.742	0.14832	0.29832	3.352
6	2.313	0.4323	8.754	0.11424	0.26424	3.784
7	2.660	0.3759	11.067	0.09036	0.24036	4.160
8	3.059	0.3269	13.727	0.07285	0.22285	4.487
9	3.518	0.2843	16.786	0.05957	0.20957	4.772
10	4.046	0.2472	20.304	0.04925	0.19925	5.019
11	4.652	0.2149	24.349	0.04107	0.19107	5.234
12	5.350	0.1869	29.002	0.03448	0.18448	5.421
13	6.153	0.1625	34.352	0.02911	0.17911	5.583
14	7.076	0.1413	40.505	0.02469	0.17469	5.724
15	8.137	0.1229	47.580	0.02102	0.17102	5.847
16	9.358	0.1069	55.717	0.01795	0.16795	5.954
17	10.761	0.0929	65.075	0.01537	0.16537	6.047
18	12.375	0.0808	75.836	0.01319	0.16319	6.128
19	14.232	0.0703	88.212	0.01134	0.16134	6.198
20	16.367	0.0611	102.444	0.00976	0.15976	6.259
21	18.822	0.0531	118.810	0.00842	0.15842	6.312
22	21.645	0.0462	137.632	0.00727	0.15727	6.359
23	24.891	0.0402	159.276	0.00628	0.15628	6.399
24	28.625	0.0349	194.168	0.00543	0.15543	6.434
25	32.919	0.0304	212.793	0.00470	0.15470	6.464
26	37.857	0.0264	245.712	0.00407	0.15407	6.491
27	43.535	0.0230	283.569	0.00353	0.15353	6.514
28	50.066	0.0200	327.104	0.00306	0.15306	6.534
29	57.575	0.0174	377.170	0.00265	0.15265	6.551
30	66.212	0.0151	434.745	0.00230	0.15230	6.566
35	133.176	0.0075	881.170	0.00113	0.15113	6.617
40	267.864	0.0037	1779.090	0.00056	0.15056	6.642
45	538.769	0.0019	3585.128	0.00028	0.15028	6.654
50	1083.657	0.0009	7217.716	0.00014	0.15014	6.661
55	2179.622	0.0005	14524.148	0.00007	0.15007	6.664
60	4383.999	0.0002	29219.992	0.00003	0.15003	6.665
65	8817.787	0.0001	58778.583	0.00002	0.15002	6.666

Table 2-4. 20% Interest factors.

Period n	Single-payment compound-amount F/P	Single-payment present-worth P/F	Uniform series compound-amount F/A	Sinking-fund payment A/F	Capital recovery A/P	Uniform-series present-worth P/A
	Future value of $1	Present value of $1	Future value of uniform series of $1	Uniform series whose future value is $1	Uniform series with present value of $1	Present value of uniform series of $1
	$(1+i)^n$	$\frac{1}{(1+i)^n}$	$\frac{(1+i)^n-1}{i}$	$\frac{i}{(1+i)^n-1}$	$\frac{i(1+i)^n}{(1+i)^n-1}$	$\frac{(1+i)^n-1}{i(1+i)^n}$
1	1.200	0.8333	1.000	1.00000	1.20000	0.833
2	1.440	0.6944	2.200	0.45455	0.65455	1.528
3	1.728	0.5787	3.640	0.27473	0.47473	2.106
4	2.074	0.4823	5.368	0.18629	0.38629	2.589
5	2.488	0.4019	7.442	0.13438	0.33438	2.991
6	2.986	0.3349	9.930	0.10071	0.30071	3.326
7	3.583	0.2791	12.916	0.07742	0.27742	3.605
8	4.300	0.2326	16.499	0.06061	0.26061	3.837
9	5.160	0.1938	20.799	0.04808	0.24808	4.031
10	6.192	0.1615	25.959	0.03852	0.23852	4.192
11	7.430	0.1346	32.150	0.03110	0.23110	4.327
12	8.916	0.1122	39.581	0.02526	0.22526	4.439
13	10.699	0.0935	48.497	0.02062	0.22062	4.533
14	12.839	0.0779	59.196	0.01689	0.21689	4.611
15	15.407	0.0649	72.035	0.01388	0.21388	4.675
16	18.488	0.0541	87.442	0.01144	0.21144	4.730
17	22.186	0.0451	105.931	0.00944	0.20944	4.775
18	26.623	0.0376	128.117	0.00781	0.20781	4.812
19	31.948	0.0313	154.740	0.00646	0.20646	4.843
20	38.338	0.0261	186.688	0.00536	0.20536	4.870
21	46.005	0.0217	225.026	0.00444	0.20444	4.891
22	55.206	0.0181	271.031	0.00369	0.20369	4.909
23	66.247	0.0151	326.237	0.00307	0.20307	4.925
24	79.497	0.0126	392.484	0.00255	0.20255	4.937
25	95.396	0.0105	471.981	0.00212	0.20212	4.948
26	114.475	0.0087	567.377	0.00176	0.20176	4.956
27	137.371	0.0073	681.853	0.00147	0.20147	4.964
28	164.845	0.0061	819.223	0.00122	0.20122	4.970
29	197.814	0.0051	984.068	0.00102	0.20102	4.975
30	237.376	0.0042	1181.882	0.00085	0.20085	4.979
35	590.668	0.0017	2948.341	0.00034	0.20034	4.992
40	1469.772	0.0007	7343.858	0.00014	0.20014	4.997
45	3657.262	0.0003	18281.310	0.00005	0.20005	4.999
50	9100.438	0.0001	45497.191	0.00002	0.20002	4.999

Table 2-5. 25% Interest factors.

Period n	Single-payment compound-amount F/P	Single-payment present-worth P/F	Uniform series compound-amount F/A	Sinking-fund payment A/F	Capital recovery A/P	Uniform-series present-worth P/A
	Future value of $1	Present value of $1	Future value of uniform series of $1	Uniform series whose future value is $1	Uniform series with present value of $1	Present value of uniform series of $1
	$(1+i)^n$	$\frac{1}{(1+i)^n}$	$\frac{(1+i)^n-1}{i}$	$\frac{i}{(1+i)^n-1}$	$\frac{i(1+i)^n}{(1+i)^n-1}$	$\frac{(1+i)^n-1}{i(1+i)^n}$
1	1.250	0.8000	1.000	1.00000	1.25000	0.800
2	1.562	0.6400	2.250	0.44444	0.69444	1.440
3	1.953	0.5120	3.812	0.26230	0.51230	1.952
4	2.441	0.4096	5.766	0.17344	0.42344	2.362
5	3.052	0.3277	8.207	0.12185	0.37185	2.689
6	3.815	0.2621	11.259	0.08882	0.33882	2.951
7	4.768	0.2097	15.073	0.06634	0.31634	3.161
8	5.960	0.1678	19.842	0.05040	0.30040	3.329
9	7.451	0.1342	25.802	0.03876	0.28876	3.463
10	9.313	0.1074	33.253	0.03007	0.28007	3.571
11	11.642	0.0859	42.566	0.02349	0.27349	3.656
12	14.552	0.0687	54.208	0.01845	0.26845	3.725
13	18.190	0.0550	68.760	0.01454	0.26454	3.780
14	22.737	0.0440	86.949	0.01150	0.26150	3.824
15	28.422	0.0352	109.687	0.00912	0.25912	3.859
16	35.527	0.0281	138.109	0.00724	0.25724	3.887
17	44.409	0.0225	173.636	0.00576	0.25576	3.910
18	55.511	0.0180	218.045	0.00459	0.25459	3.928
19	69.389	0.0144	273.556	0.00366	0.25366	3.942
20	86.736	0.0115	342.945	0.00292	0.25292	3.954
21	108.420	0.0092	429.681	0.00233	0.25233	3.963
22	135.525	0.0074	538.101	0.00186	0.25186	3.970
23	169.407	0.0059	673.626	0.00148	0.25148	3.976
24	211.758	0.0047	843.033	0.00119	0.25119	3.981
25	264.698	0.0038	1054.791	0.00095	0.25095	3.985
26	330.872	0.0030	1319.489	0.00076	0.25076	3.988
27	413.590	0.0024	1650.361	0.00061	0.25061	3.990
28	516.988	0.0019	2063.952	0.00048	0.25048	3.992
29	646.235	0.0015	2580.939	0.00039	0.25039	3.994
30	807.794	0.0012	3227.174	0.00031	0.25031	3.995
35	2465.190	0.0004	9856.761	0.00010	0.25010	3.998
40	7523.164	0.0001	30088.655	0.00003	0.25003	3.999

Table 2-6. 30% Interest factors.

Period n	Single-payment compound-amount F/P	Single-payment present-worth P/F	Uniform series compound-amount F/A	Sinking-fund payment A/F	Capital recovery A/P	Uniform-series present-worth P/A
	Future value of $1	Present value of $1	Future value of uniform series of $1	Uniform series whose future value is $1	Uniform series with present value of $1	Present value of uniform series of $1
	$(1+i)^n$	$\frac{1}{(1+i)^n}$	$\frac{(1+i)^n-1}{i}$	$\frac{i}{(1+i)^n-1}$	$\frac{i(1+i)^n}{(1+i)^n-1}$	$\frac{(1+i)^n-1}{i(1+i)^n}$
1	1.300	0.7692	1.000	1.00000	1.30000	0.769
2	1.690	0.5917	2.300	0.43478	0.73478	1.361
3	2.197	0.4552	3.990	0.25063	0.55063	1.816
4	2.856	0.3501	6.187	0.16163	0.46163	2.166
5	3.713	0.2693	9.043	0.11058	0.41058	2.436
6	4.827	0.2072	12.756	0.07839	0.37839	2.643
7	6.275	0.1594	17.583	0.05687	0.35687	2.802
8	8.157	0.1226	23.858	0.04192	0.34192	2.925
9	10.604	0.0943	32.015	0.03124	0.33124	3.019
10	13.786	0.0725	42.619	0.02346	0.32346	3.092
11	17.922	0.0558	56.405	0.01773	0.31773	3.147
12	23.298	0.0429	74.327	0.01345	0.31345	3.190
13	30.288	0.0330	97.625	0.01024	0.31024	3.223
14	39.374	0.0254	127.913	0.00782	0.30782	3.249
15	51.186	0.0195	167.286	0.00598	0.30598	3.268
16	66.542	0.0150	218.472	0.00458	0.30458	3.283
17	86.504	0.0116	285.014	0.00351	0.30351	3.295
18	112.455	0.0089	371.518	0.00269	0.30269	3.304
19	146.192	0.0068	483.973	0.00207	0.30207	3.311
20	190.050	0.0053	630.165	0.00159	0.30159	3.316
21	247.065	0.0040	820.215	0.00122	0.30122	3.320
22	321.194	0.0031	1067.280	0.00094	0.30094	3.323
23	417.539	0.0024	1388.464	0.00072	0.30072	3.325
24	542.801	0.0018	1806.003	0.00055	0.30055	3.327
25	705.641	0.0014	2348.803	0.00043	0.30043	3.329
26	917.333	0.0011	3054.444	0.00033	0.30033	3.330
27	1192.533	0.0008	3971.778	0.00025	0.30025	3.331
28	1550.293	0.0006	5164.311	0.00019	0.30019	3.331
29	2015.381	0.0005	6714.604	0.00015	0.30015	3.332
30	2619.996	0.0004	8729.985	0.00011	0.30011	3.332
35	9727.8060	0.0001	32422.868	0.00003	0.30003	3.333

Table 2-7. 40% Interest factors.

Period n	Single-payment compound-amount F/P	Single-payment present-worth P/F	Uniform series compound-amount F/A	Sinking-fund payment A/F	Capital recovery A/P	Uniform-series present-worth P/A
	Future value of $1	Present value of $1	Future value of uniform series of $1	Uniform series whose future value is $1	Uniform series with present value of $1	Present value of uniform series of $1
	$(1+i)^n$	$\frac{1}{(1+i)^n}$	$\frac{(1+i)^n-1}{i}$	$\frac{i}{(1+i)^n-1}$	$\frac{i(1+i)^n}{(1+i)^n-1}$	$\frac{(1+i)^n-1}{i(1+i)^n}$
1	1.400	0.7143	1.000	1.00000	1.40000	0.714
2	1.960	0.5102	2.400	0.41667	0.81667	1.224
3	2.744	0.3644	4.360	0.22936	0.62936	1.589
4	3.842	0.2603	7.104	0.14077	0.54077	1.849
5	5.378	0.1859	10.946	0.09136	0.49136	2.035
6	7.530	0.1328	16.324	0.06126	0.46126	2.168
7	10.541	0.0949	23.853	0.04192	0.44192	2.263
8	14.758	0.0678	34.395	0.02907	0.42907	2.331
9	20.661	0.0484	49.153	0.02034	0.42034	2.379
10	28.925	0.0346	69.814	0.01432	0.41432	2.414
11	40.496	0.0247	98.739	0.01013	0.41013	2.438
12	56.694	0.0176	139.235	0.00718	0.40718	2.456
13	79.371	0.0126	195.929	0.00510	0.40510	2.469
14	111.120	0.0090	275.300	0.00363	0.40363	2.478
15	155.568	0.0064	386.420	0.00259	0.40259	2.484
16	217.795	0.0046	541.988	0.00185	0.40185	2.489
17	304.913	0.0033	759.784	0.00132	0.40132	2.492
18	426.879	0.0023	1064.697	0.00094	0.40094	2.494
19	597.630	0.0017	1491.576	0.00067	0.40067	2.496
20	836.683	0.0012	2089.206	0.00048	0.40048	2.497
21	1171.356	0.0009	2925.889	0.00034	0.40034	2.498
22	1639.898	0.0006	4097.245	0.00024	0.40024	2.498
23	2295.857	0.0004	5737.142	0.00017	0.40017	2.499
24	3214.200	0.0003	8032.999	0.00012	0.40012	2.499
25	4499.880	0.0002	11247.199	0.00009	0.40009	2.499
26	6299.831	0.0002	15747.079	0.00006	0.40006	2.500
27	8819.764	0.0001	22046.910	0.00005	0.40005	2.500

Table 2-8. 50% Interest factors.

Period n	Single-payment compound-amount F/P	Single-payment present-worth P/F	Uniform series compound-amount F/A	Sinking-fund payment A/F	Capital recovery A/P	Uniform-series present-worth P/A
	Future value of $1	Present value of $1	Future value of uniform series of $1	Uniform series whose future value is $1	Uniform series with present value of $1	Present value of uniform series of $1
	$(1+i)^n$	$\frac{1}{(1+i)^n}$	$\frac{(1+i)^n-1}{i}$	$\frac{i}{(1+i)^n-1}$	$\frac{i(1+i)^n}{(1+i)^n-1}$	$\frac{(1+i)^n-1}{i(1+i)^n}$
1	1.500	0.6667	1.000	1.00000	1.50000	0.667
2	2.250	0.4444	2.500	0.40000	0.90000	1.111
3	3.375	0.2963	4.750	0.21053	0.71053	1.407
4	5.062	0.1975	8.125	0.12308	0.62308	1.605
5	7.594	0.1317	13.188	0.07583	0.57583	1.737
6	11.391	0.0878	20.781	0.04812	0.54812	1.824
7	17.086	0.0585	32.172	0.03108	0.53108	1.883
8	25.629	0.0390	49.258	0.02030	0.52030	1.922
9	38.443	0.0260	74.887	0.01335	0.51335	1.948
10	57.665	0.0173	113.330	0.00882	0.50882	1.965
11	86.498	0.0116	170.995	0.00585	0.50585	1.977
12	129.746	0.0077	257.493	0.00388	0.50388	1.985
13	194.620	0.0051	387.239	0.00258	0.50258	1.990
14	291.929	0.0034	581.859	0.00172	0.50172	1.993
15	437.894	0.0023	873.788	0.00114	0.50114	1.995
16	656.841	0.0015	1311.682	0.00076	0.50076	1.997
17	985.261	0.0010	1968.523	0.00051	0.50051	1.998
18	1477.892	0.0007	2953.784	0.00034	0.50034	1.999
19	2216.838	0.0005	4431.676	0.00023	0.50023	1.999
20	3325.257	0.0003	6648.513	0.00015	0.50015	1.999
21	4987.885	0.0002	9973.770	0.00010	0.50010	2.000
22	7481.828	0.0001	14961.655	0.00007	0.50007	2.000

Table 2-9. Five-year escalation table.

Present Worth of a Series of Escalating Payments Compounded Annually
Discount-Escalation Factors for n = 5 Years

Discount	Annual Escalation Rate					
Rate	0.10	0.12	0.14	0.16	0.18	0.20
0.10	5.000000	5.279234	5.572605	5.880105	6.202627	6.540569
0.11	4.866862	5.136200	5.420152	5.717603	6.029313	6.355882
0.12	4.738562	5.000000	5.274242	5.561868	5.863289	6.179066
0.13	4.615647	4.869164	5.133876	5.412404	5.704137	6.009541
0.14	4.497670	4.742953	5.000000	5.269208	5.551563	5.847029
0.15	4.384494	4.622149	4.871228	5.131703	5.404955	5.691165
0.16	4.275647	4.505953	4.747390	5.000000	5.264441	5.541511
0.17	4.171042	4.394428	4.628438	4.873699	5.129353	5.397964
0.18	4.070432	4.287089	4.513947	4.751566	5.000000	5.259749
0.19	3.973684	4.183921	4.403996	4.634350	4.875619	5.126925
0.20	3.880510	4.084577	4.298207	4.521178	4.755725	5.000000
0.21	3.790801	3.989001	4.196400	4.413341	4.640260	4.877689
0.22	3.704368	3.896891	4.098287	4.308947	4.529298	4.759649
0.23	3.621094	3.808179	4.003835	4.208479	4.422339	4.645864
0.24	3.540773	3.722628	3.912807	4.111612	4.319417	4.536517
0.25	3.463301	3.640161	3.825008	4.018249	4.220158	4.431144
0.26	3.388553	3.560586	3.740376	3.928286	4.124553	4.329514
0.27	3.316408	3.483803	3.658706	3.841442	4.032275	4.231583
0.28	3.246718	3.409649	3.579870	3.757639	3.943295	4.137057
0.29	3.179393	3.338051	3.503722	3.676771	3.857370	4.045902
0.30	3.114338	3.268861	3.430201	3.598653	3.774459	3.957921
0.31	3.051452	3.201978	3.359143	3.523171	3.694328	3.872901
0.32	2.990618	3.137327	3.290436	3.450224	3.616936	3.790808
0.33	2.939764	3.074780	3.224015	3.379722	3.542100	3.711472
0.34	2.874812	3.014281	3.159770	3.311524	3.469775	3.634758

Table 2-10. Ten-year escalation table.

Present Worth of a Series of Escalating Payments Compounded Annually
Discount-Escalation Factors for n = 10 Years

Discount	Annual Escalation Rate					
Rate	0.10	0.12	0.14	0.16	0.18	0.20
0.10	10.000000	11.056250	12.234870	13.548650	15.013550	16.646080
0.11	9.518405	10.508020	11.613440	12.844310	14.215140	15.741560
0.12	9.068870	10.000000	11.036530	12.190470	13.474590	14.903510
0.13	8.650280	9.526666	10.498990	11.582430	12.786980	14.125780
0.14	8.259741	9.084209	10.000000	11.017130	12.147890	13.403480
0.15	7.895187	8.672058	9.534301	10.490510	11.552670	12.731900
0.16	7.554141	8.286779	9.099380	10.000000	10.998720	12.106600
0.17	7.234974	7.926784	8.693151	9.542653	10.481740	11.524400
0.18	6.935890	7.589595	8.312960	9.113885	10.000000	10.980620
0.19	6.655455	7.273785	7.957330	8.713262	9.549790	10.472990
0.20	6.392080	6.977461	7.624072	8.338518	9.128122	10.000000
0.21	6.144593	6.699373	7.311519	7.987156	8.733109	9.557141
0.22	5.911755	6.437922	7.017915	7.657542	8.363208	9.141752
0.23	5.692557	6.192047	6.742093	7.348193	8.015993	8.752133
0.24	5.485921	5.960481	6.482632	7.057347	7.690163	8.387045
0.25	5.290990	5.742294	6.238276	6.783767	7.383800	8.044173
0.26	5.106956	5.536463	6.008083	6.526298	7.095769	7.721807
0.27	4.933045	5.342146	5.790929	6.283557	6.824442	7.418647
0.28	4.768518	5.158489	5.585917	6.054608	6.568835	7.133100
0.29	4.612762	4.984826	5.392166	5.838531	6.327682	6.864109
0.30	4.465205	4.820429	5.209000	5.634354	6.100129	6.610435
0.31	4.325286	4.664669	5.035615	5.441257	5.885058	6.370867
0.32	4.192478	4.517015	4.871346	5.258512	5.681746	6.144601
0.33	4.066339	4.376884	4.715648	5.085461	5.489304	5.930659
0.34	3.946452	4.243845	4.567942	4.921409	5.307107	5.728189

Table 2-11. Fifteen-year escalation table.

Present Worth of a Series of Escalating Payments Compounded Annually
Discount-Escalation Factors for $n = 15$ years

Discount Rate	Annual Escalation Rate 0.10	0.12	0.14	0.16	0.18	0.20
0.10	15.000000	17.377880	20.199780	23.549540	27.529640	32.259620
0.11	13.964150	16.126230	18.690120	21.727370	25.328490	29.601330
0.12	13.026090	15.000000	17.332040	20.090360	23.355070	27.221890
0.13	12.177030	13.981710	16.105770	18.616160	21.581750	25.087260
0.14	11.406510	13.057790	15.000000	17.287320	19.985530	23.169060
0.15	10.706220	12.220570	13.998120	16.086500	18.545150	21.442230
0.16	10.068030	11.459170	13.088900	15.000000	17.244580	19.884420
0.17	9.485654	10.766180	12.262790	14.015480	16.066830	18.477610
0.18	8.953083	10.133630	11.510270	13.118840	15.000000	17.203010
0.19	8.465335	9.555676	10.824310	12.303300	14.030830	16.047480
0.20	8.017635	9.026333	10.197550	11.560150	13.148090	15.000000
0.21	7.606115	8.540965	9.623969	10.881130	12.343120	14.046400
0.22	7.227109	8.094845	9.097863	10.259820	11.608480	13.176250
0.23	6.877548	7.684317	8.614813	9.690559	10.936240	12.381480
0.24	6.554501	7.305762	8.170423	9.167798	10.320590	11.655310
0.25	6.255518	6.956243	7.760848	8.687104	9.755424	10.990130
0.26	5.978393	6.632936	7.382943	8.244519	9.236152	10.379760
0.27	5.721101	6.333429	7.033547	7.836080	8.757889	9.819020
0.28	5.481814	6.055485	6.710042	7.458700	8.316982	9.302823
0.29	5.258970	5.797236	6.410005	7.109541	7.909701	8.827153
0.30	5.051153	5.556882	6.131433	6.785917	7.533113	8.388091
0.31	4.857052	5.332839	5.872303	6.485500	7.184156	7.982019
0.32	4.675478	5.123753	5.630905	6.206250	6.860492	7.606122
0.33	4.505413	4.928297	5.405771	5.946343	6.559743	7.257569
0.34	4.345926	4.745399	5.195502	5.704048	6.280019	6.933897

Table 2-12. Twenty-year escalation table.

Present Worth of a Series of Escalating Payments Compounded Annually
Discount-Escalation Factors for n = 20 Years

Discount	Annual Escalation Rate					
Rate	0.10	0.12	0.14	0.16	0.18	0.20
0.10	20.000000	24.295450	29.722090	36.592170	45.308970	56.383330
0.11	18.213210	22.002090	26.776150	32.799710	40.417480	50.067940
0.12	16.642370	20.000000	24.210030	29.505400	36.181240	44.614710
0.13	15.259850	18.243100	21.964990	26.634490	32.502270	39.891400
0.14	14.038630	16.694830	20.000000	24.127100	29.298170	35.789680
0.15	12.957040	15.329770	18.271200	21.929940	26.498510	32.218060
0.16	11.995640	14.121040	16.746150	20.000000	24.047720	29.098950
0.17	11.138940	13.048560	15.397670	18.300390	21.894660	26.369210
0.18	10.373120	12.093400	14.201180	16.795710	20.000000	23.970940
0.19	9.686791	11.240870	13.137510	15.463070	18.326720	21.860120
0.20	9.069737	10.477430	12.188860	14.279470	16.844020	20.000000
0.21	8.513605	9.792256	11.340570	13.224610	15.527270	18.353210
0.22	8.010912	9.175267	10.579620	12.282120	14.355520	16.890730
0.23	7.555427	8.618459	9.895583	11.438060	13.309280	15.589300
0.24	7.141531	8.114476	9.278916	10.679810	12.373300	14.429370
0.25	6.764528	7.657278	8.721467	9.997057	11.533310	13.392180
0.26	6.420316	7.241402	8.216490	9.380883	10.778020	12.462340
0.27	6.105252	6.862203	7.757722	8.823063	10.096710	11.626890
0.28	5.816151	6.515563	7.339966	8.316995	9.480940	10.874120
0.29	5.550301	6.198027	6.958601	7.856833	8.922847	10.194520
0.30	5.305312	5.906440	6.609778	7.437339	8.416060	9.579437
0.31	5.079039	5.638064	6.289875	7.054007	7.954518	9.021190
0.32	4.869585	5.390575	5.995840	6.702967	7.533406	8.513612
0.33	4.675331	5.161809	5.725066	6.380829	7.148198	8.050965
0.34	4.494838	4.949990	5.475180	6.084525	6.795200	7.628322

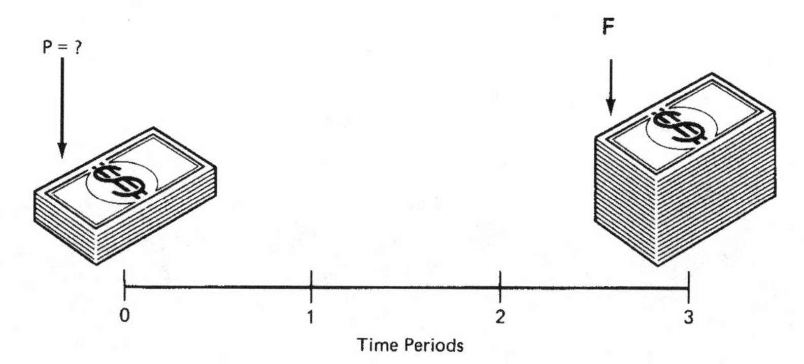

Figure 2-2. Single payment present worth (P/F).

Uniform Series Compound Amount—F/A

The F/A factor is used to determine the amount F that an equal annual payment *A* will accumulate to in *n* years at *i* percent interest. If *A* (uniform annual payment) is known, and *F* (the future worth of these payments) is required, then Equation 2-6 is used.

$$F = A \times \frac{(1+i)^n - 1}{i(1+i)^n} \quad (2\text{-}6)$$

$$F/A \times \frac{(1+i)^n - 1}{i} \quad (2\text{-}7)$$

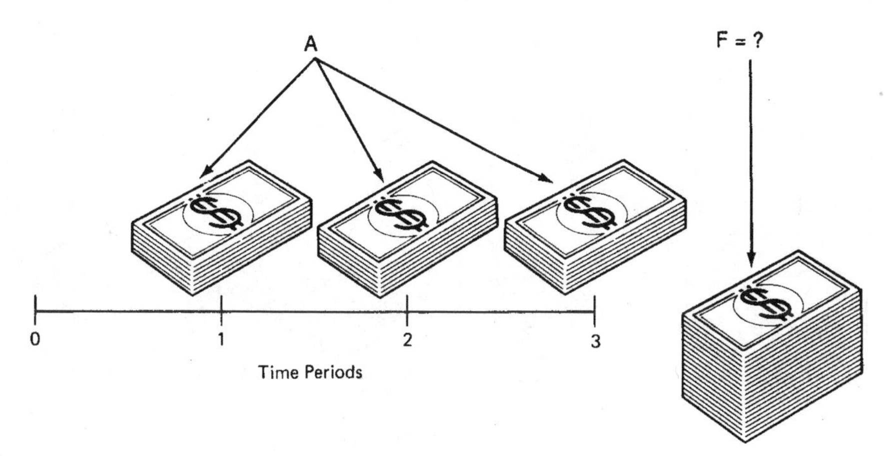

Figure 2-3. Uniform series compound amount (F/A).

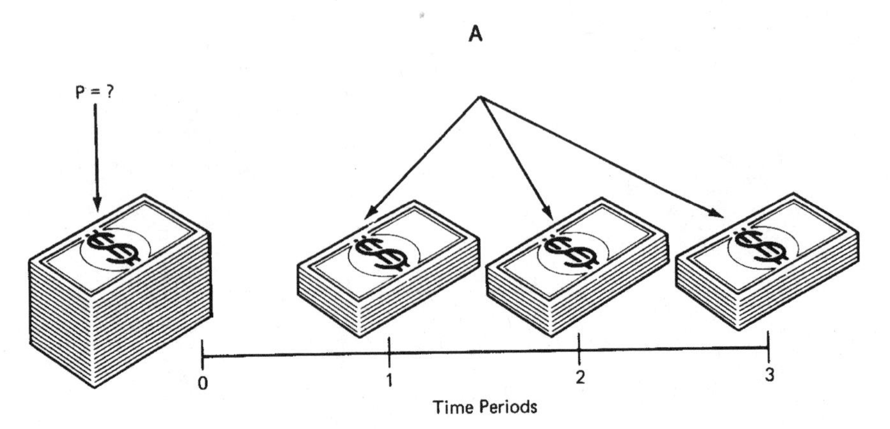

Figure 2-4. Uniform series present worth (P/A).

Uniform Series Present Worth—(P/A)

The P/A factor is used to determine the present amount P that can be paid by equal payments of A (uniform annual payment) at i percent interest, for n years. If A is known, and P is required, then Equation 2-8 is used.

$$P = A \times \frac{(1+i)^n - 1}{i(1+i)^n} \tag{2-8}$$

$$P/A = \frac{(1+i)^n - 1}{i(1+i)^n} \tag{2-9}$$

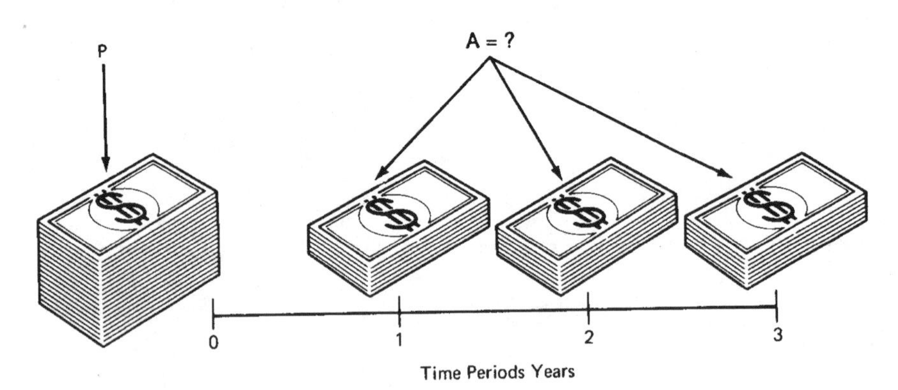

Figure 2-5. Capital recovery (A/P).

Capital Recovery—A/P

The A/P factor is used to determine an annual payment A required to pay off a present amount P at i percent interest, for n years. If the present sum of money, P, spent today is known, and the uniform payment A needed to pay back P over a stated period of time is required, then Equation 2-10 is used.

$$A = P\frac{i(1+i)^n}{(1+i)^n - 1} \tag{2-10}$$

$$A/P\frac{i(1+i)^n}{(1+i)^n - 1} \tag{2-11}$$

Sinking Fund Payment—A/F

The A/F factor is used to determine the equal annual amount R that must be invested for n years at i percent interest in order to accumulate a specified future amount. If F (the future worth of a series of annual payments) is known, and A (value of those annual payments) is required, then Equation 2-12 is used.

$$A = F \times \frac{i}{(1+i)^n - 1} \tag{2-12}$$

$$A/F = \frac{i}{(1+i)^n - 1} \tag{2-13}$$

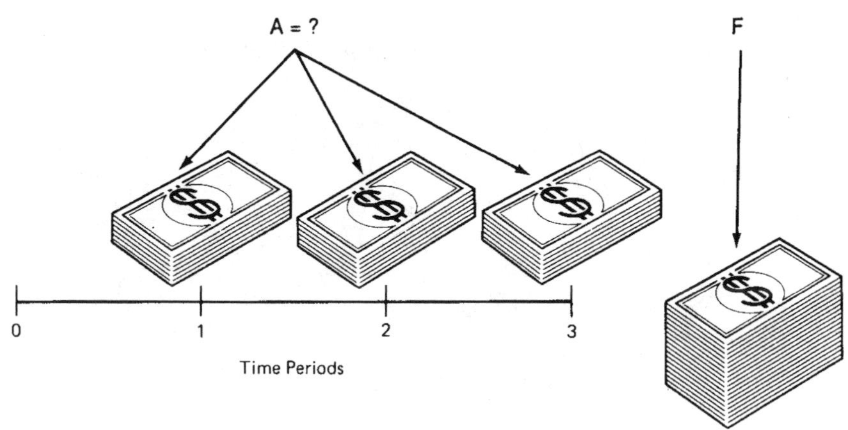

Figure 2-6. Sinking fund payment (A/F).

Gradient Present Worth—GPW

The GPW factor is used to determine the present amount P that can be paid by annual amounts A' which escalate at e percent, at i percent interest, for n years. If A' is known, and P is required, then Equation 2-14 is used. The GPW factor is a relatively new term which has gained in importance due to the impact of inflation.

$$P = A \times (\text{GPW})_n \qquad (2\text{-}14)$$

$$P/A' = GWP = \frac{\frac{1+e}{1+i}\left[1-\left(\frac{1+e}{1+i}\right)^n\right]}{1-\frac{1+e}{1+i}} \qquad (2\text{-}15)$$

The three most commonly used methods in life cycle costing are the annual cost, present worth and rate-of-return analysis.

In the present worth method a minimum rate of return (i) is stipulated. All future expenditures are converted to present values using the interest factors. The alternative with lowest effective first cost is the most desirable.

A similar procedure is implemented in the annual cost method. The difference is that the first cost is converted to an annual expenditure. The alternative with lowest effective annual cost is the most desirable.

In the rate-of-return method, a trial-and-error procedure is usually required. Interpolation from the interest tables can determine what rate of return (i) will give an interest factor which will make the overall cash

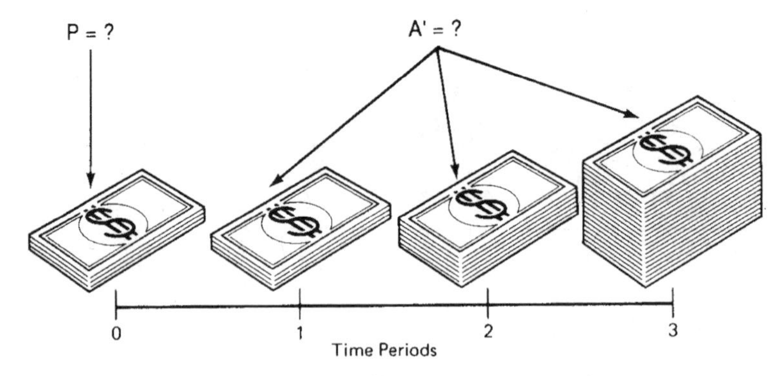

Figure 2-7. Gradient present worth.

flow balance. The rate-of-return analysis gives a good indication of the overall ranking of independent alternates.

The effect of escalation in fuel costs can influence greatly the final decision. When an annual cost grows at a steady rate it may be treated as a gradient and the gradient present worth factor can be used.

Special thanks are given to Rudolph R. Yanuck and Dr. Robert Brown for the use of their specially designed interest and escalation tables used in this text.

When life cycle costing is used to compare several alternatives the differences between costs are important. For example, if one alternate forces additional maintenance or an operating expense to occur, then these factors as well as energy costs need to be included. Remember, what was previously spent for the item to be replaced is irrelevant. The only factor to be considered is whether the new cost can be justified based on projected savings over its useful life.

THE JOB SIMULATION EXPERIENCE

Throughout the text you will experience job situations and problems. Each simulation experience is denoted by SIM. The answer will be given below the problem. Cover the answers, then you can "play the game."

SIM 2-1

An evaluation needs to be made to replace all 40-watt fluorescent lamps with a new lamp that saves 12 percent or 4.8 watts and gives the same output. The cost of each lamp is $2.80.

Assuming a rate of return before taxes of 25 percent is required, can the immediate replacement be justified? Hours of operation are 5800 and the lamp life is two years. Electricity costs 7.0¢/kWh.

ANSWER

$$A = 5800 \times 4.8 \times 0.070/1000 = \$1.94$$
$$A/P = 1.94/2.80 = .69$$

From Table 2-5 a rate of return of 25 percent is obtained. When analyzing energy conservation measures, never look at what was previously spent or the life remaining. Just determine if the new expenditure will pay for itself.

SIM 2-2

An electrical energy audit indicates electrical motor consumption is 4×10^6 kWh per year. By upgrading the motor spares with high efficiency motors a 10% savings can be realized. The additional cost for these motors is estimated at $80,000. Assuming an 8¢ per kWh energy charge and 20-year life, is the expenditure justified based on a minimum rate of return of 20% before taxes? Solve the problem using the present worth, annual cost, and rate-of-return methods.

Analysis

Present Worth Method

	Alternate 1 Present Method	*Alternate 2 Use High Efficiency Motor Spares*
(1) First Cost (P)	—	$80,000
(2) Annual Cost (*A*)	4 × 106 ×.08 = $320,000	.9 × $320,000 = $288,000
P/A (Table 2-4)	4.87	4.87
(2) *A* × 4.87 =	$1,558,400	$1,402,560
Present Worth (1)+(3)	$1,558,400	$1,482,560 Choose Alternate with Lowest First Cost

Annual Cost Method

	Alternate 1	*Alternate 2*
(1) First Cost (P)	—	$80,000
(2) Annual Cost (*A*)	$320,000	$288,000
A/P (Table 2-4)	.2	.2
(3) P × .2	—	$16,000
Annual Cost (2)+(3)	$320,000	$304,000 Choose Alternate with Lowest First Cost

Rate of Return Method

$$P = (\$320{,}000 \quad \$288{,}000)$$

$$P/A = \frac{80{,}000}{32{,}000} = 2.5$$

What value of *i* will make P/A = 2.5? *i* = 40% (Table 2-7).

SIM 2-3

Show the effect of 10 percent escalation on the rate of return analysis given the

Energy equipment investment	= \$20,000
After-tax savings	= \$2,600
Equipment life (n)	= 15 years

ANSWER

Without escalation:

$$\frac{A}{P} = \frac{2{,}600}{20{,}000} = 0.13$$

From Table 2-1, the rate of return is 10 percent. With 10 percent escalation assumed:

$$\frac{P}{A} = \frac{20{,}000}{2{,}600} = 7.69$$

From Table 2-11, the rate of return is 21 percent.

Thus we see that taking into account a modest escalation rate can dramatically affect the justification of the project.

MAKING DECISIONS FOR ALTERNATE INVESTMENTS

There are several methods for determining which energy conservation alternative is the most economical. Probably the most familiar and trusted method is the annual cost method.

When evaluating replacement of processes or equipment *do not* consider what was previously spent. The decision will be based on whether the new process or equipment proves to save substantially enough in operating costs to justify the expenditure.

Equation 2-16 is used to convert the lump sum investment P into the annual cost. In the case where the asset has a value after the end of its useful life, the annual cost becomes:

$$AC = (P - L) * \mathrm{A/P} + iL \qquad (2\text{-}16)$$

where

AC is the annual cost

L is the net sum of money that can be realized for a piece of equipment, over and above its removal cost, when it is returned at the end of the service life. L is referred to as the salvage value.

As a practical point, the salvage value is usually small and can be neglected, considering the accuracy of future costs. The annual cost technique can be implemented by using the following format:

	Alternate 1	Alternate 2
1. First cost (P)		
2. Estimated life (n)		
3. Estimated salvage value at end of life (L)		
4. Annual disbursements, including energy costs & maintenance (E)		
5. Minimum acceptable return *before* taxes (i)		
6. A/P n, i		
7. $(P - L) * A/P$		
8. Li		
9. $AC = (P - L) * A/P + Li + E$		

Choose alternate with lowest AC

The alternative with the lowest annual cost is the desired choice.

SIM 2-4

A new water line must be constructed from an existing pumping station to a reservoir. Estimates of construction and pumping costs for each pipe size have been made.

The annual cost is based on a 16-year life and a desired return on investment before taxes of 10 percent. Which is the most economical pipe size for pumping 4,000 hours/year?

Pipe Size	Estimated Construction Costs	Cost/Hour for Pumping
8″	$80,000	$4.00
10″	$100,000	$3.00
12″	$160,000	$1.50

ANSWER

	8″ Pipe	10″ Pipe		12″ Pipe
P	$80,000	$100,000		$160,000
n	16	16		16
E	16,000	12,000		6,000
i	10%	10%		10%
A/P = 0.127	—	—		—
(*P* – *L*) A/P	10,160	12,700		20,320
Li	———	———		———
AC	$26,160	$24,700	(*Choice*)	$26,320

DEPRECIATION, TAXES, AND THE TAX CREDIT

Depreciation

Depreciation affects the "accounting procedure" for determining profits and losses and the income tax of a company. In other words, for tax purposes the expenditure for an asset such as a pump or motor cannot be fully expensed in its first year. The original investment must be charged off for tax purposes over the useful life of the asset. A company usually wishes to expense an item as quickly as possible.

The Internal Revenue Service allows several methods for determining the annual depreciation rate.

Straight-line Depreciation. The simplest method is referred to as a straight-line depreciation and is defined as:

$$D = \frac{P - L}{n} \qquad (2\text{-}17)$$

where

- D is the annual depreciation rate
- L is the value of equipment at the end of its useful life, commonly referred to as salvage value
- n is the life of the equipment, which is determined by Internal Revenue Service guidelines
- P is the initial expenditure.

Sum-of-Years Digits. Another method is referred to as the sum-of-years digits. In this method the depreciation rate is determined by finding the sum of digits using the following formula,

$$N = n\,\frac{(n+1)}{2} \qquad (2\text{-}18)$$

where n is the life of equipment.

Each year's depreciation rate is determined as follows

First year
$$D = \frac{n}{N}(P-L) \qquad (2\text{-}19)$$

Second year
$$D = \frac{n-1}{N}(P-L) \qquad (\text{-}20)$$

n year
$$D = \frac{1}{N}(P-L) \qquad (2\text{-}21)$$

Declining-Balance Depreciation. The declining-balance method allows for larger depreciation charges in the early years which is sometimes referred to as fast write-off.

The rate is calculated by taking a constant percentage of the declining undepreciated balance. The most common method used to calculate the declining balance is to predetermine the depreciation rate. Under certain circumstances a rate equal to 200 percent of the straight-line depreciation rate may be used. Under other circumstances the rate is limited to 1-1/2 or 1/4 times as great as straight-line depreciation. In this method the salvage value or undepreciated book value is established once the depreciation rate is pre-established.

To calculate the undepreciated book value, Equation 2-22 is used.

$$D = 1\left(\frac{L}{P}\right)^{1/N} \tag{2-22}$$

where

D is the annual depreciation rate
L is the salvage value
P is the first cost.

The Tax Reform Act of 1986 (hereafter referred to as the "Act") represented true tax reform, as it made sweeping changes in many basic federal tax code provisions for both individuals and corporations. The Act has had significant impact on financing for cogeneration, alternative energy and energy efficiency transactions, due to substantial modifications in provisions concerning depreciation, investment and energy tax credits, tax-exempt financing, tax rates, the corporate minimum tax and tax shelters generally.

The Act lengthened the recovery periods for most depreciable assets. The Act also repealed the 10 percent investment tax credit ("ITC") for property placed in service on or after January 1, 1986, subject to the transition rules.

Tax Considerations

Tax-deductible expenses such as maintenance, energy, operating costs, insurance, and property taxes reduce the income subject to taxes.

For the after-tax life cycle analysis and payback analysis the actual incurred and annual savings is given as follows.

$$AS = (1 - I)\,E + ID \tag{2-23}$$

where

AS is the yearly annual after-tax savings (excluding effect of tax credit)
E is the yearly annual energy savings (difference between original expenses and expenses after modification)
D is the annual depreciation rate
I is the income tax bracket.

Equation 2-23 takes into account that the yearly annual energy savings are partially offset by additional taxes which must be paid due to reduced operating expenses. On the other hand, the depreciation allowance reduces taxes directly.

After-tax Analysis

To compute a rate of return which accounts for taxes, depreciation, escalation, and tax credits, a cash-flow analysis is usually required. This method analyzes all transactions including first and operating costs. To determine the after-tax rate of return a trial-and-error or computer analysis is required.

All money is converted to the present assuming an interest rate. The summation of all present dollars should equal zero when the correct interest rate is selected, as illustrated in Figure 2-8.

This analysis can be made assuming a fuel escalation rate by using the gradient present worth interest of the present worth factor.

	1	2	3	4	
Year	Investment	Tax Credit	After-tax Savings (*AS*)	Single Payment Present Worth Factor	(2 + 3) × 4 Present Worth
0	$-P$				$-P$
1		$+TC$	AS	P/F_1	$+P_1$
2			AS	P/F_2	P_2
3			AS	P/F_3	P_3
4			AS	P/F_4	P_4
Total					$\sum P$

$$AS = (1 - I)\,E + ID$$

Trial-and-Error Solution:
Correct i when $\sum P = 0$

Figure 2-8. Cash flow rate of return analysis.

SIM 2-5

Develop a set of curves that indicate the capital that can be invested to give a rate of return of 15 percent after taxes for each $1,000 saved for the following conditions:

1. The effect of escalation is not considered.
2. A 5 percent fuel escalation is considered.
3. A 10 percent fuel escalation is considered.
4. A 14 percent fuel escalation is considered.
5. A 20 percent fuel escalation is considered.

Calculate for 5-, 10-, 15-, 20-year life.

Assume straight-line depreciation over useful life, 34 percent income tax bracket, and no tax credit.

ANSWER

$$AS = (1 - I)E + ID$$

$$I = 0.4, \quad E = \$1{,}000$$

$$AS = 660 + \frac{0.34P}{N}$$

Thus, the after-tax savings (AS) are comprised of two components. The first component is a uniform series of $660 escalating at e percent/year. The second component is a uniform series of $0.34P/N$.

Each component is treated individually and converted to present day values using the GPW factor and the P/A factor, respectively. The sum of these two present worth factors must equal P. In the case of no escalation, the formula is:

$$P = 660 * \mathrm{P/A} + \frac{0.34P}{N}\,\mathrm{P/A}$$

In the case of escalation:

$$P = 660\ \mathrm{GPW} + \frac{0.34P}{N} * \mathrm{P/A}$$

Since there is only one unknown, the formulas can be readily solved. The results are indicated on the next page.

	N = 5 \$$P$	N = 10 \$$P$	N = 15 \$$P$	N =20 \$$P$
e = 0	2869	4000	4459	4648
e = 10%	3753	6292	8165	9618
e = 14%	4170	7598	10,676	13,567
e = 20%	4871	10,146	16,353	23,918

Figure 2-9 illustrates the effects of escalation. This figure can be used as a quick way to determine after-tax economics of energy utilization expenditures.

SIM 2-6

It is desired to have an after-tax savings of 15 percent. Comment on the investment that can be justified if it is assumed that the fuel rate escalation should not be considered and the annual energy savings is $2,000 with an equipment economic life of 15 years.

Comment on the above, assuming a 14 percent fuel escalation.

ANSWER

From Figure 2-9, for each $1,000 energy savings, an investment of $4,400 is justified or $8,800 for a $2,000 savings when no fuel increase is accounted for.

With a 14 percent fuel escalation rate an investment of $10,600 is justified for each $1,000 energy savings, thus $21,200 can be justified for $2,000 savings. Thus, a much higher expenditure is economically justifiable and will yield the same after-tax rate of return of 15 percent when a fuel escalation of 14 percent is considered.

IMPACT OF FUEL INFLATION ON LIFE CYCLE COSTING

As illustrated by problem 2-5 a modest estimate of fuel inflation has a major impact on improving the rate of return on investment of the project. The problem facing the energy engineer is how to forecast

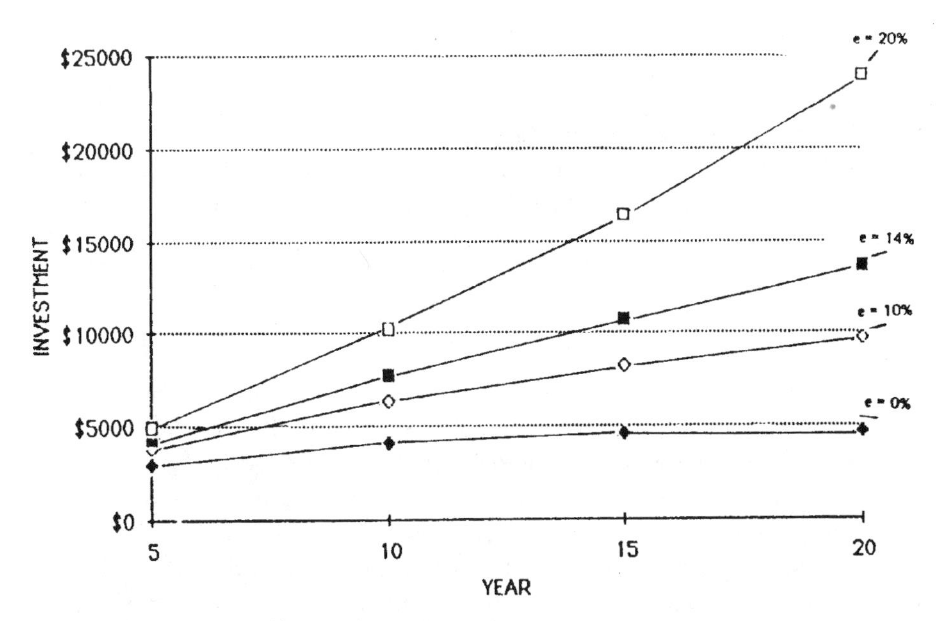

Figure 2-9. Effects of escalation on investment requirements.
Note: Maximum investment in order to attain a 15% after-tax rate of return on investment for annual savings of $1,000.

what the future of energy costs will be. All too often no fuel inflation is considered because of the difficulty of projecting the future. In making projections the following guidelines may be helpful:

- Is there a rate increase that can be forecast based on new nuclear generating capacity?

- What has been the historical rate increase for the facility? Even with fluctuations there are likely to be trends to follow,

- What events on a national or international level would impact on your costs? New state taxes, new production quotas by OPEC and other factors may effect your fuel prices.

- What do the experts say? Energy economists, forecasting services, and your local utility projections all should be taken into account.

SUMMARY OF LIFE-CYCLE COSTING

Always draw a cash flow diagram on a time basis scale. Show cash flow ins as positive and cash flow outs as negative.

In determining which interest formula to use, the following procedure may be helpful. First, put the symbols in two rows, one above the other as below:

$$\frac{\text{PAF (unknown)}}{\text{PAF (known)}}$$

The top represents the unknown values, and the bottom line represents the known. From information you have and desire, simply circle one of each line, and you have the correct factor.

For example, if you want to determine the annual saving "A" required when the cost of the energy device "P" is known, circle P on the bottom and *A* on the top. The factor A/P or capital recovery is required for this example. Table 2-11 summarizes the cash analysis for interest formulas.

Table 2-11. Cash Analysis for Interest Formulas.

GIVEN	FIND	USE
P	F	F/P
F	P	P/F
A	F	F/A
F	A	A/F
P	A	A/P
A	P	P/A

3

The Facility Survey

The survey of the facility is considered a very important part of the industrial energy audit. Chapter 1 indicated that there are many types of surveys, from a simple walk-through to a complete quantification of uses and losses. This chapter will illustrate various types of instruments that can aid in the industrial audit.

COMPARING CATALOGUE DATA WITH ACTUAL PERFORMANCE

Many energy managers are surprised when they record actual performance data of equipment and compare it with catalogue information. There is usually a great disparity between the two.

Each manufacturer has design tolerance for its equipment. For critical equipment, performance guarantees or tests should be incorporated into the initial specifications.

As part of the facility survey, nameplate data of pumps, motors, chillers, fans, etc., should be taken. The nameplate data should also be compared to actual running conditions.

The initial survey can detect motors that were over-sized. By replacing the motor with a smaller one, energy savings can be realized.

INFRARED EQUIPMENT

Some companies may have the wrong impression that infrared equipment can meet most of their instrumentation needs.

The primary use of infrared equipment in an energy utilization program is to detect building or equipment losses. Thus it is just one of the many options available.

Several energy managers find infrared in use in their plant prior to the energy utilization program. Infrared equipment, in many instances, was purchased by the electrical department and used to detect electrical hot spots.

Infrared energy is an invisible part of the electromagnetic spectrum. It exists naturally and can be measured by remote heat-sensing equipment. In recent years lightweight portable infrared systems became available to help determine energy losses. Differences in the infrared emissions from the surface of objects cause color variations to appear on the scanner. The hotter the object, the more infrared radiated. With the aid of an isotherm circuit the intensity of these radiation levels can be accurately measured and quantified. In essence the infrared scanning device is a diagnostic tool which can be used to determine building heat losses. Equipment costs range from $400 to $50,000.

An overview energy scan of the plant can be made through an aerial survey using infrared equipment. Several companies offer aerial scan services starting at $1,500. Aerial scans can determine underground stream pipe leaks, hot gas discharges, leaks, etc.

Since IR detection and measurement equipment have gained increased importance in the energy audit process, a summary of the fundamentals is reviewed in this section.

The visible portion of the spectrum runs from .4 to .75 micrometers (μm). The infrared or thermal radiation begins at this point and extends to approximately 1,000 μm. Objects such as people, plants, or buildings will emit radiation with wavelengths around 10 μm.

Infrared instruments are required to detect and measure the thermal radiation. To calibrate the instrument a special "black body" radiator is used. A black body radiator absorbs all the radiation that impinges on it and has an absorbing efficiency or emissivity of 1.

The accuracy of temperature measurements by infrared instruments depends on the three processes which are responsible for an object acting like a black body. These processes—absorbed, reflected, and transmitted radiation—are responsible for the total radiation reaching an infrared scanner.

The real temperature of the object is dependent only upon its emitted radiation.

Corrections to apparent temperatures are made by knowing the emissivity of an object at a specified temperature.

The heart of the infrared instrument is the infrared detector. The

Gamma Rays	X-Rays	UV	Visible	Infrared	Microwave	Radio Wave
10^{-6}	10^{-5}	10^{-2}	.4	.75	10^{3}	10^{6}

high energy radiation short wavelength — low energy radiation long wavelength

Figure 3-1. Electromagnetic spectrum.

detector absorbs infrared energy and converts it into electrical voltage or current. The two principal types of detectors are the thermal and photo type. Figure 3-2 shows a thermal camera with LCD display. The thermal detector generally requires a given period of time to develop an image on photographic film. The photo detectors are more sensitive and have a higher response time. Television-like displays on a cathode ray tube permit studies of dynamic thermal events on moving objects in real time.

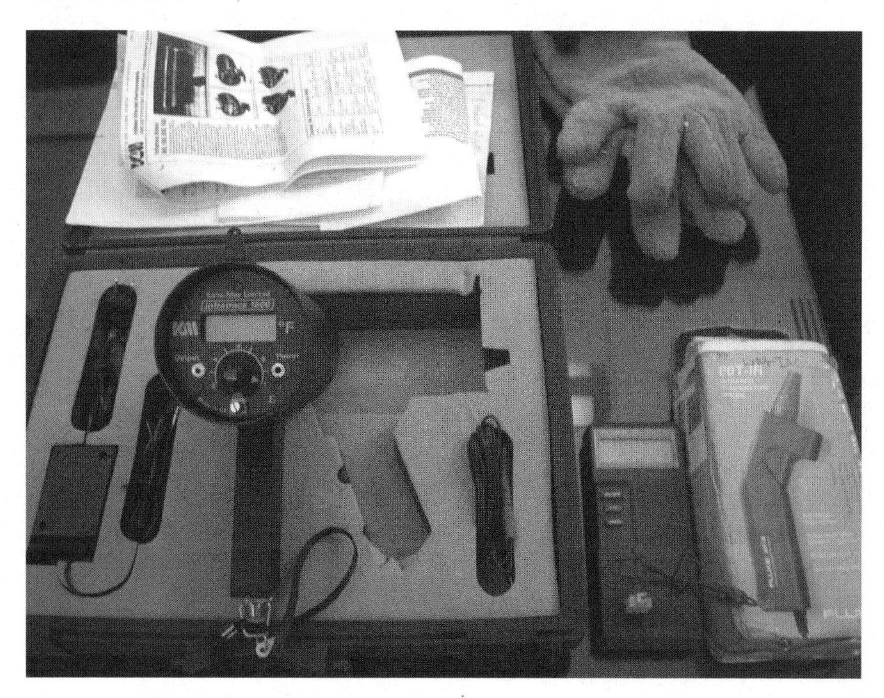

Figure 3-2. Temperature

There are various ways of displaying signals produced by infrared detectors. One way is by use of an isotherm contour. The lightest areas of the picture represent the warmest areas of the subject and the darkest areas represent the coolest portions. These instruments can show thermal variations of less than 0.1°C and can cover a range of –30°C to over 2,000°C.

The isotherm can be calibrated by means of a black body radiator so that a specific temperature is known. The scanner can then be moved and the temperatures of the various parts of the subject can be made.

MEASURING ELECTRICAL SYSTEM PERFORMANCE

The ammeter, voltmeter, wattmeter, power factor meter, and footcandle meter are usually required to do an electrical survey. These instruments are described below.

Ammeter and Voltmeter

To measure electrical currents, ammeters are used. For most audits, alternating currents are measured. Ammeters used in audits are portable and are designed to be easily attached and removed. Figure 3-3 shows a fluke 23 ammeter.

There are many brands and styles of snap-on ammeters commonly available that can read up to 1000 amperes continuously. This range can be extended to 4000 amperes continuously for some models with an accessory step-down current transformer.

The snap-on ammeters can be either indicating or recording with a printout. After attachment, the recording ammeter can keep recording current variations for as long as a full month on one roll of recording paper. This allows studying current variations in a conductor for extended periods without constant operator attention.

The ammeter supplies a direct measurement of electrical current which is one of the parameters needed to calculate electrical energy. The second parameter required to calculate energy is voltage, and it is measured by a voltmeter.

Several types of electrical meters can read the voltage or current. A voltmeter measures the difference in electrical potential between two points in an electrical circuit.

In series with the probes are the galvanometer and a fixed resis-

Figure 3-3. Clamp-on Amp Meter

tance (which determine the voltage scale). The current through this fixed resistance circuit is then proportional to the voltage and the galvanometer deflects in proportion to the voltage.

The voltage drops measured in many instances are fairly constant and need only be performed once. If there are appreciable fluctuations, additional readings or the use of a recording voltmeter may be indicated.

Most voltages measured in practice are under 600 volts and there are many portable voltmeter/ammeter clamp-ons available for this and lower ranges. Figure 3-4 shows a fluke 87 mltimeter.

Wattmeter and Power Factor Meter

The portable wattmeter can be used to indicate by direct reading electrical energy in watts. It can also be calculated by measuring voltage, current and the angle between them (power factor angle).

The basic wattmeter consists of three voltage probes and a snap-on current coil which feeds the wattmeter movement.

The typical operating limits are 300 kilowatts, 650 volts, and 600 amperes. It can be used on both one- and three-phase circuits.

The portable power factor meter is primarily a three-phase instrument. One of its three voltage probes is attached to each conductor

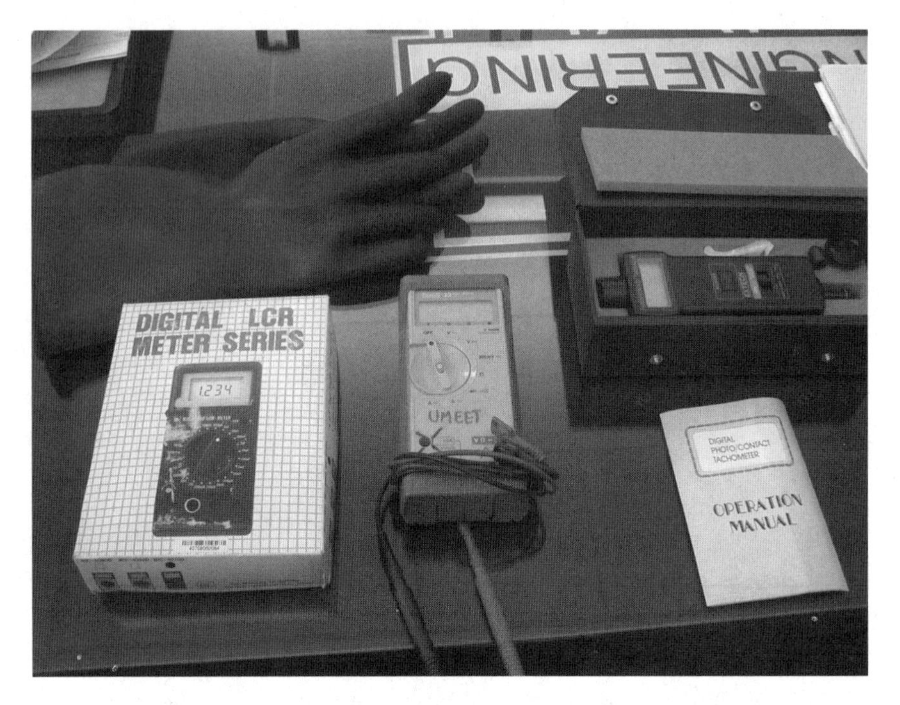

Figure 3-4. Fluke 87 multimeter.

phase and a snap-on jaw is placed about one of the phases. By disconnecting the wattmeter circuitry, it will directly read the power factor of the circuit to which it is attached.

It can measure power factor over a range of 1.0 leading to 1.0 lagging with "ampacities" up to 1500 amperes at 600 volts. This range covers the large bulk of the applications found in light industry and commerce.

The power factor is a basic parameter whose value must be known to calculate electric energy usage. Diagnostically it is a useful instrument to determine the sources of poor power factor in a facility.

Portable digital kWh and kW demand units are now available. Figure 3-5 shows a portable load logger.

Digital read-outs of energy usage in both kWh and kW demand or in dollars and cents, including instantaneous usage, accumulated usage, projected usage for a particular billing period, alarms when over-target levels are desired for usage, and control-outputs for load shedding and cycling are possible.

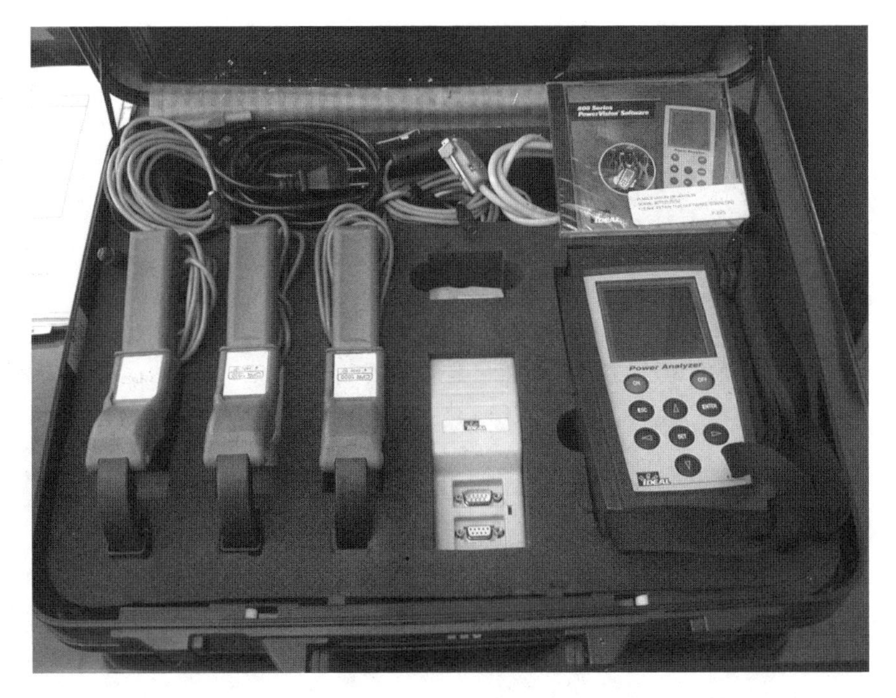

Figure 3-5. Portable load logger.

Continuous displays or intermittent alternating displays are available at the touch of a button of any information needed such as the cost of operating a production machine for one shift, one hour or one week.

Foot-candle Meter

Foot-candle meters measure illumination in units of foot-candles through light-sensitive barrier layers of cells contained within them. They are usually pocket size and portable and are meant to be used as field instruments to survey levels of illumination. Foot-candle meters differ from conventional photographic lightmeters in that they are color and cosine corrected. Figure 3-6 shows a light meter.

TEMPERATURE MEASUREMENTS

To maximize system performance, knowledge of the temperature of a fluid, surface, etc., is essential. Several types of temperature devices are described in this section.

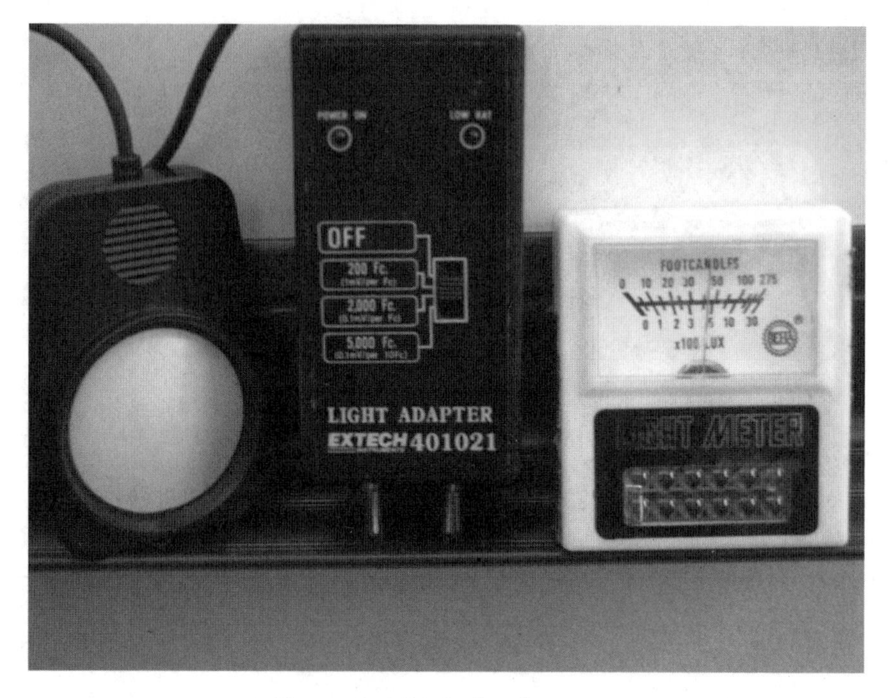

Figure 3-6. Light level meter.

Thermometer

There are many types of thermometers that can be used in an energy audit. The choice of what to use is usually dictated by cost, durability, and application.

For air-conditioning, ventilation and hot-water service applications (temperature ranges 50°F to 250°F) a multipurpose portable battery-operated thermometer is used. Three separate probes are usually provided to measure liquid, air or surface temperatures. Figure 3-7 shows a typical thermometer used in HVAC applications.

For boiler and oven stacks (1000°F) a dial thermometer is used. Thermocouples are used for measurements above 1000°F.

Surface Pyrometer

Surface pyrometers are instruments which measure the temperature of surfaces. They are somewhat more complex than other temperature instruments because their probe must make intimate contact with the surface being measured.

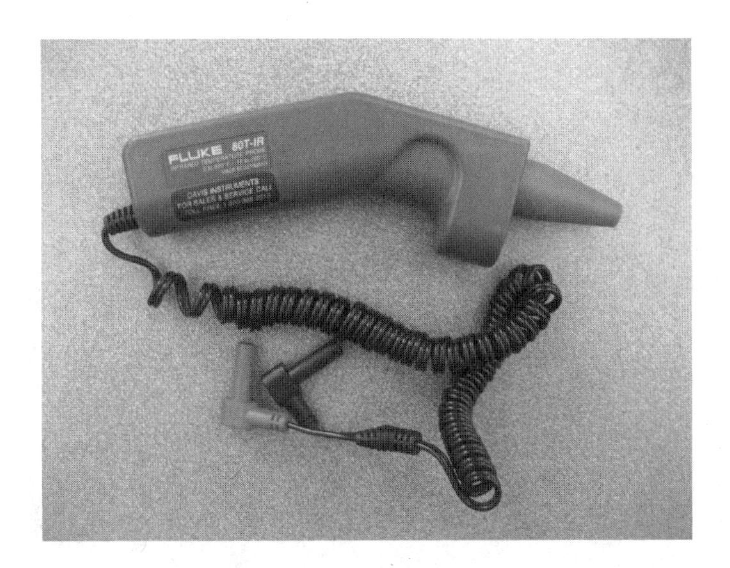

Figure 3-7. High Temperature Probe

Surface pyrometers are of immense help in assessing heat losses through walls and also for testing steam traps.

They may be divided into two classes: low-temperature (up to 250°F) and high-temperature (up to 600°F to 700°F). The low-temperature unit is usually part of the multipurpose thermometer kit. The high-temperature unit is more specialized, but needed for evaluating fired units and general steam service. Figure 3-8 shows a surface pyrometer.

There are also noncontact surface pyrometers which measure infrared radiation from surfaces in terms of temperature. These are suitable for general work and also for measuring surfaces which are visually but not physically accessible.

A more specialized instrument is the optical pyrometer. This is for high-temperature work (above 1500°F) because it measures the temperature of bodies which are incandescent because of their temperature.

Psychrometer

A psychrometer is an instrument which measures relative humidity based on the relation of the dry-bulb temperature and the wet-bulb temperature.

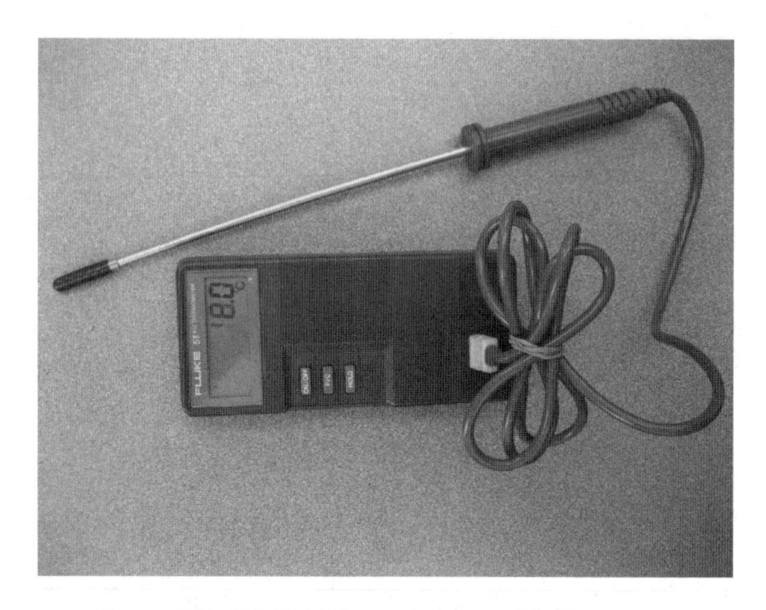

Figure 3-8. Digital Thermometer with Pyrometer

Relative humidity is of prime importance in HVAC and drying operations. Recording psychrometers are also available. Above 200°F humidity studies constitute a specialized field of endeavor.

Portable Electronic Thermometer

The portable electronic thermometer is an adaptable temperature measurement tool. The battery-powered basic instrument, when housed in a carrying case, is suitable for laboratory or industrial use. Figure 3-9 shows an electronic thermometer.

A pocket-size digital, battery-operated thermometer is especially convenient for spot checks or where a number of rapid readings of process temperatures need to be taken.

Thermocouple Probe

No matter what sort of indicating instrument is employed, the thermocouple used should be carefully selected to match the application and properly positioned if a representative temperature is to be measured. The same care is needed for all sensing devices—thermocouple, bimetals, resistance elements, fluid expansion, and vapor pressure bulbs.

Suction Pyrometer

Errors arise if a normal sheathed thermocouple is used to measure gas temperatures, especially high ones. The suction pyrometer overcomes these by shielding the thermocouple from wall radiation and drawing gases over it at high velocity to ensure good convective heat transfer. The thermocouple thus produces a reading which approaches the true temperature at the sampling point rather than a temperature between that of the walls and the gases.

MEASURING COMBUSTION SYSTEMS

To maximize combustion efficiency it is necessary to know the composition of the flue gas. By obtaining a good air-fuel ratio substantial energy will be saved.

Combustion Tester

Combustion testing consists of determining the concentrations of the products of combustion in a stack gas. The products of combustion usually considered are carbon dioxide and carbon monoxide. Oxygen is tested to assure proper excess air levels.

The definitive test for these constituents is an Orsat apparatus. This test consists of taking a measured volume of stack gas and measuring successive volumes after intimate contact with selective absorbing solutions. The reduction in volume after each absorption is the measure of each constituent.

The Orsat has a number of disadvantages. The main ones are that it requires considerable time to set up and use and its operator must have a good degree of dexterity and be in constant practice.

Instead of an Orsat, there are portable and easy-to-use absorbing instruments which can easily determine the concentrations of the constituents of interest on an individual basis. Setup and operating times are minimal and just about anyone can learn to use them.

The typical range of concentrations are CO_2: 0-20%, O_2: 0-21%, and CO: 0-0.5%. The CO_2 or O_2 content, along with knowledge of flue gas temperature and fuel type, allows the flue gas loss to be determined off standard charts.

Boiler Test Kit

The boiler test kit contains the following:

CO_2 Gas analyzer

O_2 Gas analyzer
Inclined monometer

CO Gas analyzer.

The purpose of the components of the kit is to help evaluate fireside boiler operation. Good combustion usually means high carbon dioxide (CO_2), low oxygen (O_2), and little or no trace of carbon monoxide (CO).

Gas Analyzers

The gas analyzers are usually of the Fyrite type. The Fyrite type differs from the Orsat apparatus in that it is more limited in application and less accurate. The chief advantages of the Fyrite are that it is simple and easy to use and is inexpensive. This device is often used in an energy audit. Three readings using the Fyrite analyzer should be made and the results averaged.

Smoke Tester

To measure combustion completeness the smoke detector is used. Smoke is unburned carbon, which wastes fuel, causes air pollution, and fouls heat-exchanger surfaces. To use the instrument, a measured volume of flue gas is drawn through filter paper with the probe. The smoke spot is compared visually with a standard scale and a measure of smoke density is determined.

Combustion Analyzer

The combustion electronic analyzer permits fast, close adjustments. The unit contains digital displays. A standard sampler assembly with probe allows for stack measurements through a single stack or breaching hole. Figure 3-9 shows a typical combustion analyzer.

MEASURING HEATING, VENTILATION AND AIR-CONDITIONING (HVAC) SYSTEM PERFORMANCE

Air Velocity, Pressurization & Leakage Rates

- *Blower Door*—a device containing a fan, controller and several pressure gauges and a frame which fits into a doorway of a building. This device is used to determine the pressurization and leakage rates of a building and its air distribution system under varying pressure conditions

- *Smoke Generators*—useful in determining airflow characteristics in buildings, and air distribution systems

- *Anemometers*—two types, the vane and hot-wire, are used to measure air velocity

- *Airflow hoods*—contain an airspeed integrating manifold which averages the velocity across the opening and reads out the airflow volume

Temperature Measurement

The temperature devices most commonly used are as follows:

Figure 9. Combustion analyzer.

- *Glass thermometers*—considered to be the most useful to temperature measuring instruments—accurate and convenient but fragile.

- *Resistance thermometers*—considered to be very useful for A/C testing. Accuracy is good and they are reliable and convenient to use.

- *Thermocouples*—similar to resistance thermocouple, but do not require battery power source. Chrome-Alum or iron types are the most useful and have satisfactory accuracy and repeatability.

- *Pressure bulb thermometers*—more suitable for permanent installation.

- *Optical pyrometers*—only suitable for furnace settings and therefore limited in use.

- *Radiation pyrometers*—limited in use for A/C work. These are better suited for spot measurements in hard-to-reach spots. Figure 3-11 shows a radiation pyrometer.

- *Thermographs*—use for recording room or space temperature and gives a chart indicating variations over a 12- or 168-hour period. Reasonably accurate.

Pressure Measurement (Absolute and Differential)

Common devices used for measuring pressure in HVAC applications are as follows:

- *Micromanometer*—not usually portable, but suitable for fixed measurement of pressure differentials across filter, coils, etc.

- *Draft gauges*—can be portable and used for either direct pressure or pressure differential.

- *Manometers*—can be portable. Used for direct pressure reading and with Pitot tubes for air flows. Very useful.

- *Swing vane gauges*—can be portable. Usually used for air flow.

- *Bourdon tube gauges*—very useful for measuring all forms of system fluid pressures from 5 psi up. Special types for refrigeration plants.

Humidity Measurement

The data given below indicate the type of instruments available for humidity measurement. The following indicates equipment suitable for HVAC applications:

- *Psychrometers*—basically these are wet and dry bulb thermometers. They can be fixed on a portable stand or mounted in a frame with a handle for revolving in air.

- *Dimensional change*—device usually consists of a "hair," which changes in length proportionally with humidity changes. Not usually portable, fragile, and only suitable for limited temperature and humidity ranges.

- *Electrical conductivity*—can be compact and portable but of a higher cost. Very convenient to use.

- *Electrolytic*—as above. But for very low temperature ranges. Therefore unsuitable for HVAC test work.

- *Gravimeter*—not suitable.

4

Electrical System Optimization

APPLYING PROVEN TECHNIQUES TO REDUCE THE ELECTRICAL BILL

Electrical bills can be reduced by up to 30 percent by knowing the utility rate structure, by improving the plant power factor, by reducing peak loads, and by the efficient use of lighting. This chapter will illustrate these aspects as they apply to the energy utilization program.

WHY THE PLANT MANAGER SHOULD UNDERSTAND THE ELECTRIC RATE STRUCTURE

Each plant manager should understand how the plant is billed. Utility companies usually have several rate structures offered to customers. By understanding the electrical characteristics of the plant, the best rate structure for the plant is determined. Understanding the rate structure also enables the plant manager to avoid the penalties the utility company incorporates into its rates.

ELECTRICAL RATE TARIFF

Customer Charge—a fixed monthly charge that covers the utility's expenses for customer metering, postage for billing, etc.

Energy Charge—The energy charge is charge for the number of kWh consumed that month. This covers the utility's cost for fuel, operation of equipment and maintenance expenses.

Demand Charge—The demand charge is a function of the peak kW consumed at the customer's facility during the month. This number may represent the actual peak kW consumed or it may be a percentage of the maximum kW measured during the last eleven months if the utility maintains a demand ratchet. A demand ratchet may exist because a utility must maintain reserve capacity to meet client demands. Thus, a ratchet allows the utility to pass on part of the cost of maintaining reserve capacity to the client.

Power Factor Charge or Reactive Demand Charge—The power factor is a measurement of the ratio of resistive power (kW) consumed by a load to the apparent power (kVA) consumed. In an industrial facility, the electrical load consists of significant reactive load (kVAR) due to the magnetism necessary to operate motors and transformers. In a residential load, the apparent power consumed is almost total resistive so the client is only billed for kW. If the utility is required to consume fuel to generate reactive energy (kVAR) then it must be allowed to recover those costs. The power factor measurement allows the utility to determine if the client load consumes too much reactive power in relation to its overall load demand. A typical threshold value is when the load power factor is below 0.85.

Time-of-use

Larger facilities usually purchase electricity from utilities under some form of time-of-use rate schedules. A time-of-use rate schedule basically divides the 24-hour day into three periods of four to twelve hours: On-peak, partial peak (mid-peak) and off-peak. Some rate schedules have winter on-peak periods, some have summer peak periods. This varies regionally due to HVAC loads.

Different rates are charged for each time period. Also maximum demand charges are applied for the on-peak and partial peak periods on a monthly basis.

REAL TIME PRICING (RTP)

The concept behind real time pricing is that the client will be charged for the "real" cost of the energy being consumed based upon the market value for that energy at that point in time. The market value fluctuates as a function of available generation, temperature, weather, etc. While value fluctuates on a daily basis, the value is strongly tied to both season and time of day. Pricing is provided typically on a "day ahead" basis though some contracts use "week ahead" pricing.

Customer Base Line (CBL)—Typically, a client establishing real time pricing with a utility negotiates a customer base line. The CBL defines a "best guess" for the client's consumption for the upcoming year. It is often based upon the client demand for the past year and modified to reflect predicted changes. The CBL is a key point in the contract negotiation and requires significant forethought to optimize cost savings. Once the CBL is defined, RTP prices are a function of marginal daily price variance. Savings can be achieved by moving energy use from periods of high marginal costs to periods of low energy costs.

RTP prices may be set on a day-ahead or week-ahead schedule. RTP prices are developed from daily system cost information and vary depending on such system conditions as weather and demand. For example, electricity is less expensive to produce when hydroelectric power is plentiful, or during times of low demand, such as late at night or on weekends.

Time of day—discounts are allowed for electrical usage during off-peak hours.

Ratchet rate—the billing demand is based on 80 to 90 percent of peak demand for any one month. The billing demand will remain at that ratchet for 12 months even though the actual demand for the succeeding months may be less.

POWER BASICS—THE KEY TO ELECTRICAL ENERGY REDUCTION

By understanding power basics, one can reduce the electrical bill. The total power requirement is comprised of two components, as illustrated in the power triangle, Figure 4-1. This diagram shows the resistive portion or kilowatt (kW), 90° out of phase with the reactive portion, kilovolt ampere reactive (kvar). The reactive current is necessary to build up the flux for the magnetic field of inductive devices, but otherwise it is non-usable. The resistive portion is also known as the active power which is directly converted to useful work. The hypotenuse of the power triangle is referred to as the kilovolt ampere or apparent power (kVa). The angle between kW and kVa is the power factor angle.

$$\text{kW} = \text{kVa} \cos \theta \tag{4-1a}$$

$$\text{kVa} = \text{kW} / \cos \theta \tag{4-1b}$$

$$\text{kvar} = \text{kVa} \sin \theta \tag{4-1c}$$

$$\text{P.F.} = \cos \theta \tag{4-1d}$$

where P.F. is referred to as the power factor.

Note: Only power portions in phase with each other can be combined. For example: resistive portions of one load can be added to resistive portions of another. The same will hold for reactive loads.

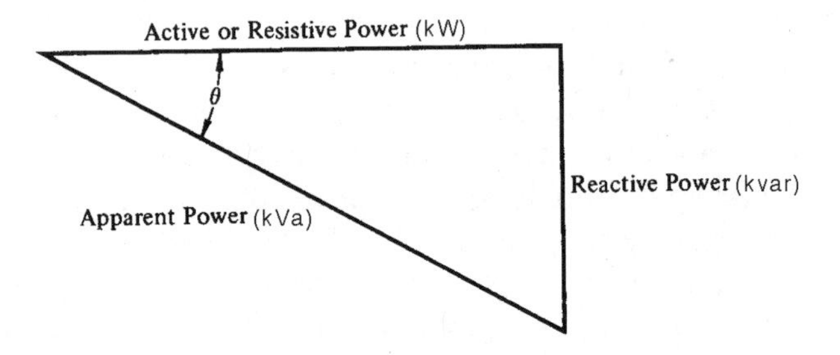

Figure 4-1. The power triangle.

RELATIONSHIPS BETWEEN POWER, VOLTAGE, AND CURRENT

For a balanced 3-phase load,

$$\text{Power watts} = \sqrt{3}\, V_L\, I_L \cos\theta \qquad (4\text{-}2a)$$

or

$$P = 3\, V_{LN}\, I_L \cos\theta$$

For a balanced 1-phase load,

$$\text{Power} = V_{LN}\, I_L \cos\theta \qquad (4\text{-}2b)$$

where

V_L = Voltage between hot legs
V_{LN} = Voltage from hot leg to neutral
I_L = Line current.

MOTOR LOADS

Each electrical load in a system has an inherent power factor. Motor loads are usually specified by horsepower ratings. These may be converted to kVa, by use of Equation 4-3.

$$\text{kVa} = \frac{\text{hp} \times 0.746}{\text{n} \times \text{P.F.}} \tag{4-3}$$

where

n = Motor efficiency
P.F. = Motor power factor
hp = Motor horsepower.

Most motor manufacturers can supply information on motor efficiencies and power factors. Typical values are illustrated in Table 4-1. From this table, it is evident that smaller motors running partly loaded are the least efficient and have the poorest power factor.

Table 4-1. Typical motor horsepowers & efficiencies for 1800 RPM, "T" frame, totally enclosed fan-cooled (TEFC) motors.

	1/2 Load		3/4 Load		Full Load	
Motor hp-Range	*n*	P.F.	*n*	P.F.	*n*	P.F.
3-30	83.3	70.1	85.8	79.2	86.2	83.5
40-100	89.2	79.2	90.7	85.4	90.9	87.7

WHAT ARE THE ADVANTAGES OF POWER FACTOR CORRECTION?

Several advantages that usually offset the cost of correcting the power factor are indicated below:

1 The monthly electric bill is lowered due to the utility company power rate structure.
2 The plant system capacity is increased since the transformer load can be increased.
3 Electrical system losses are decreased and voltage regulation is improved.

SIM 4-1

A plant load is comprised of the following:

50 kW lighting at unity power factor
Ten 30 hp motors running at full load
Two 40 hp motors running at full load.

What is the overall power factor resulting from these loads?

ANSWER

10-30 hp motors

Assume that efficiency is 86.2% at full load and PF is .835

$$n = 0.862 \qquad \cos\theta = 0.835$$

$$\text{PkW} = \frac{\text{hp} \times 0.746}{n} = \frac{10 \times 30 \times 0.746}{0.862} = 259.6 \text{ kW}$$

$$\theta = 33°$$

$$\tan\theta = 0.65$$

$$\text{kvar} = \text{kW} \tan\theta = 259.6 \times 0.65 = 168.7 \text{ kvar}$$

40 hp motors – $\theta = 29°$

Assume that efficiency at full load = 90% and PF = .88

$$\text{PkW} = \frac{2 \times 40 \times 0.746}{0.90} = 66.3 \text{ kW}$$

$$\tan\theta = 0.55$$

$$\text{kvar} = 66.3 \times 0.55 = 36.4$$

Load Total:

$$\text{kW} = 50 + 259.6 + 66.3 = 375.9$$

$$\text{kvar} = 168.7 + 36.4 = 205.1$$

$$\tan\theta = \frac{\text{kvar}}{\text{kW}} = \frac{205.1}{375.9} = .546$$

$$\theta = 28°$$

$$\cos\theta = 0.88 = \text{overall power factor}$$

It should be noted that the lighting load at a P.F. of 1 improves the overall power factor. The composite load is illustrated in Figure 4-2.

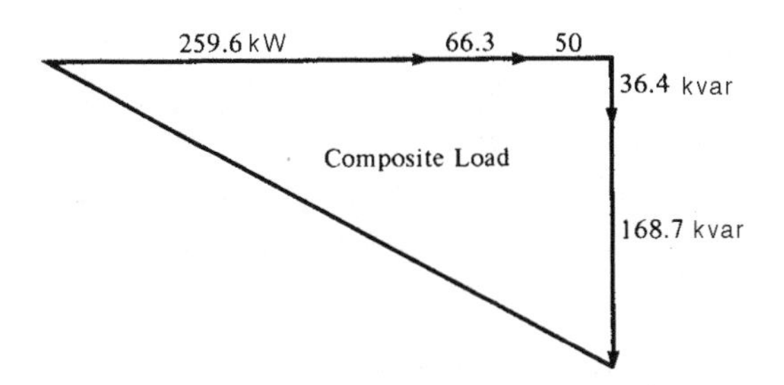

Figure 4-2. The composite load.

HOW TO IMPROVE THE PLANT POWER FACTOR

The plant power factor is improved by:

1. Reducing inefficient loadings; motors running at full load have a significantly better power factor.

2. Providing external capacitors at the motor or at the distribution equipment.

3. Use of energy-efficient motors.

4. Using synchronous motors instead of induction motors. [The main applications of synchronous motors are in plants that require large slow speed motor drives (1,200 rpm and below).]

How Capacitors Improve Power Factor

Capacitors supply the reactive kilovars or magnetizing power required for reactive loads. Thus, the kilovars required from the generating source decreases. This is illustrated in Figure 4-3.

SIM 4-2

Specify the required capacitor kilovars to improve the power factor of SIM 4-2 to 0.95. Indicate the reduction in kVa and line current at 460 volts due to power factor correction.

ANSWER

The kW load of SIM 4-2 is fixed at 375.9 kW. The desired power factor is 0.95 or θ= 18°.

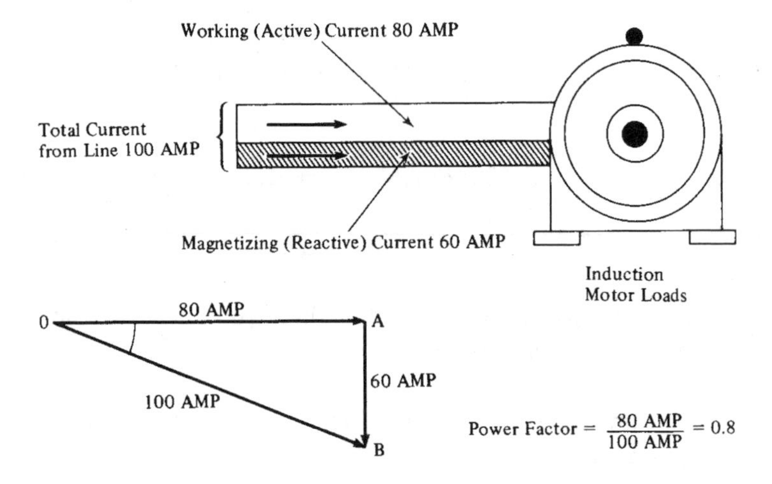

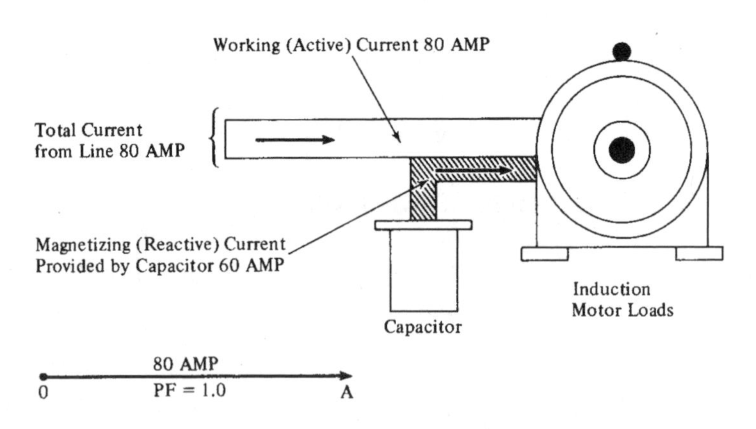

Figure 4-3. The effect of capacitors. (Reprinted by permission of *Specifying Engineer.*)

Thus: kvar = kW tan θ= 375.9 × 0.32 = 120 kvar. The required capacitor kilovars is 205.1 – 120 = 85.1 kvar. The load analysis is illustrated in Figure 4-4.

The kVa is reduced from: $kVa_1 = kW / \cos\theta = 375.9 / 0.88 = 427.1$ kVa, to: $kVa_2 = 375.9 / 0.95 = 395.6$. To calculate the line current use Eq. 4-2a.

Thus,

I (Before) $= 427.1 \times 10^3 / \sqrt{3} \times 460 = 536$ amps

I (After) $= 395.6 \times 10^3 / \sqrt{3} \times 460 = 497$ amps.

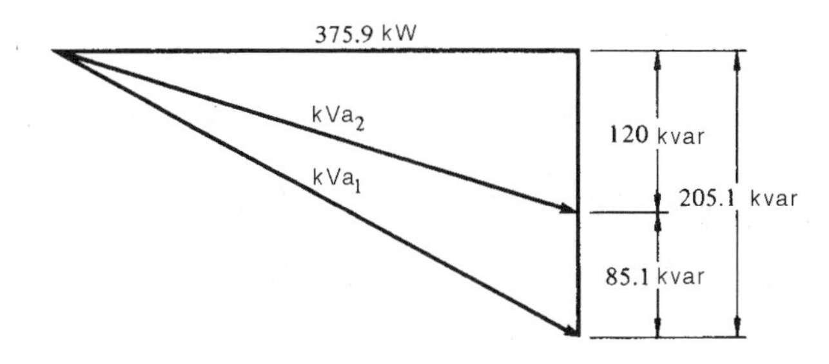

Figure 4-4. The effect of power factor correction on power triangle.

Shortcut Methods

A handy shortcut table which can be used to find the value of the capacitor required to improve the plant power factor is illustrated by Table 4-2. Figure 4-5 illustrates the additional capacity available when capacitors are used.

WHERE TO LOCATE CAPACITORS

As indicated, the primary purpose of capacitors is to reduce the power consumption. Additional benefits are derived by capacitor location. Figure 4-6 indicates typical capacitor locations. Maximum benefit of capacitors is derived by locating them as close as possible to the load. At this location, its kilovars are confined to the smallest possible segment, decreasing the load current. This, in turn, will reduce power losses of the system substantially. Power losses are proportional to the square of the current. When power losses are reduced, voltage at the motor increases; thus, motor performance may also increase.

Table 4-2. Shortcut method—power factor correction.

kW Multipliers for Determining Capacitor Kilovars

Desired Power Factor in Percentage

Original Power Factor in Percentage	80	81	82	83	84	85	86	87	88	89	90	91	92	93	94	95	96	97	98	99	100
50	.982	1.008	1.034	1.060	1.086	1.112	1.139	1.165	1.192	1.220	1.248	1.276	1.303	1.337	1.369	1.403	1.441	1.481	1.529	1.590	1.732
51	.936	.962	.988	1.014	1.040	1.066	1.093	1.119	1.146	1.174	1.202	1.230	1.257	1.291	1.323	1.357	1.395	1.435	1.483	1.544	1.688
52	.894	.920	.946	.972	.998	1.024	1.051	1.077	1.104	1.132	1.160	1.188	1.215	1.249	1.281	1.315	1.353	1.393	1.441	1.502	1.644
53	.850	.876	.902	.928	.954	.980	1.007	1.033	1.060	1.088	1.116	1.144	1.171	1.205	1.237	1.271	1.309	1.349	1.397	1.458	1.600
54	.809	.835	.861	.887	.913	.939	.966	.992	1.019	1.047	1.075	1.103	1.130	1.164	1.196	1.230	1.268	1.308	1.356	1.417	1.559
55	.769	.795	.821	.847	.873	.899	.926	.952	.979	1.007	1.035	1.063	1.090	1.124	1.156	1.190	1.228	1.268	1.316	1.377	1.519
56	.730	.756	.782	.808	.834	.860	.887	.913	.940	.968	.996	1.024	1.051	1.085	1.117	1.151	1.189	1.229	1.277	1.338	1.480
57	.692	.718	.744	.770	.796	.822	.849	.875	.902	.930	.958	.986	1.013	1.047	1.079	1.113	1.151	1.191	1.239	1.300	1.442
58	.655	.681	.707	.733	.759	.785	.812	.838	.865	.893	.921	.949	.976	1.010	1.042	1.076	1.114	1.154	1.202	1.263	1.405
59	.618	.644	.670	.696	.722	.748	.775	.801	.828	.856	.884	.912	.939	.973	1.005	1.039	1.077	1.117	1.165	1.226	1.368
60	.584	.610	.636	.662	.688	.714	.741	.767	.794	.822	.849	.878	.905	.939	.971	1.005	1.043	1.083	1.131	1.192	1.334
61	.549	.575	.601	.627	.653	.679	.706	.732	.759	.787	.815	.843	.870	.904	.936	.970	1.008	1.048	1.096	1.157	1.299
62	.515	.541	.567	.593	.619	.645	.672	.698	.725	.753	.781	.809	.836	.870	.902	.936	.974	1.014	1.062	1.123	1.265
63	.483	.509	.535	.561	.587	.613	.640	.666	.693	.721	.749	.777	.804	.838	.870	.904	.942	.982	1.030	1.091	1.233
64	.450	.476	.502	.528	.554	.580	.607	.633	.660	.688	.716	.744	.771	.805	.837	.871	.909	.949	.997	1.058	1.200
65	.419	.445	.471	.497	.523	.549	.576	.602	.629	.657	.685	.713	.740	.774	.806	.840	.878	.918	.966	1.027	1.169
66	.388	.414	.440	.466	.492	.518	.545	.571	.598	.626	.654	.682	.709	.743	.775	.809	.847	.887	.935	.996	1.138
67	.358	.384	.410	.436	.462	.488	.515	.541	.568	.596	.624	.652	.679	.713	.745	.779	.817	.857	.905	.966	1.108
68	.329	.355	.381	.407	.433	.459	.486	.512	.539	.567	.595	.623	.650	.684	.716	.750	.788	.828	.876	.937	1.079
69	.299	.325	.351	.377	.403	.429	.456	.482	.509	.537	.565	.593	.620	.654	.686	.720	.758	.798	.840	.907	1.049
70	.270	.296	.322	.348	.374	.400	.427	.453	.480	.508	.536	.564	.591	.625	.657	.691	.729	.769	.811	.878	1.020
71	.242	.268	.294	.320	.346	.372	.399	.425	.452	.480	.508	.536	.563	.597	.629	.683	.701	.741	.783	.850	.992
72	.213	.239	.265	.291	.317	.343	.370	.396	.423	.451	.479	.507	.534	.568	.600	.634	.672	.712	.754	.821	.963
73	.186	.212	.238	.264	.290	.316	.343	.369	.396	.424	.452	.480	.507	.541	.573	.607	.645	.685	.727	.794	.936
74	.159	.185	.211	237	.263	.289	.316	.342	.369	.397	.425	.453	.480	.514	.546	.580	.618	.658	.700	.767	.909
75	.132	.158	.184	.210	.236	.262	.289	.315	.342	.370	.398	.426	.453	.487	.519	.553	.591	.631	.673	.740	.882
76	.105	.131	.157	.183	.209	.235	.262	.288	.315	.343	.371	.399	.426	.460	.492	.526	.564	.604	.652	.713	.855
77	.079	.105	.131	.157	.183	.209	.236	.262	.289	.317	.345	.373	.400	.434	.466	.500	.538	.578	.620	.687	.829
78	.053	.079	.105	.131	.157	.183	.210	.236	.263	.291	.319	.347	.374	.408	.440	.474	.512	.552	.594	.661	.803
79	.026	.052	.078	.104	.130	.156	.183	.209	.236	.264	.292	.320	.347	.381	.413	.447	.485	.525	.567	.634	.776
80	.000	026	.052	.078	.104	.130	.157	.183	.210	.238	.266	.294	.321	.355	.387	.421	.450	.499	.541	.608	.750
81	—	.000	.026	.052	.078	.104	.131	.157	.184	.212	.240	.268	.295	.329	.361	.395	.433	.473	.515	.582	.724
82	—	—	.000	.026	.052	.078	.105	.131	.158	.186	.214	.242	.269	.303	.335	.369	.407	.447	.489	.556	.698
83	—	—	—	.000	.026	.052	.079	.105	.132	.160	.188	.216	.243	.277	.309	.343	.381	.421	.463	.530	.672
84	—	—	—	—	.000	.026	.053	.079	.106	.134	.162	.190	.217	.251	.283	.317	.355	.395	.437	.504	.645
85	—	—	—	—	—	.000	.027	.053	.080	.108	.136	.164	.191	.225	.257	.291	.329	.369	.417	.478	.620

Example: Total kW input of load from wattmeter reading 100 kW at a power factor of 60%. The leading reactive kvar necessary to raise the power factor to 90% is found by multiplying the 100 kW by the factor found in the table, which is .849. Then 100 kW × 0.849 = 84.9 kvar. Use 85 kvar.

Reprinted by permission of Federal Pacific Electric Company.

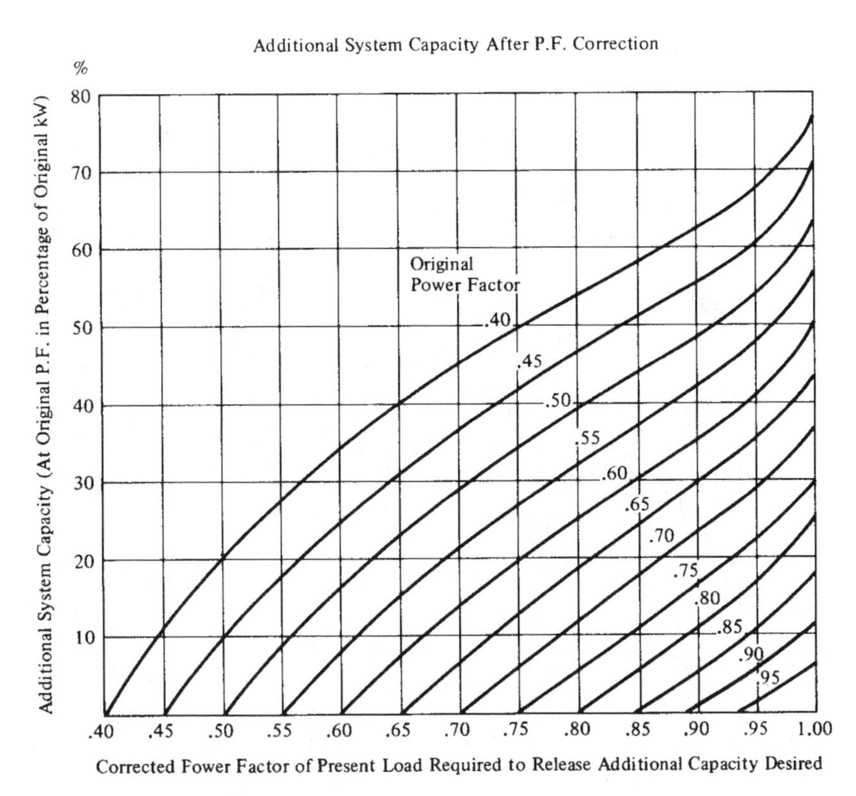

Example: Ten percent of additional capacity is required; the plant power factor is 0.85. What value should the Power Factor be corrected to in order to release the additional capacity?

Answer: From the Figure above, the Power Factor should be corrected to 0.95.

Figure 4-5. Shortcut method to determine released capacity by power factor correction. Adapted from *Specifying Engineer*.

Locations C1A, C1B and C1C of Figure 4-6 indicate three different arrangements at the load. Note that in all three locations, extra switches are not required, since the capacitor is either switched with the motor starter or the breaker before the starter. Case C1A is recommended for new installation, since the maximum benefit is derived and the size of the motor thermal protector is reduced. In Case C1B, as in Case ClA, the capacitor is energized only when the motor is in operation. Case C1B is recommended in cases where the installation is existing and the thermal protector does

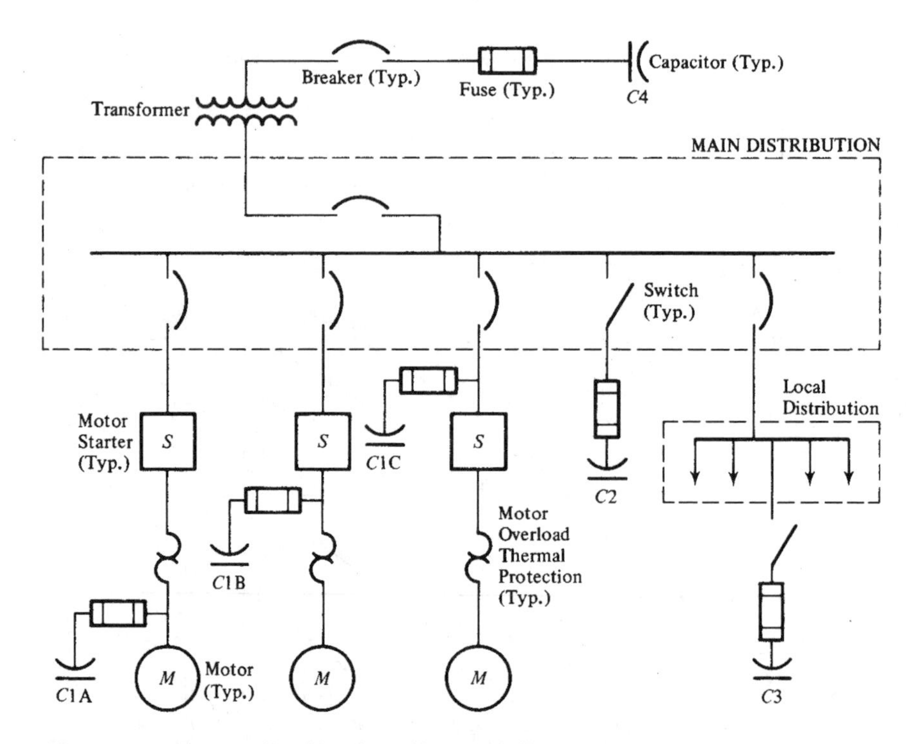

Figure 4-6. Power distribution diagram illustrating capacitor locations.

not need to be re-sized. In position C1C, the capacitor is permanently connected to the circuit, but does not require a separate switch, since it can be disconnected by the breaker before the starter.

It should be noted that the rating of the capacitor should *not* be greater than the no-load magnetizing kvar of the motor. If this condition exists, damaging over-voltage or transient torques can occur. This is why most motor manufacturers specify maximum capacitor ratings to be applied to specific motors.

The next preference for capacitor locations as illustrated by Figure 4-6 is at locations C_2 and C_3. In these locations, a breaker or switch will be required. Location C_4 requires a high voltage breaker. The advantage of locating capacitors at power centers or feeders is that they can be grouped together. When several motors are running intermittently, the capacitors are permitted to be on line all the time, reducing the total power regardless of load. Figures 4-7 and 4-8 illustrate typical capacitor installations.

Figure 4-7. Installation of capacitors in central area. Machine shop and welding plant. 75 kvar installed capacitors.

EFFICIENT MOTORS

Another method to improve the plant or building power factor is to use energy efficient motors. Energy efficient motors are available from several manufacturers. Energy efficient motors are slightly more expensive than their standard counterpart. Based on the energy cost it can be determined if the added investment is justified. With the emphasis on energy conservation, new lines of energy efficient motors are being introduced. Figures 4-9 and 4-10 illustrate a typical comparison between energy efficient and standard motors.

SYNCHRONOUS MOTORS AND POWER FACTOR CORRECTION

Synchronous motors find applications when constant speed operation is essential. A synchronous motor, unlike its induction

Figure 4-8. Installation of capacitors at motor. Blacktop and gravel plant. 45 kvar installed capacitors.

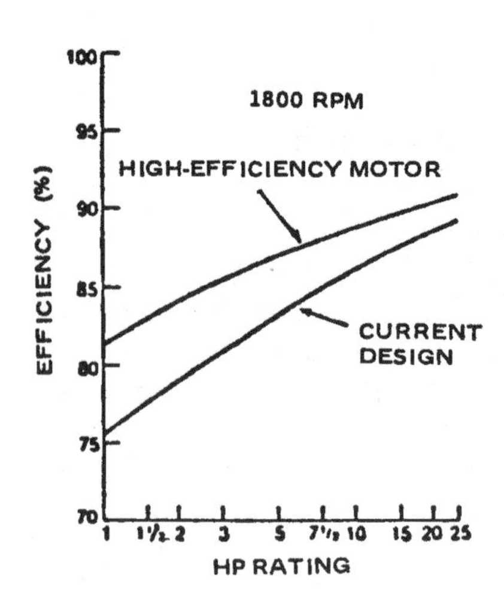

Figure 4-9. Efficiency vs. horsepower rating (drip-proof motors).

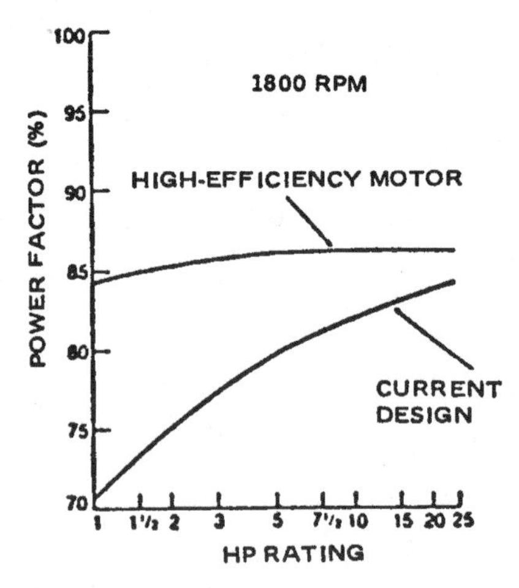

Figure 4-10. Power factor vs. horsepower rating (drip-proof motors).

motor counterpart, requires D.C. power as well as A.C. power. Many synchronous motors are self-excited; thus, the A.C. power to the motor is the only requirement. The D.C. for the field windings is generated intrinsic to the motor. Synchronous motors in ratings above 300 hp and speed below 1200 rpm are often cheaper than induction motors. Another factor for synchronous motor selection is that power factor is improved with their use.

Figure 4-11 illustrates the use of different synchronous motors. The 0.8 power factor synchronous motor delivers leading kilovars similar to that of a capacitor, while the unity power factor synchronous motor only delivers leading kilovars when operating at reduced loads. Both types of synchronous motors improve plant power factor.

What Method Should be Used to Improve the Plant Power Factor?

Today plant engineers/managers must improve the overall power quality of their facility. Poor application of capacitors such as installation with a variable speed drive could cause harmonic distortions. Historically, plants with poor power factors and utility rates with a power factor penalty have chosen the installation of capacitors to correct the problem.

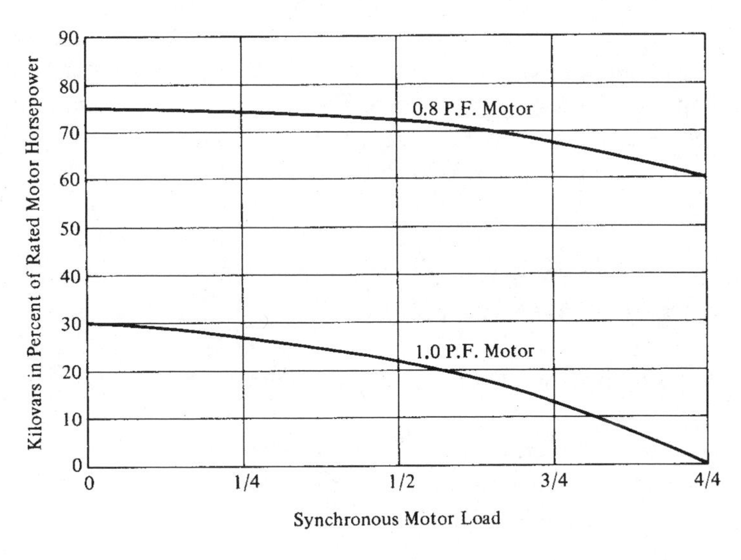

Figure 4-11. Kilovars supplied by synchronous motors.

In today's market, replacing older, standard efficient motors with carefully selected high-efficient, high power factor motors is a better alternative to power factor correction for small-sized motors (5-20 hp). This alternative usually results in a better rate of return on investment and does not impact negatively on power quality. One exception is small motors with a very low load factor. Capacitors in this case are usually more economical.

Correcting the power factor of a larger motor (20 hp or greater) is still usually accomplished with a capacitor. The economics of power factor correction indicate that correcting beyond 0.95 is seldom justified. Since power factor correction using capacitors usually has a payback period of 3 years or less, it should have a high priority in the overall program.

Power factor correction is basic to electrical design. It has been rediscovered due to increased energy cost. This is one example of rediscovery; if you look hard enough you will find many more.

WHAT IS LOAD MANAGEMENT?

Load management is an umbrella term that describes the methods and technologies a utility can use to control the timing and peak of customer power use. Its objective is to reduce the demand for electricity during peak use periods and increase the demand during off-peak periods. The utility rate structure is the vehicle used to meet this objective. The utility rate structure penalizes a customer for peak power demands. Other structures have penalties to discourage the use of electricity during certain times of the day. Load management enables the utility company to use its power generating equipment more efficiently.

WHAT HAVE BEEN SOME OF THE RESULTS OF LOAD MANAGEMENT?

Some of the results of load management programs are as follows:

1 A large glass manufacturer in Tipton, Pennsylvania, rescheduled the start-up of its tempering ovens to night-time hours—*Result:* Peak demand reduced by 1,250 kilowatts.

2 A Flemington, New Jersey, plant rescheduled the time for recharging

lift truck batteries—*Result:* Peak daytime demand reduced by 100 kilowatts.

3 A tool manufacturing company in Meadville, Pennsylvania, installed demand-limiting equipment and started using waste heat from some of its process operations to supplement heat to its building—*Result:* Overall power demand dropped by 1,000 kilowatts.

4 In Hamburg, Pennsylvania, a metal coating company rescheduled furnace loading from its first shift to its second shift—*Result:* Demand reduced by 1000 kilowatts during peak use periods. Expected savings are $35,000 to $40,000 per year.

From the above, two load management techniques can be seen. One is to reschedule energy-related activities to non-peak hour times; the other is to automatically shed loads by use of a load demand controller or computer. In either case, the first step is to make a load assessment. A load assessment indicates the electrical energy characteristics for the billing period. The electric characteristics are compared with plant operations to determine what has caused the peak power usages.

APPLICATION OF AUTOMATIC LOAD SHEDDING

Load shedding should be considered when power usage demands fluctuate substantially and load leveling is feasible because of substantial non-essential or controllable loads. Load shedding has been used widely in the steel industry, but the principles of load shedding can be applied to any large industrial or commercial user. The first step in applying load shedding is to establish a target demand. The target demand is based on actual load readings or on a load analysis. The second step is to identify controllable loads which can be shut off to obtain the desired limit. Examples of controllable loads are electric furnaces, electric boilers, compressors, snow melters, air conditioners, heating and ventilating fans, comfort cooling, and noncritical "batch processes." Ideally, if controllable loads matched intermittent peak uses, the plant under load demand control would use a preset amount of kilowatts in a fixed amount of time. Figure 4-12 shows the effect of load shedding on peak power demands.

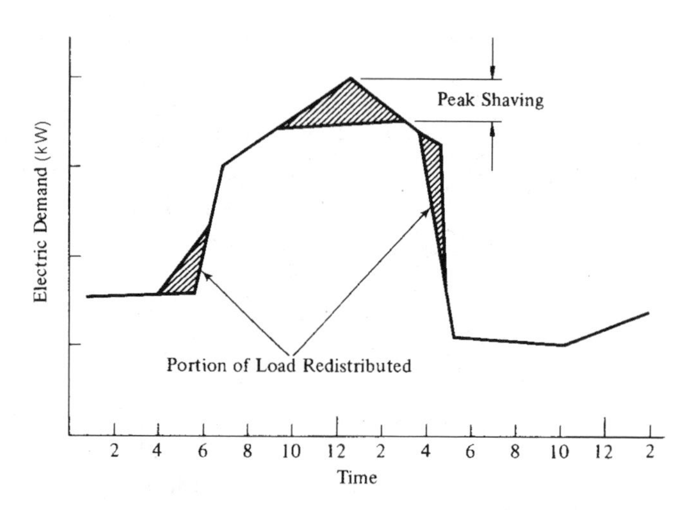

Figure 4-12. Load shedding.

HOW DOES LOAD DEMAND CONTROL WORK?

Figure 4-13 is the block diagram of a demand control system. The demand controller is essentially a computer. It compares the consumers' actual rate of energy consumption to a predetermined ideal rate of energy consumption, during any demand interval. Let's look at each aspect.

Inputs

The same metering which the utility company uses for billing is used to supply information to the Load Demand control. The Watt Hour Meter supplies information on the kilowatt hours used. The information supplied is in the form of pulses. The Demand Meter supplies information on the end of the demand interval. This period of time is usually 15, 30, or 60 minutes.

Logic

The logic elements compare input data to a predetermined ideal rate. Signals to shed load are activated when the actual usage rate indicates that the present demand will be exceeded. Signals to restore loads are activated whenever a new demand is started. Loads are shed within the last few minutes of the demand interval in order to avoid unnecessary control action. Refer to Figure 4-14.

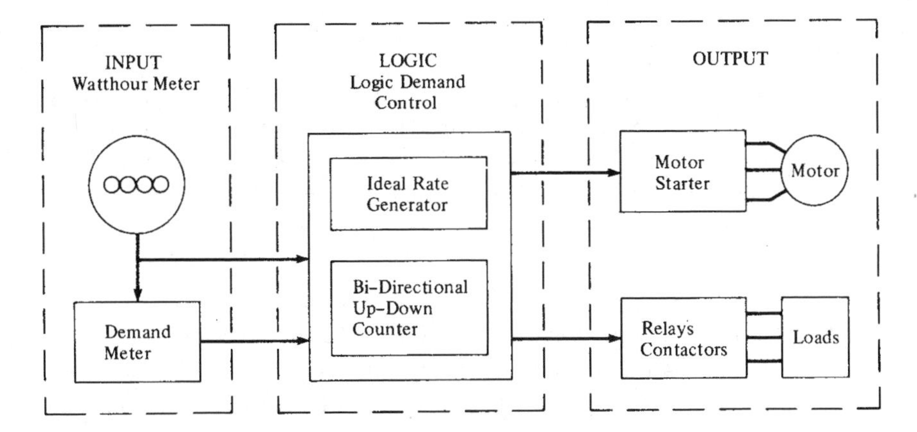

Figure 4-13. Block diagram of load demand controller. (Reprinted by permission of *Electrical Consultant Magazine.*)

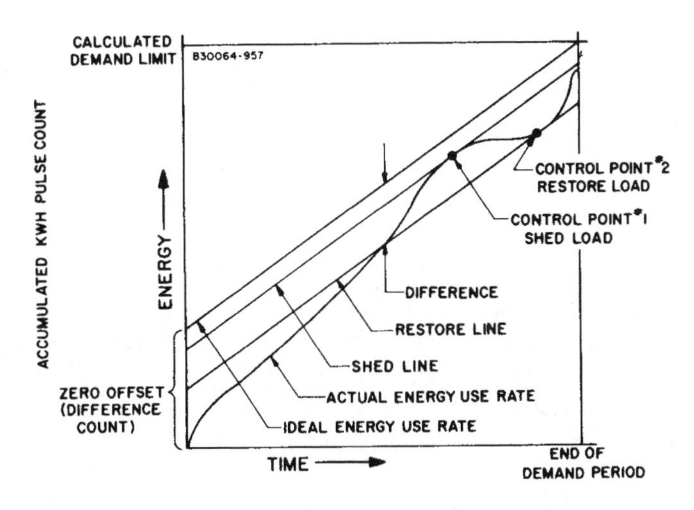

Figure 4-14. Control curves for load demand controller. (Reprinted by permission of Square D Company.)

Outputs

Signals from the logic elements activate relays, contractors, or motor starters to shed or restore loads.

Caution

In order to accomplish load shedding, be careful to avoid cycling of equipment for short periods of time, i.e., five minutes. For example, turning motors on and off lowers the life of the equipment and could severely damage it. Always check the specifications of the controllables to insure that many starts in a short period of time do not damage the equipment.

Is a Programmable Logic Controller (PLC) Required to Shed Loads?

Before buying a standard PLC to automatically shed loads, consider some of the packaged solid state load demand controllers. These controllers are relatively inexpensive and several models are available to meet most needs. If PLC is already available, then programmable control of load shedding becomes more attractive. In either case, an analysis of specific electrical usage should be made by an independent consultant in order to determine which system is best.

THE CONFUSION OVER ENERGY MANAGEMENT SYSTEMS

With the myriad electrical management systems on the market, it is no wonder why the energy manager is confused. First, there is no standardization of equipment specifications. Second, the term *energy management* system itself is confusing.

An energy management system could be one or all of the following:

1 Simple time clock.

2 Simple duty cycle device.

3 Peak load shedding device.

4 Heating, ventilation, and air conditioning controllers (enthalpy controller, temperature switch, etc.).

The main reason the controller is specified in the first place is to reduce peak electrical demand. Once the controller is constructed for that purpose, features such as a time clock and duty cycle control are economically added.

The terminology usually changes from a simple peak load shedding device to an energy management system when the heating and ventilation control features are added.

The size of the unit to be purchased depends on the number of loads that must be cycled to reduce peak demand.

LIGHTING BASICS—THE KEY TO REDUCING LIGHTING WASTES

By understanding the basics of lighting design, several ways to improve the efficiency of lighting systems will become apparent.

There are two common lighting methods used: One is called the "Lumen" method, while the other is the "Point by Point" method. The Lumen method assumes an equal footcandle level throughout the area. This method is used frequently by lighting designers since it is simplest; however, it wastes energy, since it is the light "at the task" which must be maintained and not the light in the surrounding areas. The "Point by Point" method calculates the lighting requirements for the task in question.

The methods are illustrated by Eqs. 4-4, 4-5, 4-6, and 4-7.

Lumen Method

$$N = \frac{F_1 \times A}{Lu \times L_1 \times L_2 \times Cu} \tag{4-4}$$

where

N is the number of lamps required.

F_1 is the required foot-candle level at the task. A foot-candle is a measure of illumination: one standard candle power measured one foot away.

A is the area of the room in square feet.

Lu is the Lumen output per lamp. A Lumen is a measure of lamp intensity: its value is found in the manufacturer's catalogue.

Cu is the coefficient of utilization. It represents the ratio of the Lumens reaching the working plane to the total Lumens generated by the lamp. The coefficient of utilization makes allowances for light absorbed or reflected by walls, ceilings, and the fixture itself. Its values are found in the manufacturer's catalogue.

L_1 is the lamp depreciation factor. It takes into account that the lamp Lumen depreciates with time. Its value is found in the manufacturer's catalogue.

L_2 is the luminaire (fixture) dirt depreciation factor. It takes into account the effect of dirt on a luminaire, and varies with type of luminaire and the atmosphere in which it is operated.

Point by Point Methods

There are three commonly used lighting formulas associated with this method. Eq. 4-5 is used for infinite length illumination sources, such as a row of non-louvered industrial fixtures. Eqs. 4-6 and 4-7 are used for point sources. An compact fluorescent lamp or mercury vapor luminaire is treated as a point source. These equations omit inter reflections; thus, the total measured illumination will be greater than the calculated values. Inter reflections can be taken into account by referring to a lighting handbook.

$$F_2 = \frac{0.35 \times CP}{D} \qquad (4\text{-}5)$$

$$F_3 = \frac{CP \times \cos\theta}{D^2} \qquad (4\text{-}6)$$

$$F_4 = \frac{CP \times \sin\theta}{D^2} \qquad (4\text{-}7)$$

where

F_2 is the illumination produced at a point on a plane directly parallel to and directly under the source.

F_3 is the illumination on a horizontal plane.

F_4 is the illumination on a vertical plane.

CP *is the candle power of the source in the particular direction. Its value is found in a manufacturer's catalogue.*

D is the distance in feet to the point of illumination.

θ is the angle between "D" and the direct component. Refer to Figure 4-15.

Analysis of Methods

From the two methods of computing lighting requirements, the following conclusions are made:

- The "point by point" method should be used with "at the task" lighting levels.
- Efficient lamps with high lumen and candlepower output should be used.
- Luminaires (fixtures) should be chosen based on a high coefficient of utilization for the application.
- Luminaires should be chosen on the basis of the environment; i.e., in a dirty environment, luminaires should prevent dust buildup.

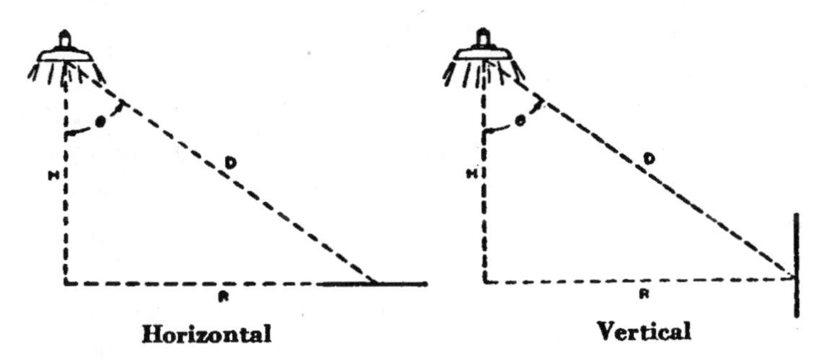

Figure 4-15. Point-by-point method of illumination.

- Lamps with good lamp lumen depreciation characteristics should be used.

LIGHTING ILLUMINATION REQUIREMENTS

The levels of illumination specified by the Illuminating Engineering Society (IES) in the *IES Lighting Handbook* are recommended levels at the *Task Surface*. For years, lighting designers have been using IES task lighting values as the criteria for the total space. An option available to lighting designers is to use lighting levels below Task values for areas surrounding the task location. The level of the surrounding area should be no more than 1/3 of the weighted average of the foot-candle levels for task areas with a 20 foot-candle minimum. (10 foot-candle minimum can be used for non-critical general areas.)

THE EFFICIENT USE OF LAMPS

Figure 4-16 illustrates lumen outputs of various types of lamps.

EFFICIENT TYPES OF INCANDESCENTS FOR LIMITED USE

Attempts to increase the efficiency of incandescent lighting while maintaining good color rendition have led to the manufacture of a number of energy-saving incandescent lamps for limited residential use.

Tungsten Halogen—These lamps vary from the standard incandescent by the addition of halogen gases to the bulb. Halogen gases keep the glass bulb from darkening by preventing the filament from evaporating, and thereby increase lifetime up to four times that of a standard bulb. The lumen-per-watt rating is approximately the same for both types of incandescents, but tungsten halogen lamps average 94% efficiency throughout their extended lifetime, offering significant energy and operating cost savings. However, tungsten halogen lamps require special fixtures, and during operation the surface of the bulb reaches very high temperatures, so they are not commonly used in the home.

Reflector or R-Lamps—Reflector lamps (R-lamps) are incandescents with an interior coating of aluminum that directs the light to the front of the bulb. Certain incandescent light fixtures, such as recessed or directional fixtures, trap light inside. Reflector lamps project a cone of light out of the

Lighting Comparison Chart

Lighting Type	*Efficacy (lumens/watt)*	*Lifetime (hours)*	*Color Rendition Index (CRI)*	*Color Temperature (K)*	*Indoors/ Outdoors*
Incandescent					
Standard "A" bulb	10-17	750-2500	98-100 (excellent)	2700-2800 (warm)	Indoors/ Outdoors
Tungsten halogen	12-22	2000-4000	98-100 (excellent)	2900-3200 (warm to neutral)	Indoors/ Outdoors
Reflector	12-19	2000-3000	98-100 (excellent)	2800 (warm)	Indoors/ Outdoors
Fluorescent					
Straight tube	30-110	7000-24,000	50-90 (fair to to good)	2700-6500 (warm to to cold	Indoors/ Outdoors
Compact fluorescent lamp (CFL)	50-70	10,000	65-88 (good)	2700-6500 (warm to cold)	Indoors/ Outdoors
Circline	40-50	12,000			Indoors
High-Intensity Discharge					
Mercury vapor	25-60	16,000-24,000	50 (poor to fair)	3200-7000 (warm to cold)	Indoors/ Outdoors
Metal halide	70-115	5000-20,000	70 (fair)	3700 (cold)	Indoors/ Outdoors
High-pressure Sodium	50-140	16,000-24,000	25 (poor)	2100 (warm)	Outdoors
Low-Pressure Sodium	60-150	12,000-18,000	-44 (very poor)		Outdoors

Figure 4-16. Efficiency of various light sources.

fixture and into the room, so that more light is delivered where it is needed. In these fixtures, a 50-watt reflector bulb will provide better lighting and use less energy when substituted for a 100-watt standard incandescent bulb.

Reflector lamps may be appropriate for task lighting, because they directly illuminate a work area, and for accent lighting. Reflector lamps are manufactured in sizes from 30 to 1,500 watts and in various light distributions. While they have a lower initial efficiency (lumens per watt) than regular incandescents, they direct light more effectively, so that more light is actually delivered than with regular incandescents.

PAR Lamps—Parabolic aluminized reflector (PAR) lamps are reflector lamps with a lens of heavy, durable glass, which makes them an appropriate choice for outdoor flood and spot lighting. They are available in 75, 150, and 250 watts. They have longer lifetimes with less depreciation than standard incandescents.

ER Lamps—Ellipsoidal reflector (ER) lamps are ideally suited for recessed fixtures, because the beam of light produced is focused two inches ahead of the lamp to reduce the amount of light trapped in the fixture. In a directional fixture, a 75-watt ellipsoidal reflector lamp delivers more light than a 150-watt R-lamp.

Fluorescent Lamps

Fluorescent lamps have made dramatic advances in the last 20 years. From the introduction of reduced wattage lamps in the mid-1970s, to the marketing of several styles of low wattage, compact lamps recently, there has been a steady parade of new products. The range of colors is more complete than mercury vapor and lamp manufacturers have recently made significant progress in developing fluorescent and metal halide lamps which have much more consistent color rendering properties allowing greater flexibility in mixing these two sources without creating disturbing color mismatches. The wide range of compact fluorescent lamps open up a whole new market for fluorescent sources. These lamps permit design of much smaller luminaires which can compete with incandescent and mercury vapor in the low cost, square or round fixture market which the incandescent and mercury sources have dominated for so long.

Energy Efficient Fluorescents

The energy efficient fluorescents represent the second generation of improved fluorescent lighting.

Typical energy savings are about 6 W per lamp for the popular 4-ft 40 W replacement and 15 W per lamp for the popular 8-ft slimline 75 W replacement. Savings in energy cost normally pay back the new lamp cost in a year at typical power rates and lamp costs.

While switching from T-12 standard fixtures to T-8 bulbs with electronic ballasts creates savings in the office environment, T-5 fluorescent bulbs offer a competitive solution with high bay applications. For example, a 6 bulb, T-5 high output fixture can yield 25% energy savings over a 400

Watt metal halide fixture. Another benefit is the T-5 fixtures do not have a long warm-up or re-strike delay time associated with them.

Light Emitting Diode (LED) Fixtures

Retailers have found LED lighting offers a myriad of light color options while creating significant energy savings over conventional display lights. Prior to the development of LED lighting, ambient lighting could only be provided by specialized, expensive, incandescent lights. LED lamps offer significant energy savings and much longer life than their incandescent counterparts. They work better in environments where vibrations are present because they are not as fragile since they have no filament. Their color rendering index is superior to compact fluorescent lamps and their light is much more directed improving transfer efficiency.

Compact Fluorescent Lamps (CFL)

A recently developed arc-discharge lamp, called compact fluorescent lamp, can replace an incandescent light source in particular applications. Three configurations are possible for the installation of compact fluorescent lamps: dedicated, self-ballasted, and modular. Dedicated compact fluorescent lamp systems are similar to full-size fluorescent lighting systems in which a ballast is hard-wired to lamp holders within a luminaire. Self-ballasted and modular compact fluorescent lamp products have screwbases designed for installation in medium screw base sockets; they typically replace incandescent lamps. A self-ballasted compact fluorescent lamp contains a lamp and ballast as an inseparable unit. A modular compact fluorescent lamp product consists of a screwbase ballast with a replaceable lamp. The ballast and lamp connect together using a socket-and-base design that ensures compatibility of lamps and ballasts. While most of the modular types are operated in the preheat mode, the electronic ballasted lamps are operated in the rapid start mode and in principle could be dimmed.

High Intensity Discharge (HID) Lamps

High intensity discharge lamps are electric discharge sources. The basic difference from fluorescent lamps is that HID lamps operate at a much higher arc pressure. Spectral characteristics differ from those of fluorescent lamps because the higher pressure arc emits a large portion of its visible light. HID lamps produce full light output only at full operating pressure usually several minutes after starting. Most HID lamps contain both an inner and an outer bulb. The inner bulb is made of quartz or polycrystalline aluminum; the outer bulb is generally made of thermal shock-resistant glass. HID lamps require current limiting devices, which consume 10% to 20% additional watts. HID lamps include mercury, metal halide, high-pressure sodium, and low-pressure sodium lamps:

Mercury Lamps

These are low in efficacy compared to other HID sources, and are obsolescent for most industrial lighting applications. They are available with either "clear" or phosphor-coated bulbs of 40 to 1,000 W, and in various sizes and shapes. Typical efficacy ranges from 30 to 63 lumens per watt, not including ballast loss. "Clear" mercury lamps produce light rich in yellow and green tones while lacking in red. Phosphor-coated lamps provide improved color. Special types include semi-reflector, reflectorized, and self-ballasted lamps.

Metal Halide (MH) Lamps

These are similar in construction to mercury lamps. The difference is in the arc tube, which contains various metal halides in addition to mercury. They are available in either clear or phosphor-coated bulbs from 32 to 1,500 W. Present efficacies range from 70 to 125 lumens per watt, not including ballast power loss. Color improvement is achieved by the metal halide additives.

Metal-halide lamps generally have fairly good color rendering qualities. While this lamp displays some very desirable qualities, it also has some distinct drawbacks including relatively short life for an HID lamp, long restrike time to restart after the lamp has been shut off (about 15-20 minutes at 70°F) and a pronounced tendency to shift colors as the lamp ages.

High-pressure Sodium (HPS) Lamps

Light is produced by electricity passing through sodium vapor. They are presently available in sizes of 35 to 1,000 W. Typical initial efficacies are about twice that of mercury lamps: from 80 to 140 lumens per watt, not including ballast power loss. Normally with clear outer envelopes, they may also be obtained with coatings that improve diffusion. The color of light produced is golden white.

The lamp's primary drawback is the rendering of some colors. The lamp produces a high percentage of light in the yellow range of the spectrum. This tends to accentuate colors in the yellow region.

Rendering of reds and greens shows a pronounced color shift. In areas where color selection, matching and discrimination are necessary, high pressure sodium should not be used as the only source of light. It is possible to gain quite satisfactory color rendering by mixing high pressure sodium and metal halide in the proper proportions. Since both sources have relatively high efficacies, there is not a significant loss in energy efficiency by making this compromise.

Recently lamp manufacturers have introduced high pressure sodium lamps with improved color rendering qualities. However, as with most things in this world, the improvement in color rendering was not gained without cost—the efficacy of the color-improved lamps is somewhat lower, approximately 90 lumens per watt.

Low-pressure Sodium Lamps

Low-pressure sodium lamps provide the highest efficacy of any of the sources for general lighting with values ranging up to 180 lumens per watt. Low pressure sodium produces an almost pure yellow light with very high efficacy and renders all colors gray except yellow or near yellow. The effect of this is there can be no color discrimination under low pressure sodium lighting and it is suitable for use in a very limited number of applications. It is an acceptable source for warehouse lighting where it is only necessary to read labels but not to choose items by color. This source has application for either indoor or outdoor safety or security lighting, again as long as color rendering is not important.

Control Equipment

Table 4-3 lists various types of equipment that can be components of a lighting control system, with a description of the predominant characteristic of each type of equipment. Static equipment can alter light levels semipermanently. Dynamic equipment can alter light levels automatically over short intervals to correspond to the activities in a space. Different sets of components can be used to form various lighting control systems in order to accomplish different combinations of control strategies.

Solid-state Ballasts

After more than 10 years of development and 5 years of manufacturing experience, operating fluorescent lamps at high frequency

Table 4-3. Lighting control equipment.

System	Remarks
STATIC:	
Delamping	Method for reducing light level 50%.
Impedance Monitors	Method for reducing light level 30, 50%.
DYNAMIC:	
Light Controllers	
Switches/Relays	Method for on-off switching of large banks of lamps.
Voltage/Phase Control	Method for controlling light level continuously 100 to 50%.
Solid-State Dimming Ballasts	Ballasts that operate fluorescent lamps efficiently and can dim them continuously (100 to 10%) with low voltage.
SENSORS:	
Clocks	System to regulate the illumination distribution as a function of time.
Personnel	Sensor that detects whether a space is occupied by sensing the motion of an occupant.
Photocell	Sensor that measures illumination level of a designated area.
COMMUNICATION:	
Computer/Micro-processor	Method for automatically communicating instructions and/or input from sensors to commands to the light controllers.
Power-Line Carrier	Method for carrying information over existing power lines rather than dedicated hard-wired communication lines.

(20 to 30 kHz) with solid-state ballasts has achieved credibility. The fact that all of the major ballast manufacturers offer solid-state ballasts and the major lamp companies have designed new lamps to be operated at high frequency is evidence that the solid-state high frequency ballast is now state-of-the-art.

It has been shown that fluorescent lamps operated at high frequency are 10 to 15% more efficacious than 60 Hz operation. In addition, the solid-state ballast is more efficient than conventional ballasts in conditioning the input power for the lamps, such that the total system efficacy increase is between 20 and 25 percent. That is, for a standard two-lamp 40 watt F40 T-12 rapid start system, overall efficacy is increased from 63 lm/w to over 80 lm/w.

SIM 4-3

The waste treatment plant has an average 2,000-kW connected load at a PF of 0.8. Comment on the yearly savings before taxes for installing capacitors to improve the PF to 0.9. The demand charge that accounts in this billing structure for poor PF is $9/kVa/month. The total installation cost of the capacitors is $25/kvar.

An analysis of the electrical demand indicates that a peak of 2,500 kW can be reduced by automatically turning off decorative lighting and lighting for non-essential uses. Comment on adding a package load demand controller to reduce the peak demand to 2,000 kW at a PF of 0.9. The installed cost of the unit is $5,000.

ANSWER

Power Factor Improvement

First cost: From Table 4-1, the multiplier is 0.266. The required correction is 2,000kW × 0.266 = 532 kvar. The capacitor cost is $40/kvar × 532 kvar =$21,280.

Annual savings:

With correction—

Billing kVa = 2,000/0.8 = 2,500 kVa

Without correction—

Billing kVa = 2,000/0.9 = 2,222 kVa

Savings = (2,500—2,222) × \$9/kVa/month × 12 = \$30,024
Since the payback period is less than one year, investment is justified.

Load Management

In load shedding, the peak demand is reduced (2,500/0.9—2,000/0.9) without considering energy usage savings; the net savings are:

$$\$9(2{,}777 - 2{,}222) \times 12 = \$54{,}000$$

With a payback period of a fraction of the year, investment is justified.

5

Utility and Process System Optimization

The energy manager should analyze the total utility needs and the process for energy utilization opportunities. The overall heat and material balance and process flow diagram are very important tools. Each subprocess must also be analyzed in detail.

In this chapter, waste heat recovery, boiler operation, utility, and process systems will be discussed.

BASIS OF THERMODYNAMICS

Thermodynamics deals with the relationships between heat and work. It is based on two basic laws of nature: the first and second laws of thermodynamics. The principles are used in the design of equipment such as steam engines, turbines, pumps, and refrigerators, and in practically every process involving a flow of heat or a chemical equilibrium.

First Law: The first law states that energy can neither be created nor destroyed, thus, it is referred to as the law of conservation of energy. Equation 5-4 expresses the first law for the steady state condition.

$$E_2 - E_1 = Q - W \tag{5-1}$$

where

$E_2 - E_1$ is the change in stored energy at the boundary states 1 and 2 of the system

Q is the heat added to the system

W is the work done by the system

Figure 5-1 illustrates a thermodynamic process where mass enters and leaves the system. The potential energy (Z) and the kinetic energy ($V^2/64.2$) plus the enthalpy represents the stored energy of the mass. Note, "Z" is the elevation above the reference point in feet, and "V" is the velocity of the mass in ft/sec. In the case of the steam turbine, the change in Z, V, and Q are small in comparison to the change in enthalpy. Thus, the energy equation reduces to

$$W/778 = h_1 - h_2 \quad (5\text{-}2)$$

where

W is the work done in ft • lb/lb
h_1 is the enthalpy of the entering steam Btu/lb
h_2 is the enthalpy of the exhaust steam, Btu/lb
And 1 Btu equals 778 ft • lb.

Second Law: The second law qualifies the first law by discussing the conversion between heat and work. All forms of energy, including work, can be converted to heat, but the converse is not generally true. The Kelvin-Planck statement of the second law of thermodynamics says

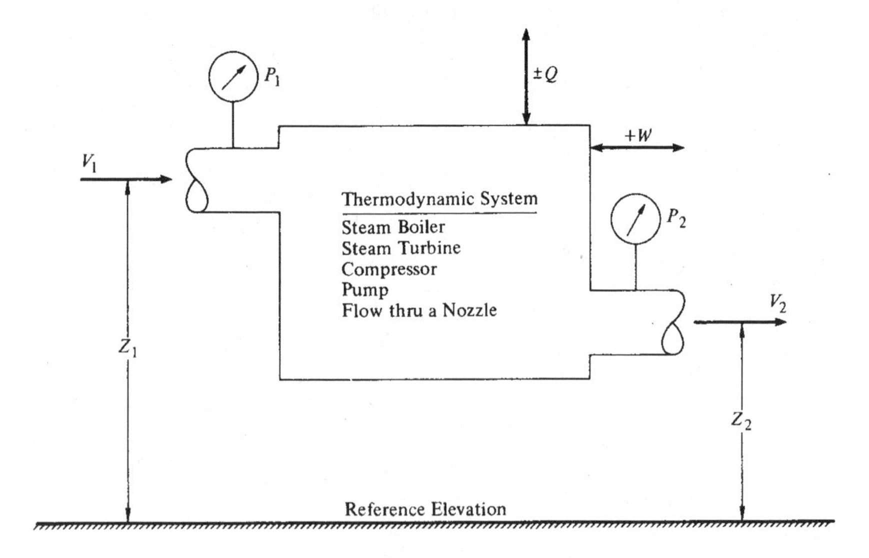

Figure 5-1. System illustrating conservation of energy.

essentially the following: Only a portion of the heat from a heat work cycle, such as a steam power plant, can be converted to work. The remaining heat must be rejected as heat to a sink of lower temperature; to the atmosphere, for instance.

The Clausius statement, which also deals with the second law, states that heat, in the absence of some form of external assistance, can only flow from a hotter to a colder body.

THE CARNOT CYCLE

The Carnot cycle is of interest because it is used as a comparison of the efficiency of equipment performance. The Carnot cycle offers the maximum thermal efficiency attainable between any given temperatures of heat source and sink. A thermodynamic cycle is a series of processes forming a closed curve on any system of thermodynamic coordinates. The Carnot cycle is illustrated on a temperature-entropy diagram, Figure 5-2A, and on the Mollier Diagram for super-heated steam, Figure 5-2B.

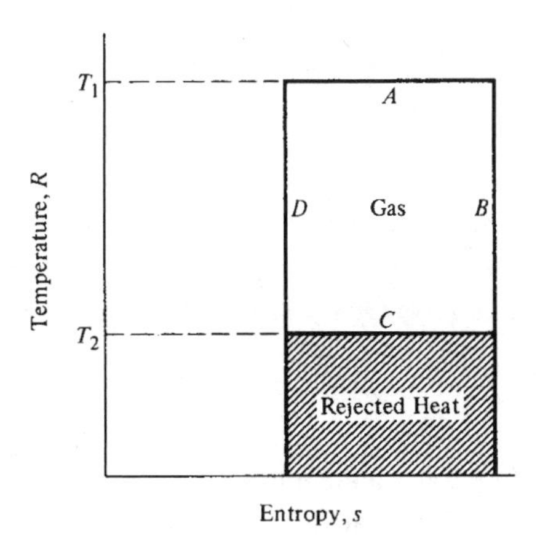

A. Temperature—entropy diagram; gas.

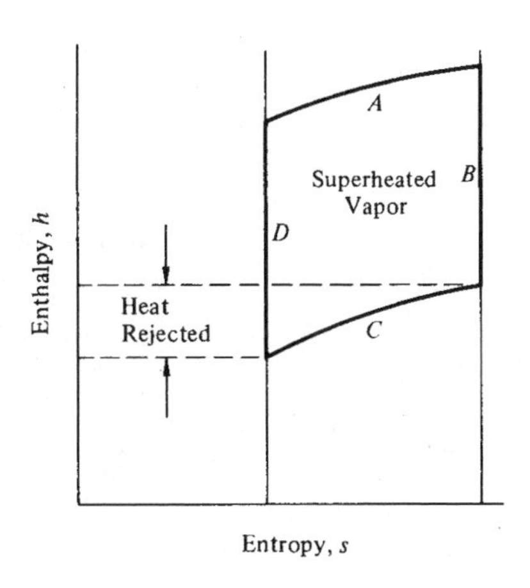

B. Mollier Diagram; superheated vapor.

Figure 5-2. Carnot cycles.

The cycle consists of the following:

1. Heat addition at constant temperature, resulting in expansion work and changes in enthalpy.

2. Adiabatic isentropic expansion (change in entropy is zero) with expansion work and an equivalent decrease in enthalpy.

3. Constant temperature heat rejection to the surroundings, equal to the compression work and any changes in enthalpy.

4. Adiabatic isentropic compression returning to the starting temperature with compression work and an equivalent increase in enthalpy.

The Carnot cycle is an example of a reversible process and has no counterpart in practice. Nevertheless, this cycle illustrates the principles of thermodynamics. The thermal efficiency for the Carnot cycle is illustrated by Eq. 5-3.

$$\text{Thermal efficiency} = \frac{T_1 T_2}{T_1} \tag{5-3}$$

where

T_1 = Absolute temperature of heat source, °R (Rankine)
T_2 = Absolute temperature of heat sink, °R

and absolute temperature is given by Eq. 5-4.

$$\text{Absolute temperature} = 460 + \text{temperature in Fahrenheit.} \tag{5-4}$$

T_2 is usually based on atmospheric temperature, which is taken as 500°R.

SIM 5-1

Compute the ideal thermal efficiency for a steam engine, based on the Carnot cycle and a saturated steam temperature of 540°F.

ANSWER

$$\text{Thermal efficiency} = \frac{(540 + 460) - 500}{(540 + 460)}$$

Properties of Steam Pressure and Temperature

Water boils at 212°F when it is in an open vessel under atmospheric pressure equal to 14.7 psia (pounds per square inch, absolute). Absolute pressure is the amount of pressure exerted by a system on its boundaries and is used to differentiate it from gage pressure. A pressure gage indicates the difference between the pressure of the system and atmospheric pressure.

$$\text{psia} = \text{psig} + \text{atmospheric pressure in psia} \qquad (5\text{-}5)$$

Changing the pressure of water changes the boiling temperature. Thus, water can be vaporized at 170°F, at 300°F, or any other temperature, as long as the applied pressure corresponds to that boiling point.

Solid, Liquid, and Vapor States of a Liquid

Water, as well as other liquids, can exist in three states: solid, liquid, and vapor. In order to change the state from ice to water or from water to steam, heat must be added. The heat required to change a solid to liquid is called the *latent heat of fusion.* The heat required to change a liquid to a vapor is called the *latent heat of vaporization.*

In condensing steam, heat must be removed. The quantity is exactly equal to the latent heat that went into the water to change it to steam.

Heat supplied to a fluid, during the change of state to a vapor, will not cause the temperature to rise; thus, it is referred to as the latent heat of vaporization. Heat given off by a substance when it condenses from steam to a liquid is called *sensible* heat. Physical properties of water, such as the latent heat of vaporization, also change with variations in pressure.

Steam properties are given in the ASME steam tables.

USE OF THE SPECIFIC HEAT CONCEPT

Another physical property of a material is the *specific heat.* The specific heat is defined as the amount of heat in Btu required to raise one pound of a substance one degree F. For water, it can be seen from the previous examples that one Btu of heat is required to raise one lb water 1°F; thus, the specific heat of water $C_p = 1$. Specific heats for other materials are illustrated in Table 5-1. The following three equations are useful in heat recovery problems:

$$q = wC_p \Delta T \tag{5-6}$$

where

q = quantity of heat, Btu
w = weight of substance, lb
C_p = specific heat of substance, Btu/lb-°F
ΔT = temperature change of substance, °F.

$$q = MC_p\Delta T \tag{5-7}$$

Table 5-1. Specific heat of various substances.

SUBSTANCE	SPECIFIC HEAT	SUBSTANCE	SPECIFIC HEAT
SOLIDS		LIQUIDS	
ALUMINUM	0.230	ALCOHOL	0.600
ASBESTOS	0.195	AMMONIA	1.100
BRASS	0.086	BRINE, CALCIUM (20% SOLUTION)	0.730
BRICK	0.220	BRINE, SODIUM (20% SOLUTION)	0.810
BRONZE	0.086	CARBON TETRACHLORIDE	0.200
CHALK	0.215	CHLOROFORM	0.230
CONCRETE	0.270	ETHER	0.530
COPPER	0.093	GASOLINE	0.700
CORK	0.485	GLYCERINE	0.576
GLASS, CROWN	0.161	KEROSENE	0.500
GLASS, FLINT	0.117	MACHINE OIL	0.400
GLASS, THERMOMETER	0.199	MERCURY	0.033
GOLD	0.030	PETROLEUM	0.500
GRANITE	0.192	SULPHURIC ACID	0.336
GYPSUM	0.259	TURPENTINE	0.470
ICE	0.480	WATER	1.000
IRON, CAST	0.130	WATER, SEA	0.940
IRON, WROUGHT	0.114		
LEAD	0.031	GASES	
LEATHER	0.360	AIR	0.240
LIMESTONE	0.216	AMMONIA	0.520
MARBLE	0.210	BROMINE	0.056
MONEL METAL	0.128	CARBON DIOXIDE	0.200
PORCELAIN	0.255	CARBON MONOXIDE	0.243
RUBBER	0.481	CHLOROFORM	0.144
SILVER	0.055	ETHER	0.428
STEEL	0.118	HYDROGEN	3.410
TIN	0.045	METHANE	0.593
WOOD	0.330	NITROGEN	0.240
ZINC	0.092	OXYGEN	0.220
		SULPHUR DIOXIDE	0.154
		STEAM (SUPERHEATED, 1 PSI)	0.450

Reprinted by permission of The Trane Company.

where

q = quantity of heat, Btu/hr (Btu/hr is sometimes abbreviated as Btuh)
M = Flow rate, lb/hr
C_p = Specific heat, Btu/lb°F
ΔT = temperature change of substance °F.

$$q = M\Delta h \tag{5-8}$$

where
Δh = change in enthalpy of the fluid
q and M are defined above.

PRACTICAL APPLICATIONS FOR ENERGY CONSERVATION

The Steam Balance

The first step in evaluating energy conservation measures is to compile a flowsheet and steam balance. The steam balance indicates ways in which steam usage can be minimized. To start a balance, the following is evaluated:

1. Requirements of users.
2. Steam pressure levels to satisfy process needs.
3. Turbine drive operating pressures.
4. Pressure ratings for piping valves and fittings.

Figure 5-3 illustrates a simple block heat balance diagram for a steam generating process. To conserve energy, losses must be minimized and furnace efficiency must be maximized.

In Chapter 6, heat losses will be treated in detail. The remaining portion of this chapter will indicate ways to reduce steam consumption and apply systems efficiently.

Using the Steam Turbine

Compared to other prime movers, the steam turbine remains a flexible component for most steam systems. Sizes vary from single stage units of 50 to 600 hp for driving pumps, fans, and compressors,

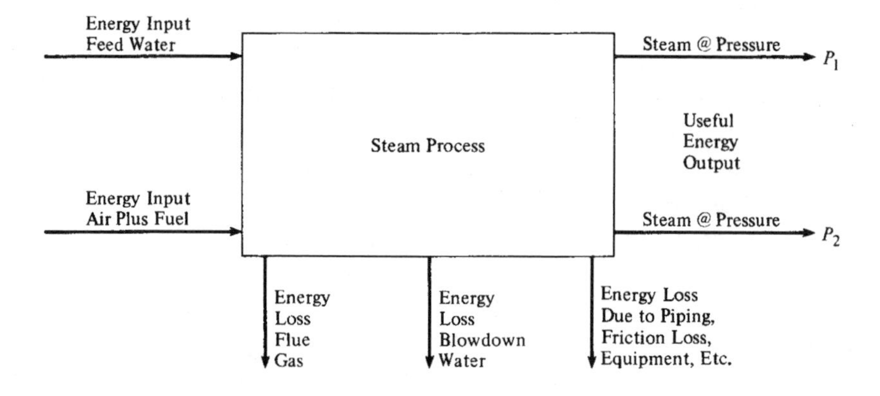

Figure 5-3. Heat balance for steam process.

to multi-stage units rated as high as 50,000 hp. The large units can drive huge process compressor trains. Turbines operate in the following way: Steam enters the turbine through steam inlet valves and is expanded through the nozzles. It then impinges on the blades of the rotor and exits through exhaust connections. The process of expanding the steam through the nozzles changes the heat energy into mechanical energy.

Turbines can operate with inlet steam from 200 psig to over 900 psig. The throttling of the inlet steam occurs under an adiabatic (no heat loss) process. A throttling process reduces the pressure while the enthalpy remains constant. The mechanical energy can drive process equipment or a generator to produce electricity.

Some of the applications of a steam turbine are as follows:

1. Used as a pressure reducing valve, but instead of wasting energy, output is used to create electricity.

2. Permits generation of high pressure and high temperature steam.

3. Used as a variable speed drive for fans and equipment.

4. Used for reliability applications where standby power is required.

Returning Condensate to the Boiler

When condensate is returned to the boiler plant, the amount of fuel used for steam generation is reduced by 10 to 30%. Condensate returned to the boiler plant reduces water pollution and saves:

1. Energy and chemicals used for water treating.
2. Treated makeup boiler feed water.

Returning condensate is a common practice in new plant design, but many existing plants have not put into practice this valuable energy conservation technique.

SIM 5-2

An 80% efficient boiler uses No. 2 fuel oil to generate steam. What is the yearly (8000 h/yr) savings, excluding piping amortization, of returning 20,000 lb/h of 25 psig condensate to the boiler plant, instead of using 70°F make-up water? Assume the fuel costs $6.00 per million Btu.

ANSWER

Assume atmospheric pressure is 15 psia and adds it to gage pressure of 25 psig.

h_f (40 psia) = 236 Btu/lb

h_f (70°) = 38 Btu/lb

$q = 20{,}000\ (236\text{-}38) = 3.96 \times 10^6$ Btuh

$$(q)\,(\text{Fuel oil requirement}) = \frac{3.96 \times 10^6}{0.8} = 4.95 \times 10^6\text{Btuh}$$

Yearly energy savings before taxes = 4.95 × 8000 × $6.00 = $237,600.

Flashing Condensate to Lower Pressure

High pressure condensate can be flashed to lower pressure steam. This is often desirable when remote locations discourage returning condensate to the boiler plant and low pressure steam is required.

Figure 5-4 should be used to calculate the amount of low pressure steam that is produced from higher pressure condensate. To use the figure, simply take the difference between the percent condensate flashed to 0 psig for the steam and condensate pressures in question. Multiply this percent by the pounds per hour of high pressure condensate available.

SIM 5-3

Compute the amount of 25 psig steam which is generated by flashing 10,000 lb/hr of 125 psig condensate. What is the yearly (8,000 h/yr)

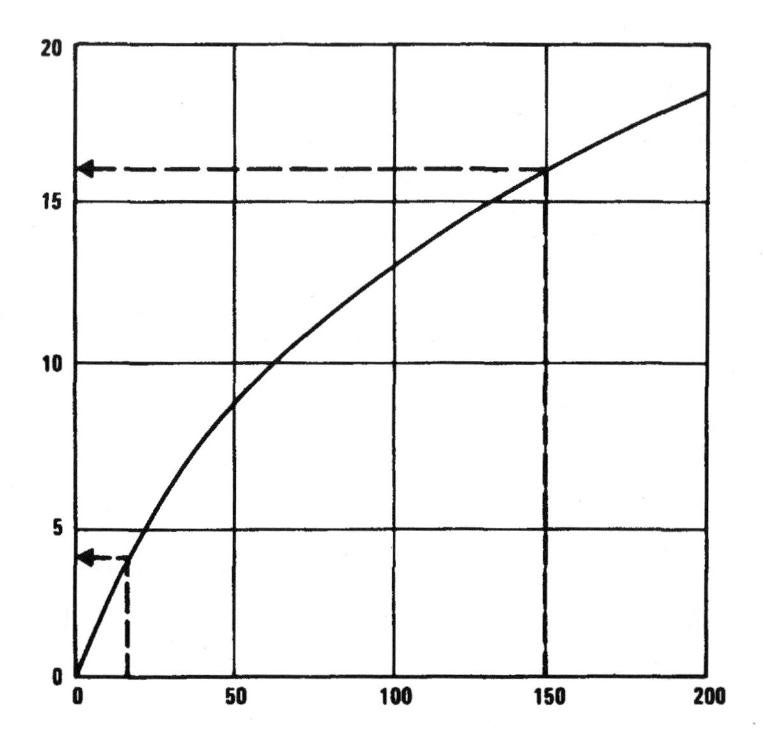

Figure 5-4. Steam condensate flashing (calculated from steam tables). (Adapted from the *NBS Handbook 115.*)

savings, excluding piping amortization, for this process, as opposed to using 70°F water? The boiler is 80% efficient and uses No. 2 fuel oil. Assume fuel costs $6.00/million Btu.

ANSWER

From Figure 5-4:

At 125 psig, condensate flashed to 0 psig	15%
At 25 psig, condensate flashed to 0 psig	5.5%
Percent of 25 psig steam available	= 9.5%

The potential amount of 25 psig steam available from flashing the 125 psig condensate is:

$$10,000 \text{ lb/hr} \times .095 = 950 \text{ lb/hr.}$$

The annual energy savings:

hg (40 psia) = 1169.8 Btu/lb
hf (70°) = 38 Btu/lb
$q = 950 \times (1169.8 - 38)$
$q = 1.075 \times 10^6$ Btuh

$$\text{Fuel oil requirement q} = \frac{1.075 \times 10^6}{0.8} = 1.34 \times 10^6\text{Btuh}$$

Yearly energy savings $= 1.34 \times 8000 \times \$6.00$
$= \$64,320.$

FURNACE EFFICIENCY

In order to produce steam, a boiler requires a source of heat at a sufficient temperature level. Fossil fuels, such as coal, oil, and gas are generally burned for this purpose, in the furnace of a boiler. The combustion of fossil fuels is defined as the rapid chemical combination of oxygen with the combustible elements of the fuel. The three combustible elements of fossil fuels are carbon, hydrogen, and sulfur. Sulfur is a major source of pollution and corrosion but is a minor source of heat.

Air is usually the source of oxygen for boiler furnaces.

The combustion reactions release about 61,000 Btu/lb of hydrogen burned and 14,100 Btu/lb of carbon burned.

Good combustion releases all of this heat while minimizing losses from excess air and combustion imperfections.

Unburned fuel, leaving carbon in the ash, or incompletely burned carbon represents a loss. A greater loss is the heat loss up the stack. To assure complete combustion, it is necessary to use more than theoretical air requirements. For an "ideal" union of gas and air, excess air would not be required. In order to hold down stack losses, it is necessary to keep excess air to a minimum. Air not used in fuel combustion leaves the unit at stack temperature. When this air is heated from room temperature to stack temperature, the heat required serves no useful purpose and is therefore lost heat. A checklist of items to consider for maximum furnace efficiency is illustrated in Table 5-2.

Table 5-2. Furnace efficiency checklist.

1. Excess air should be monitored to keep it as low as is practical.

2. Portable or permanent oxygen analyzers should be used on furnaces to help minimize oxygen in flue gas.

3. Fuel oil temperature should be monitored and checked against manufacturer's recommendations for the fuel specification used. Proper fuel oil temperature insures good atomization.

4. Atomizing steam should be dry.

5. If fuel gas is at dew point, then a knock-out drum at the furnace with steam tracing to the burners should be used.

6. Atomizing steam rate should be at the minimum required for acceptable combustion.

7. Furnace systems should be evaluated for heat recovery, i.e., installing air preheaters or waste heat boilers.

8. Furnace systems should be evaluated for using preheated combustion air from gas-turbine exhausts.

9. Furnace systems should be evaluated for using high pressure gas for atomization instead of steam.

10. Furnace system should incorporate automated controls and recorders to maximize fuel usage and help the operator manage his system.

11. In particular, for existing furnace installations, the following should be checked:
    (a) Reduce all air leaks (infrared photography can be used to check for leaks).
    (b) Burners and soot blowers should be inspected and maintained on a scheduled basis.
    (c) Consider replacing burners with more efficient models.

The Effect of Flue Gas and Combustion Air Temperature

Two common energy wastes in furnaces, boiler plants, and other heat processing equipment are high flue gas temperature and nonpreheated combustion air.

A common heat recovery system that saves energy is illustrated in Figure 5-5. A portion of the flue gas heat is recovered through the use of a heat exchanger, recouper, regenerator, or similar equipment. The heat of the flue gas is used to preheat the combustion air. As an example, without combustion air preheat, a furnace operating at 2,000°F uses almost 40 percent of its fuel just to heat the combustion air to the burning temperature.

Figure 5-6 illustrates heat savings using preheated combustion air.

SIM 5-4

The plant engineer is studying the savings by adding a combustion air preheater to a furnace operating on heavy oil at 1,500°F. Comment on the savings of preheating the combustion air to 750°F.

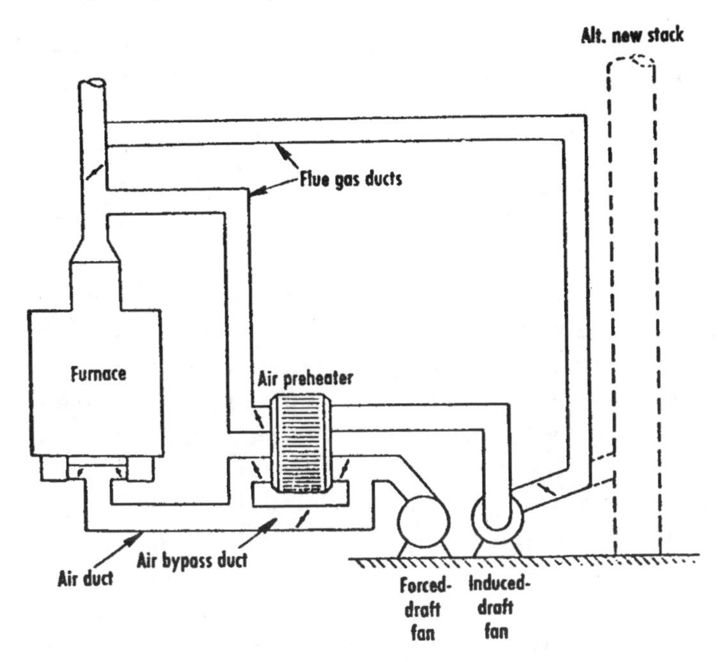

Figure 5-5. Application of air preheater. (Reprinted by permission of *Oil and Gas Journal*.)

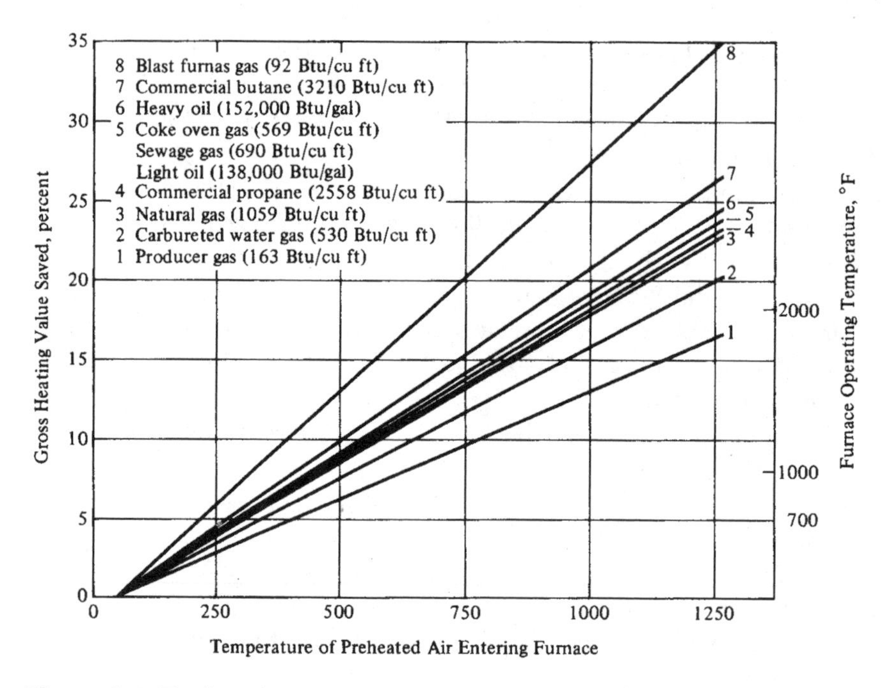

Figure 5-6. Fuel savings resulting from use of preheated combustion air. (Figure adapted with permission from *Plant Engineering*.)

ANSWER

From Figure 5-6, the net savings in fuel input is 13%.

Reducing Flue Gas Temperature

Even with a heat recovery system for preheating combustion air, flue gas temperatures should be maintained as low as possible. The amount of combustion air directly affects the oxygen available, which in turn increases the flue gas temperature and causes energy to be wasted. Figure 5-7 illustrates the fuel savings which is achieved by reducing the excess air or oxygen content from the operating condition to 2% oxygen, corresponding to 10% excess air. New plant design should incorporate combustion control systems and oxygen analyzers to accomplish low excess air requirements.

To use Figure 5-7, first find the oxygen (O_2) content in the flue gas. Note that oxygen content and percent excess air are related by curve "A." To find the savings that will result from reducing the O_2 content to 2%, simply find the intersection of the original O_2 level with the flue

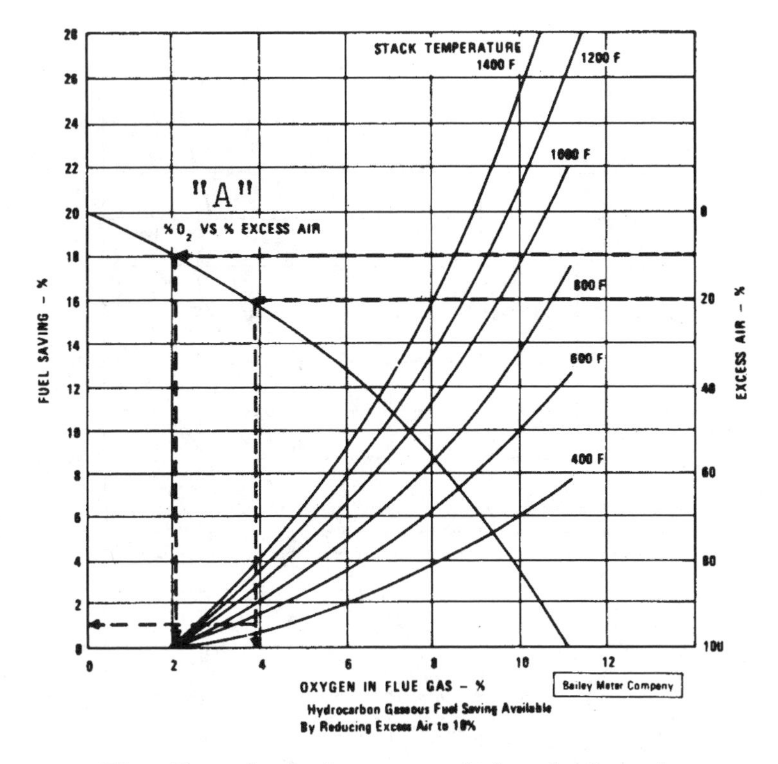

Figure 5-7. The effect of reducing excess air for a hydrocarbon gaseous fuel. (Adapted from the *NBS Handbook 115.*)

gas stack temperature. The fuel savings is read directly to the left. This curve is typical for hydrocarbon gaseous fuels such as natural gas.

Figure 5-8 is a similar curve for liquid petroleum fuels, where the oxygen content can be reduced to 4%, corresponding to 20% excess air.

STEAM TRACING

Steam tracing is commonly used to protect piping and equipment from freeze-ups and for process fluid requirements. It consists of a small tube in contact with the piping or equipment, through which steam is used as the heating medium. For systems requiring fifteen or more traces, an ethylene glycol solution can be used economically in place of steam. One such solution, SR-1, which is manufactured by Dow, freezes at –34°F . The ethylene glycol solution is heated via a heat exchanger and pumped

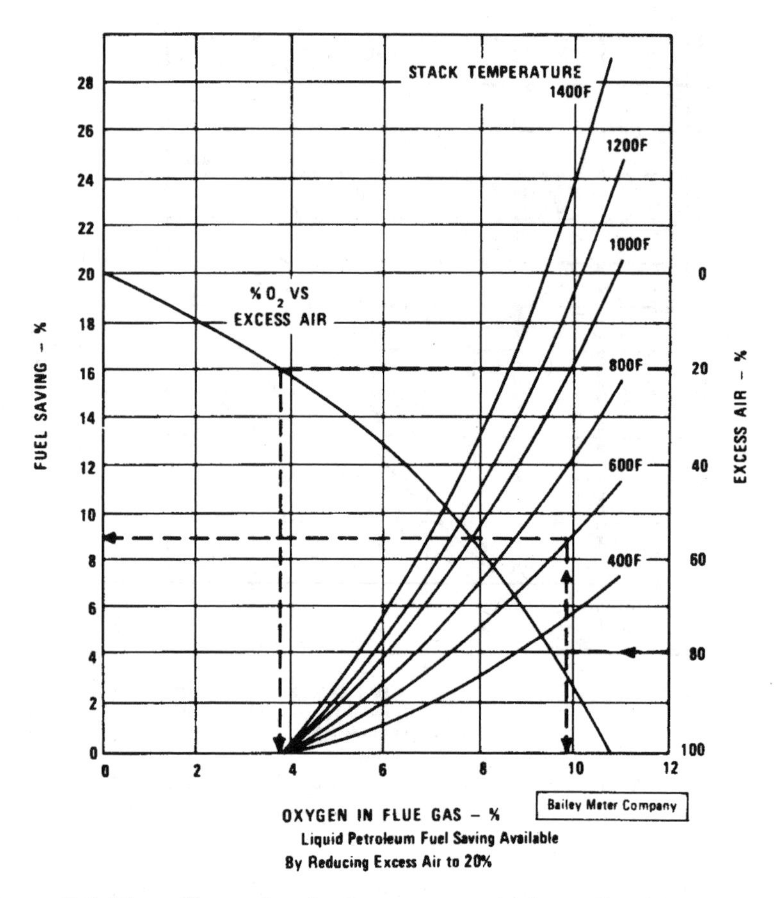

Figure 5-8 The effect of reducing excess air for a liquid petroleum fuel. (Adapted from the *NBS Handbook 115*.)

through the various tracers. The net savings in steam can be a thousand pounds per foot of tracer per year and achieve an overall energy saving of 40%. A caution should be noted that too low an excess air value, especially below 10% for gaseous fuels and 20% for liquid fuels, can cause excessive corrosion of furnace tubes.

HEAT RECOVERY

Waste heat has been defined as heat that is rejected from a process at a temperature enough above the ambient temperature to permit the manager or engineer to extract additional value from it. Sources of waste energy can be divided according to temperature into three temperature ranges. The high temperature range refers to temperatures above 1,200°F. The medium temperature range is between 450°F and 1,200°F, and the low temperature range is below 450°F.

High- and medium-temperature waste heat can be used to produce process steam. If one has high-temperature waste heat, instead of producing steam directly, one should consider the possibility of using the high-temperature energy to do useful work before the waste heat is extracted. Both gas and steam turbines are useful and fully developed heat engines.

In the low temperature range, waste energy that would be otherwise useless can sometimes be made useful by application of mechanical work through a device called the heat pump. An interesting application of this is in petroleum distillation, where the working fluid of the heat pump can be the liquid being distilled. (This application was developed by the British Petroleum Co.).

Sources of Waste Heat

The combustion of hydrocarbon fuels produces product gases in the high temperature range. The maximum theoretical temperature possible in atmospheric combustors is somewhat under 3,500°F, while measured flame temperatures in practical combustors are just under 3,000°F. Secondary air or some other dilutant is often admitted to the combustor to lower the temperature of the products to the required process temperature, for example to protect equipment, thus lowering the practical waste heat temperature.

Table 5-3 gives temperatures of waste gases from industrial process equipment in the high temperature range. All of these result from direct fuel fired processes.

Table 5-4 gives the temperatures of waste gases from process equipment in the medium temperature range. Most of the waste heat in this temperature range comes from the exhausts of directly fired process units. Medium temperature waste heat is still hot enough to allow consideration of the extraction of mechanical work from the waste heat, by a steam or gas turbine. Gas turbines can be economically utilized in some cases at inlet pressures in the range of 15 to 30 lb/in.^{2}g. Steam can be generated at almost any desired pressure and steam turbines used when economical.

Table 5-3.

Type of Device	Temperature, °F
Nickel refining furnace	2500-3000
Aluminum refining furnace	1200-1400
Zinc refining furnace	1400-2000
Copper refining furnace	1400-1500
Steel heating furnace	1700-1900
Copper reverberatory furnace	1650-2000
Open hearth furnace	1200-1300
Cement kiln (Dry process)	1150-1350
Glass melting furnace	1800-2800
Hydrogen plant	1200-1800
Solid waste incinerator	1200-1800
Fume incinerator	1200-2600

Table 5-4.

Type of Device	Temperature, °F
Steam boiler exhaust	450-900
Gas turbine exhaust	700-1000
Reciprocating engine exhaust	600-1100
Reciprocating engine exhaust (turbocharged)	450-700
Heat treating furnace	800-1200
Drying and baking oven	450-1100
Catalytic cracker	800-1200
Annealing furnace cooling system	800-1200

Table 5-5 lists some heat sources in the low temperature range. In this range it is usually not practicable to extract work from the source, though steam production may not be completely excluded if there is a need for low-pressure steam. Low-temperature waste heat may be useful in a supplementary way for preheating purposes. Taking a common example, it is possible to use economically the energy from an air conditioning condenser operating at around 90°F to heat the domestic water supply. Since the hot water must be heated to about 160°F, obviously the air conditioner waste heat is not hot enough. However, since the cold water enters the domestic water system at about 50°F, energy interchange can take place raising the water to something less than 90°F. Depending upon the relative air conditioning lead and hot water requirements, any excess condenser heat can be rejected and the additional energy required by the hot water provided by the usual electrical or fired heater.

Table 5-5.

Source	Temperature, °F
Process steam condensate	130-190
Cooling water from:	
Furnace doors	90-130
Bearings	90-190
Welding machines	90-190
Injection molding machines	90-190
Annealing furnaces	150-450
Forming dies	80-190
Air compressors	80-120
Pumps	80-190
Internal combustion engines	150-250
Air conditioning and refrigeration condensers	90-110
Liquid still condensers	90-190
Drying, baking and curing ovens	200-450
Hot processed liquids	90-450
Hot processed solids	200-450

How to Use Waste Heat

To use waste heat from sources such as those above, one often wishes to transfer the heat in one fluid stream to another (e.g., from flue gas to feedwater or combustion air). The device that accomplishes the transfer is called a heat exchanger. In the discussion immediately below is a listing of common uses for waste heat energy and in some cases, the name of the heat exchanger that would normally be applied in each particular case.

The equipment that is used to recover waste heat can range from something as simple as a pipe or duct to something as complex as a waste heat boiler. Here we categorize and describe some waste recovery systems that are available commercially, suitable for retrofitting in existing plants, with lists of potential applications for each of the described devices. These are developed technologies which have been employed for years in some industries.

1. Medium to high temperature exhaust gases can be used to preheat the combustion air for:
   Boilers using air-preheaters
   Furnaces using recuperators
   Ovens using recuperators
   Gas turbines using regenerators.

2. Low to medium temperature exhaust gases can be used to preheat boiler feedwater or boiler makeup water using *economizers,* which are simply gas-to-liquid water heating devices.

3. Exhaust gases and cooling water from condensers can be used to preheat liquid and/or solid feedstocks in industrial processes. Finned tubes and tube-in-shell *heat exchangers* are used.

4. Exhaust gases can be used to generate steam in *waste heat boilers* to produce electrical power, mechanical power, process steam, and any combination of above.

5. Waste heat may be transferred to liquid or gaseous process units directly through pipes and ducts or indirectly through a secondary fluid such as steam or oil.

6. Waste heat may be transferred to an intermediate fluid by heat exchangers or waste heat boilers, or it may be used by circulating the hot exit gas through pipes or ducts. Waste heat can be used to operate an absorption cooling unit for air conditioning or refrigeration.

Waste Heat Recovery Equipment

Industrial heat exchangers have many pseudonyms. They are sometimes called recuperators, regenerators, waste heat steam generators, condensers, heat wheels, temperature and moisture exchangers, etc. Whatever name they may have, they all perform one basic function: the transfer of heat.

Heat exchangers are characterized as single or multipass gas to gas, liquid to gas, liquid to liquid, evaporator, condenser, parallel flow, counter flow, or cross flow. The terms *single* or *multipass* refer to the heating or cooling media passing over the heat transfer surface once or a number of times. Multipass flow involves the use of internal baffles. The next three terms refer to the two fluids between which heat is transferred in the heat exchanger, and imply that no phase changes occur in those fluids. Here the term *fluid* is used in the most general sense. Thus, we can say that these terms apply to nonevaporator and noncondensing heat exchangers. The term *evaporator* applies to a heat exchanger in which heat is transferred to an evaporating (boiling) liquid, while a *condenser* is a heat exchanger in which heat is removed from a condensing vapor. A parallel flow heat exchanger is one in which both fluids flow in approximately the same direction whereas in counterflow the two fluids move in opposite directions. When the two fluids move at right angles to each other, the heat exchanger is considered to be of the crossflow type.

The principal methods of reclaiming waste heat in industrial plants make use of heat exchangers. The heat exchanger is a system which separates the stream containing waste heat and the medium which is to absorb it, but allows the flow of heat across the separation boundaries. Reasons for separating the two streams may be any of the following:

1. A pressure difference may exist between the two streams of fluid. The rigid boundaries of the heat exchanger can be designed to withstand the pressure difference.

2. In many, if not most, cases the one stream would contaminate the other, if they were permitted to mix. The heat exchanger prevents mixing.

3. Heat exchangers permit the use of an intermediate fluid better suited than either of the principal exchange media for transporting waste heat through long distances. The secondary fluid is often steam, but another substance may be selected for special properties.

4. Certain types of heat exchangers, specifically the heat wheel, are capable of transferring liquids as well as heat. Vapors being cooled in the gases are condensed in the wheel and later re-evaporated into the gas being heated. This can result in improved humidity and/or process control, abatement of atmospheric air pollution, and conservation of valuable resources.

The various names or designations applied to heat exchangers are partly an attempt to describe their function and partly the result of tradition within certain industries. For example, a recuperator is a heat exchanger which recovers waste heat from the exhaust gases of a furnace to heat the incoming air for combustion. This is the name used in both the steel and the glass making industries. The heat exchanger performing the same function in the steam generator of an electric power plant is termed an air preheater, and in the case of a gas turbine plant, a regenerator.

However, in the glass and steel industries, the word regenerator refers to two chambers of brick checkerwork which alternately absorb heat from the exhaust gases and then give up part of that heat to the incoming air. The flows of flue gas and of air are periodically reversed by valves so that one chamber of the regenerator is being heated by the products of combustion while the other is being cooled by the incoming air. Regenerators are often more expensive to buy and more expensive to maintain than are recuperators, and their application is primarily in glass melt tanks and in open hearth steel furnaces.

It must be pointed out, however, that although their functions are similar, the three heat exchangers mentioned above may be structurally quite different as well as different in their principal modes of heat transfer.

The specification of an industrial heat exchanger must include the heat exchange capacity, the temperatures of the fluids, the allowable pressure drop in each fluid path, and the properties and volumetric flow of the fluids entering the exchanger. These specifications will determine construction parameters and thus the cost of the heat exchanger. The final design will be a compromise between pressure drop, heat exchanger effectiveness, and cost. Decisions leading to that final design will balance out the cost of maintenance and operation of the overall system against the fixed costs in such a way as to minimize the total. Advice on selection and design of heat exchangers is available from vendors.

The essential parameters that should be known in order to make an optimum choice of waste heat recovery devices are:

- Temperature of waste heat fluid
- Flow rate of waste heat fluid
- Chemical composition of waste heat fluid
- Minimum allowable temperature of waste heat fluid
- Temperature of heated fluid
- Chemical composition of heated fluid
- Maximum allowable temperature of heated fluid
- Control temperature, if control required.

Table 5-6 presents the collation of a number of significant attributes of the most common types of industrial heat exchangers, in matrix form. This matrix allows rapid comparisons to be made in selecting competing types of heat exchangers. The characteristics given in the table for each type of heat exchanger are: allowable temperature range, ability to transfer moisture, ability to withstand large temperature differentials, availability as packaged units, suitability for retrofitting, and compactness and the allowable combinations of heat transfer fluids.

THE MOLLIER DIAGRAM

A visual tool for understanding and using the properties of steam is illustrated by the Mollier Diagram, Figure 5-9. The Mollier Diagram enables one to find the relationship between temperature, pressure, enthalpy, and entropy, for steam. Constant temperature and pressure curves illustrate the effect of various processes on steam.

For a constant temperature process (isothermal), the change in entropy is equal to the heat added (or subtracted) divided by the temperature at which the process is carried out. This is a simple way of explaining the physical meaning of entropy. The *change* in entropy is of interest. Increases in entropy are a measure of the portion of heat in a process that is unavailable for conversion to work. Entropy has a close relation to the second law of thermodynamics, discussed later in this chapter.

Another item of interest from the Mollier Diagram is the saturation line. The saturation line indicates temperature and pressure relationships corresponding to saturated steam.

Table 5-6. Operation and application characteristics of industrial heat exchangers.

Commercial Heat Transfer Equipment / Specifications for Waste Recovery Unit	Low Temperature, Sub-zero–250°F	Intermediate Temperature, 250°F–1200°F	High Temperature, 1200°F–2000°F	Recovers Moisture	Large Temperature Differentials Permitted	Packaged Units Available	Can Be Retrofit	No Cross-contamination	Compact Size	Gas-to-gas Heat Exchange	Gas-to-liquid Heat Exchanger	Liquid-to-liquid Heat Exchanger	Corrosive Gases Permitted with Special Construction
Radiation recuperator			•		•	1	•	•		•			•
Convection recuperator		•	•		•	•	•	•		•			•
Metallic heat wheel	•	•		2		•	•	3	•	•			•
Hygroscopic heat wheel	•			•		•	•	3	•	•			
Ceramic heat wheel		•	•		•	•	•		•	•			•

Passive regenerator	•	•			•	•	•	•		•			•
Finned-tube heat exchanger	•	•			•	•	•	•	•		•		4
Tube shell-and-tube exchanger	•	•			•	•	•	•	•		•	•	
Waste heat boilers	•	•				•	•	•			•		4
Heat pipes	•	•			5	•	•	•	•	•			•

[1] Off-the-shelf items available in small capacities only.
[2] Controversial subject. Some authorities claim moisture recovery. Do not advise depending on it.
[3] With a purge section added, cross-contamination can be limited to less than 1 percent by mass.
[4] Can be constructed of corrosion-resistant materials, but consider possible extensive damage to equipment caused by leaks or tube ruptures.
[5] Allowable temperatures and temperature differential limited by the phase equilibrium properties of the internal fluid.

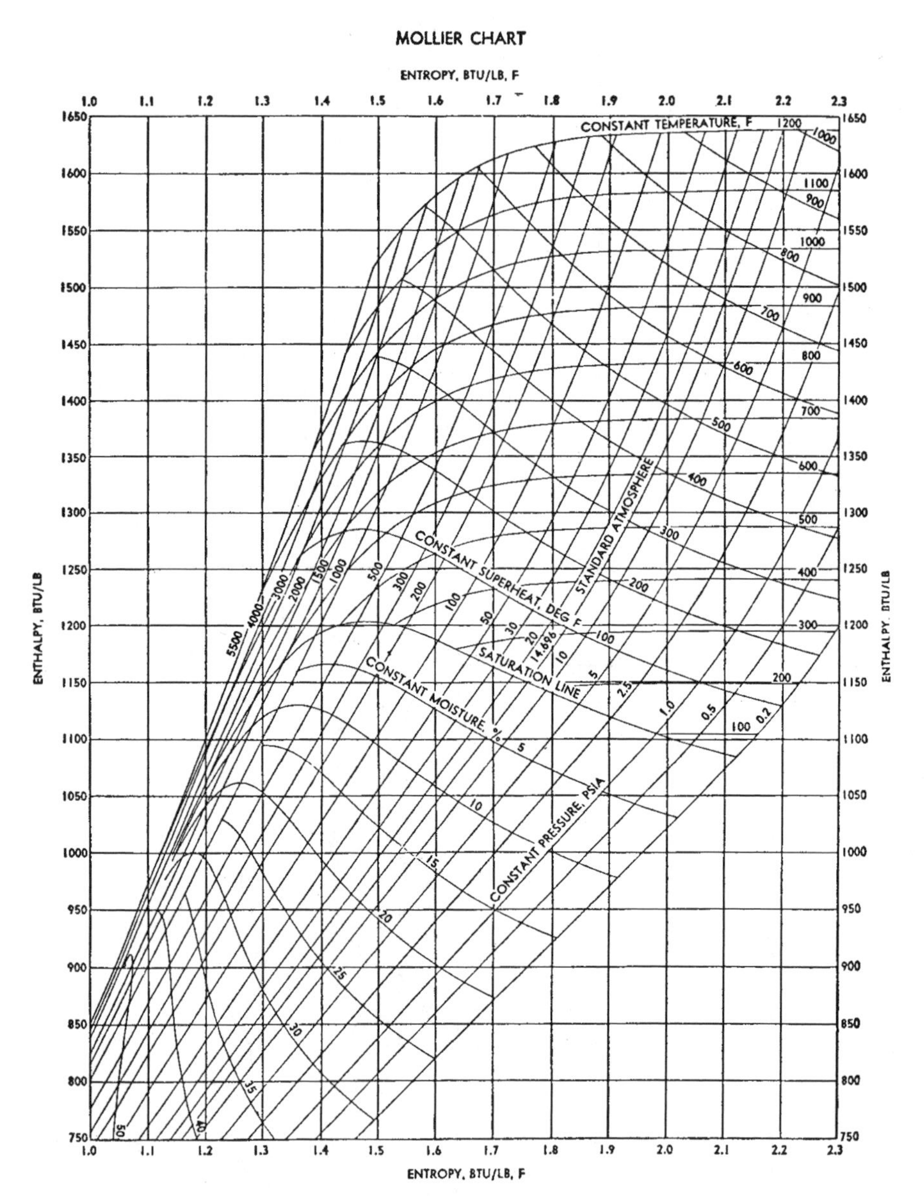

Figure 5-9. Mollier Diagram. (*Courtesy of the Babcock & Wilcox Company and ASME.*)

Below this curve, steam contains a percent moisture, as indicated by the percent moisture curves. Steam at temperatures above the saturation curve is referred to as superheated. As an example, this chart indicates that at 212°F and 14.696 psia, the enthalpy h_g is 1,150 Btu/lb.

If additional heat is added to raise the temperature of steam over the point at which it was evaporated, the steam is termed superheated. Thus, steam at the same temperature as boiling water is saturated steam. Steam at a temperature higher than boiling water, at the same pressure, is superheated steam.

Steam cannot be superheated in the presence of water because the heat supplied will only evaporate the water. Thus, the water will be evaporated prior to becoming superheated. Superheated steam is condensed by first cooling it down to the boiling point corresponding to its pressure. When the steam has been de-superheated, further removal of heat will result in its condensation.

In the generation of power, superheated steam has many uses.

STEAM GENERATION USING WASTE HEAT RECOVERY

A heat recovery system which uses the wasted exhaust from a gas turbine has proved to be very efficient. One of the largest power plants (300,000 kW) using heat recovery from gas turbines was built in 1971. The heat recovery concept works as follows:

- A heat recovery (waste heat) boiler is installed in the gas turbine hot air exhaust.
- The unfired heat recovery boiler produces steam which drives a condensing steam turbine.

Both the gas turbine and steam turbine can drive electric generators. Thus, the wasted gas turbine exhaust is used to generate up to 50% more electric power. One such process is illustrated in Figure 5-10.

SIM 5-8

The plant engineer is evaluating two alternate power generation systems for the boiler house as illustrated in Figure 5-11. An expansion to the Ajax plant requires 78,300 lb/hr of 150 psig process steam and

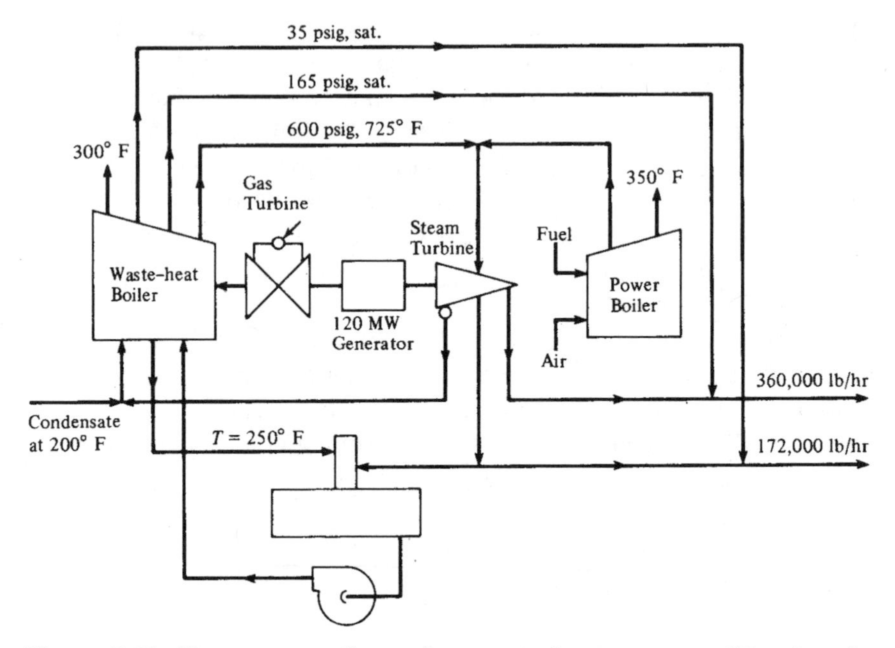

Figure 5-10. Steam generation using waste heat recovery. [Reprinted by special permission from *Chemical Engineering* (January 21, 1974) Copyright © (1974) by McGraw-Hill, New York, NY]

11,000 lb/hr of 30 psig process steam. The expansion also requires 3,000 hp of additional drivers for pumps, compressors, fans, etc.

Alternate #1 requires the generation of 150 psig saturated steam with 11,000 lb/hr to be depressured to 30 psig. The 3,000 driver hp is to be provided with electric motors.

Alternate #2 requires the generation of 600 psig 600°F superheated steam. This superheated steam is to be used to drive turbines to provide the 3000 driver hp and the turbine exhaust steam is to be used to satisfy the process steam requirements.

The following assumptions are to be made:

1. Fuel oil costs $4.00/10^6$ Btu.
2. Electricity costs $0.045/kWh.
3. Temperature of boiler make-up is 70°F.
4. Turbine efficiency is 0.5.
5. Neglect effect of heating combustion air.
6. Boiler efficiency is 0.8.

ALTERNATE #1: Steam Generation without Turbines.

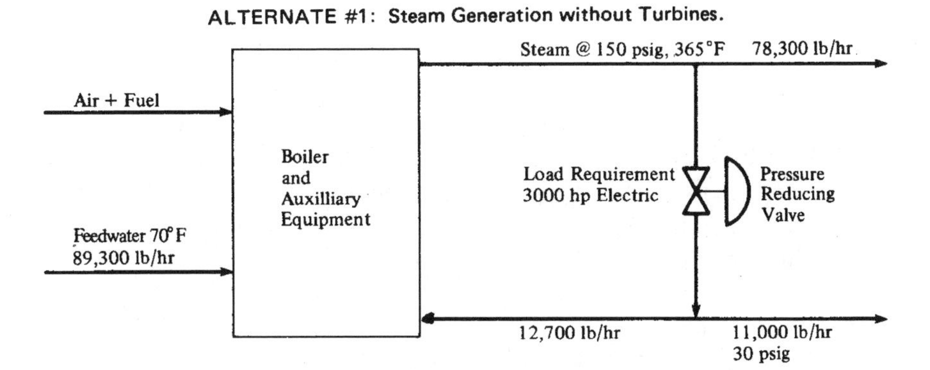

ALTERNATE #2: Steam Generation with Turbines.

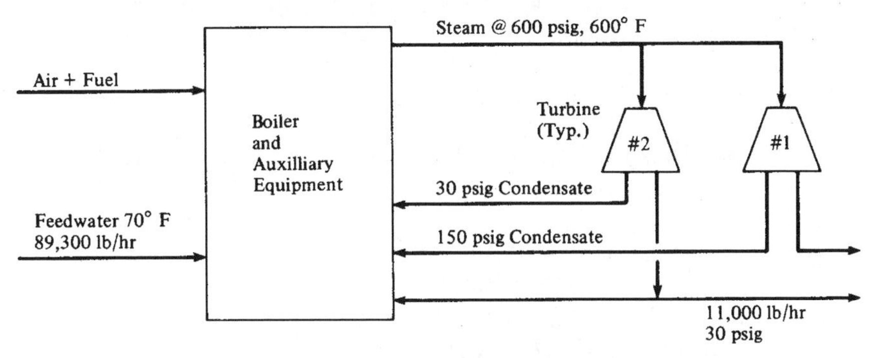

Figure 5-11. Steam generation alternatives.

7. 4,000 hours per year of operation (16 hours/day, 250 days/year).
8. Neglect costs associated with pumping boiler feed water.
9. In alternate #1, 12,700 lb/hr of 30 psig steam is returned for de-aeration of boiler make-up water.
10. In alternate #2, 30 psig condensate and 150 psig condensate from turbines, plus 30 psig steam is returned to the boiler for heat recovery and de-aeration of boiler make-up water.
11. Electrical efficiency is 0.8.

The following are required.

1. Compute a material and energy balance for each alternate system.
2. Determine the energy requirements in dollars per year for each system.

ANSWER

A material balance requires that lb/hr "in" equals the lb/hr "out." An energy balance requires the Btu/hr "in" equals the Btu/hr "out."

The steps to be taken are:

1. First find all missing material quantities.
2. Second, find the corresponding enthalpies from the Mollier Diagram (Figure 5-9).
3. Third, compute the energy balance.

For Alternate #1

Step 1: The quantity of 150 psig is 102,000 lb/hr.

Step 2: The enthalpy of 150 psig (165 psia) steam is 1,195.6 Btu/lb. The enthalpy of 30 psig (45 psia) steam is 1,172 Btu/lb. The enthalpy of water at 70°F is 38.04 Btu/lb.

Step 3: The energy balance for the system is:

$$(102{,}000 \text{ lb/hr})\ (1{,}195.6 \text{ Btu/lb}) = (89{,}300 \text{ lb/hr}\ (38.04 \text{ Btu/lb}) + (12{,}700 \text{ lb/hr})\ (1{,}172.0 \text{ Btu/lb}) + \text{input energy.}$$

Input energy = 121,951,200 Btu/hr – 3,396,972 Btu/hr – 14,884,400
= 103,669,828 Btu/hr.

The fuel required is 103,669,828 ÷ 0.8 = 129,587,285 Btu/hr.

The fuel cost is 129,587,285 × \$4.00 ÷ 10^6
= \$518.33/hr or \$2,073,353/yr.

The electrical cost is:

$$\frac{(3000 \text{ hp} \times 4000 \text{ hr/yr})(0.746 \text{ kW/hp})(\$0.045 \text{ kWh})}{0.8} = \$503{,}550/\text{yr.}$$

The total energy cost per year for alternate system #1 is \$2,073,353 + \$503,550 = \$2,576,903/yr.

For Alternate #2

Steps 1 and 2: To analyze turbine performance, use the Mollier Diagram of Figure 5-12. The first point "*A*" at 600 psig, 600°F, is located on the diagram (h = 1,290 Btu/lb). The second point "*B*" is drawn vertically from "*A*" until it intersects the 165 psia constant pressure line (h = 1,170 Btu/lb). The difference between inlet and exhaust enthalpy is

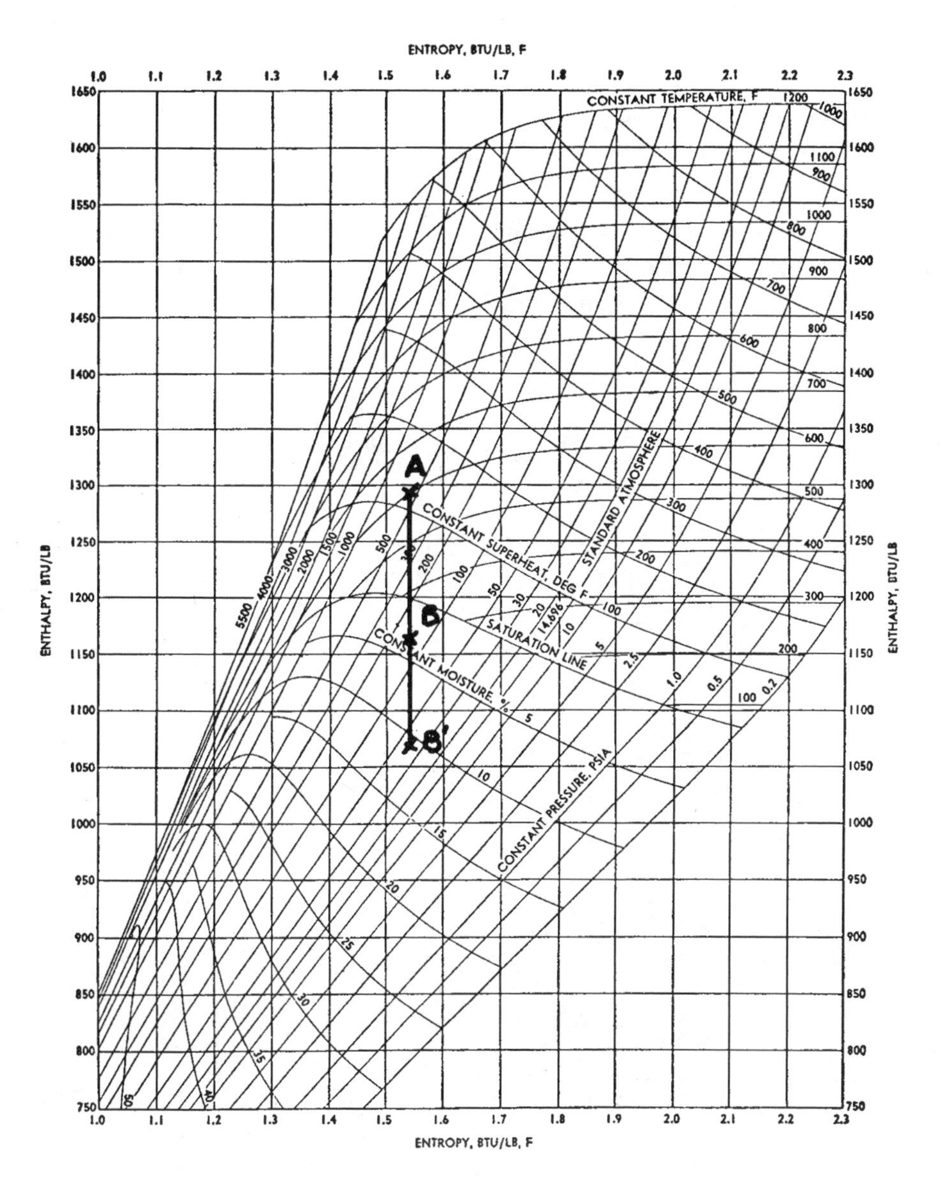

Figure 5-12. Turbine performance by use of Mollier Diagram.

the amount of energy available to the turbine to generate horsepower. The amount of condensate formed is indicated by the intersection with the constant moisture line. At point "*B*" the constant moisture is approximately 3 percent.

The material balance is then drawn for Turbine #1. Refer to Figure 5-13A.

$$X = 78{,}300 + 0.03X$$
$$0.97X = 78{,}300$$
$$X = 80{,}721 \text{ lb/hr to turbine}$$

$$150 \text{ psigcondensate} = 2{,}421 \text{ lb/hr}$$

$$hp = \frac{\text{Steam (lb/hr)} \times \Delta h \times n}{2545} \quad (5\text{-}8)$$

Equation 5-8 is based on energy Equation 5-2. Turbine efficiency (n) and conversion factor (2,545 Btu/hp) are incuded.

For Turbine No. 1

$$hp = \frac{80721 \times (1290 - 1170) \times 0.5}{2545}$$

$$hp = 1903$$

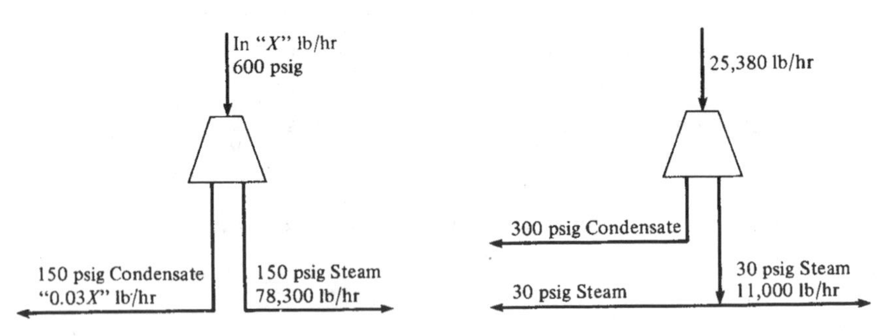

Figure 5-13. Material balances for turbines.

Material Balance for Turbine No. 2
Steps 1 & 2: Since the total horsepower requirement is 3,000, Turbine No. 2 must supply 3,000 – 1,903 = 1,097. This determines the steam rate to the turbine. To find the percent of moisture draw a vertical line from point "*A*" of Figure 5-12 until it intersects the 45 psia constant pressure line. (h = 1,070) and the constat moisture = 11%. Thus,

$$\text{hp} = \frac{X\left(1290 - 1070\right) \times 0.5}{2545} = 1097$$

$$X = 25{,}380 \text{ lb/hr}$$

A material balance for Turbine No. 2 is drawn as illustrated in Figure 5-13B.

30 psig condensate = 0.11 (25,380) = 2,791 lb/hr

30 psig steam = 25,380 – 11,000 – 2791 = 11,589 lb/hr

The total balance is illustrated in Figure 5-14.

Step 3: The energy balance of the system becomes:

$$\begin{aligned}
106{,}101\ (1290) &= 89{,}300\ (38) \\
&\quad + 2{,}791\ (243) + 11{,}589\ (1070) \\
&\quad + 2{,}421\ (338) \\
&\quad + \text{input energy} \\
\text{Input energy} &= 136.8 \times 10^6 - 3.39 \times 10^6 \\
&\quad - .67 \times 10^6 - 12.4 \times 10^6 \\
&\quad - .81 \times 10^6 \\
&= 119.53 \times 10^6 \text{ Btu/hr}
\end{aligned}$$

Fuel required is $119.53 \times 10^6 \div 0.8 = 149.4 \times 10^6$ Btu/hr.
The fuel cost = 149.4 × \$4.0 × 4000 = \$2,390,400/yr.
Savings for Alternate No. 2 without equipment cost amortization: \$2,576,903 – \$2,390,400 = \$186,503.

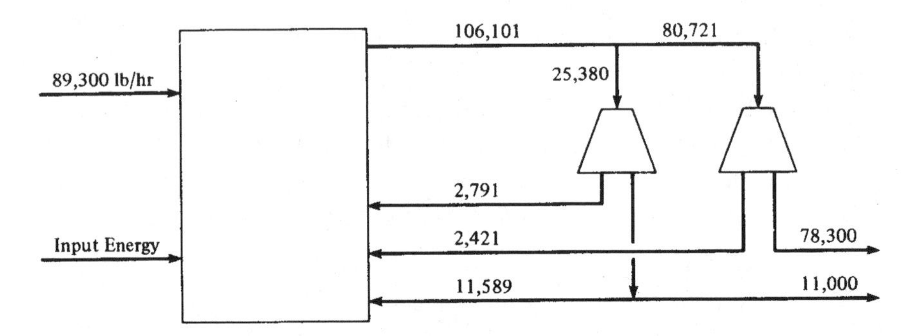

Figure 5-14. Total material balance for steam generation.

PUMPS AND PIPING SYSTEMS

Rules of thumb have governed the selection of pipe sizing for pump applications. With the cost of energy increasing, many of these rules of thumb are not valid. The basic energy balance (Equation 5-1) applies as well to fluid flow. Basically, the size of discharge line piping from the pump determines the friction loss through the pipe that the pump must overcome. The greater the line loss, the more pump horsepower required. If the line is short or has a small flow, this loss may not be significant in terms of the total system head requirements. On the other hand, if the line is long and has a large flow rate, the line loss will be significant. Therefore, an economic analysis should be made on each pumping system. Table 5-7 (p. 136-137) illustrates typical pressure drops for various flows and pipe diameters. The table is based on Crane Technical Paper 410. To compute the rake horsepower of a pump:

$$\text{Brake horsepower} = \frac{\text{GPM} \times \Delta \text{P}}{1715 \times \eta} = 1097 \tag{5-9}$$

The brake horsepower represents the actual horsepower required.

where

ΔP is the differential pressure across the pump in psi.

η is the pump efficiency

GPM is the required flow in gallons per minute. Figure 5-15 illustrates a typical pumping system and the associated pressure drops. In order to compute the differential pressure across a pump, the pressure drops need to be computed and the discharge pressure, suction pressure, and the relative pumping elevations need to be known.

Figure 5-16 illustrates a pump and piping worksheet for computing brake horsepower. The Net Positive Suction Head represents the minimum head available at the pump suction that will permit it to operate. Note: The pump efficiency and the Net Positive Suction Head of the pump should be checked against the actual pump selected.

Another equation which relaes psi to head in feet is:

$$\text{Head in feet} = \frac{\text{psi} \times 2.31}{\text{specific gravity of fluid}} \tag{5-10}$$

SIM 5-11

Due to corrosion, the pump discharge piping required for six identical processes will need to be replaced. Based on the flow rate, either 8 in., 10 in., or 12 in. piping can be used. Comment on the most economical piping system, given the following:

Process and Design Conditions	
Delivery pressure	80 psia
Flow	2,000 GMP
Static head discharge	13.4 psi
Suction pressure	15 psi
Suction static head	—
Suction line loss	1 psi
Pump efficiency	0.7
	Assume 150 hp motor will be applicable for the three schemes with an efficiency and P.F. of 0.9

Piping data
6-90° elbows
300 feet of pipe
2 gate valves

The additional costs to install the three systems are:

10″ – \$3,000 $i = 10\%$
12″ – \$6,900 $n = 10$

The cost for electricity is \$0.045/kWh.
Refer to Figure 5-15 for pump system and exclude control valve.

Table 5-7. Flow of water through schedule 40 steel pipe.

Discharge		Pressure Drop per 100 feet and Velocity in Schedule 40 Pipe for Water at 60°F.															
Gallons per Minute	Cubic Ft. per Second	Velocity Feet per Second	Press. Drop Lbs. per Sq. In.	Velocity Feet per Second	Press. Drop Lbs. per Sq. In.	Velocity Feet per Second	Press. Drop Lbs. per Sq. In.	Velocity Feet per Second	Press. Drop Lbs. per Sq. In.	Velocity Feet per Second	Press. Drop Lbs. per Sq. In.	Velocity Feet per Second	Press. Drop Lbs. per Sq. In.	Velocity Feet per Second	Press. Drop Lbs. per Sq. In.	Velocity Feet per Second	Press. Drop Lbs. per Sq. In.
		1/8"		1/4"		3/8"		1/2"									
.2	0.000446	1.13	1.86	0.616	0.359												
										3/4"							
.3	0.000668	1.69	4.22	0.924	0.903	0.504	0.159	0.317	0.061								
.4	0.000891	2.26	6.98	1.23	1.61	0.672	0.345	0.422	0.086								
.5	0.00111	2.82	10.5	1.54	2.39	0.840	0.539	0.528	0.167	0.301	0.033						
.6	0.00134	3.39	14.7	1.85	3.29	1.01	0.751	0.633	0.240	0.361	0.041						
.8	0.00178	4.52	25.0	2.46	5.44	1.34	1.25	0.844	0.408	0.481	0.102						
												1"		1¼"		1½"	
1	0.00223	5.65	37.2	3.08	8.28	1.68	1.85	1.06	0.600	0.602	0.155	0.371	0.048				
2	0.00446	11.29	134.4	6.16	30.1	3.36	6.58	2.11	2.10	1.20	0.526	0.743	0.164	0.429	0.044		
3	0.00668			9.25	64.1	5.04	13.9	3.17	4.33	1.81	1.09	1.114	0.336	0.644	0.090	0.473	0.043
4	0.00891			12.33	111.2	6.72	23.9	4.22	7.42	2.41	1.83	1.49	0.565	0.858	0.150	0.630	0.071
5	0.01114	2"				8.40	36.7	5.28	11.2	3.01	2.75	1.86	0.835	1.073	0.223	0.788	0.104
				2½"													
6	0.01337	0.574	0.044			10.08	51.9	6.33	15.8	3.61	3.84	2.23	1.17	1.29	0.309	0.946	0.145
8	0.01782	0.765	0.073			13.44	91.1	8.45	27.7	4.81	6.60	2.97	1.99	1.72	0.518	1.26	0.241
10	0.02228	0.956	0.108	0.670	0.046			10.56	42.4	6.02	9.99	3.71	2.99	2.15	0.774	1.58	0.361
15	0.03342	1.43	0.224	1.01	0.094	3"				9.03	21.6	5.57	6.36	3.22	1.63	2.37	0.755
20	0.04456	1.91	0.375	1.34	0.158	0.868	0.056	3½"		12.03	37.8	7.43	10.9	4.29	2.78	3.16	1.28
										4"							
25	0.05570	2.39	0.561	1.68	0.234	1.09	0.083	0.812	0.041			9.28	16.7	5.37	4.22	3.94	1.93
30	0.06684	2.87	0.786	2.01	0.327	1.30	0.114	0.974	0.056			11.14	23.8	6.44	5.92	4.73	2.72
35	0.07798	3.35	1.05	2.35	0.436	1.52	0.151	1.14	0.704	0.882	0.041	12.99	32.2	7.51	7.90	5.52	3.64
40	0.08912	3.83	1.35	2.68	0.556	1.74	0.192	1.30	0.095	1.01	0.052	14.85	41.5	8.59	10.24	6.30	4.65
45	0.1003	4.30	1.67	3.02	0.668	1.95	0.239	1.46	0.117	1.13	0.064			9.67	12.80	7.09	5.85
												5"					
50	0.1114	4.78	2.03	3.35	0.839	2.17	0.288	1.62	0.142	1.26	0.076			10.74	15.66	7.88	7.15
60	0.1337	5.74	2.87	4.02	1.18	2.60	0.406	1.95	0.204	1.51	0.107			12.89	22.2	9.47	10.21
70	0.1560	6.70	3.84	4.69	1.59	3.04	0.540	2.27	0.261	1.76	0.143	1.12	0.047			11.05	13.71
80	0.1782	7.65	4.97	5.36	2.03	3.47	0.687	2.60	0.334	2.02	0.180	1.28	0.060			12.62	17.59
90	0.2005	8.60	6.20	6.03	2.53	3.91	0.861	2.92	0.416	2.27	0.224	1.44	0.074	6"		14.20	22.0
100	0.2228	9.56	7.59	6.70	3.09	4.34	1.05	3.25	0.509	2.52	0.272	1.60	0.090	1.11	0.036	15.78	26.9
125	0.2785	11.97	11.76	8.38	4.71	5.43	1.61	4.06	0.769	3.15	0.415	2.01	0.135	1.39	0.055	19.72	41.4
150	0.3342	14.36	16.70	10.05	6.69	6.51	2.24	4.87	1.08	3.78	0.580	2.41	0.190	1.67	0.077		
175	0.3899	16.75	22.3	11.73	8.97	7.60	3.00	5.68	1.44	4.41	0.774	2.81	0.253	1.94	0.102	8"	
200	0.4456	19.14	28.8	13.42	11.68	8.68	3.87	6.49	1.85	5.04	0.985	3.21	0.323	2.22	0.130		
225	0.5013	...	...	15.09	14.63	9.77	4.83	7.30	2.32	5.67	1.23	3.61	0.401	2.50	0.162	1.44	0.043
250	0.557	...	...	...	...	10.85	5.93	8.12	2.84	6.30	1.46	4.01	0.495	2.78	0.195	1.60	0.051
275	0.6127	...	...	...	...	11.94	7.14	8.93	3.40	6.93	1.79	4.41	0.583	3.05	0.234	1.76	0.061
300	0.6684	...	...	...	...	13.00	8.36	9.74	4.02	7.56	2.11	4.81	0.683	3.33	0.275	1.92	0.072
325	0.7241	...	...	...	...	14.12	9.89	10.53	4.09	8.19	2.47	5.21	0.797	3.61	0.320	2.08	0.083

		1		2		3		4		5		6		7		8	
350	0.7798			...	...	...	...	11.36	**5.41**	8.82	**2.84**	5.62	**0.919**	3.89	**0.367**	2.24	**0.095**
375	0.8355			...	...	...	...	12.17	**6.18**	9.45	**3.25**	6.02	**1.05**	4.16	**0.416**	2.40	**0.108**
400	0.8912			...	...	...	...	12.98	**7.03**	10.08	**3.68**	6.42	**1.19**	4.44	**0.471**	2.56	**0.121**
425	0.9469			...	...	...	...	13.80	**7.89**	10.71	**4.12**	6.82	**1.33**	4.72	**0.529**	2.73	**0.136**
450	1.003	**10″**		...	...	...	...	14.61	**8.80**	11.34	**4.60**	7.22	**1.48**	5.00	**0.590**	2.89	**0.151**
475	1.059	1.93	**0.054**			...	...	...	...	11.97	**5.12**	7.62	**1.64**	5.27	**0.653**	3.04	**0.166**
500	1.114	2.03	**0.059**			...	...	...	...	12.60	**5.65**	8.02	**1.81**	5.55	**0.720**	3.21	**0.182**
550	1.225	2.24	**0.071**			...	...	...	...	13.85	**6.79**	8.82	**2.17**	6.11	**0.861**	3.53	**0.219**
600	1.337	2.44	**0.083**			...	...	...	...	15.12	**8.04**	9.63	**2.55**	6.66	**1.02**	3.85	**0.258**
650	1.448	2.64	**0.097**	**12″**		...	...	...	...	...	...	10.43	**2.98**	7.22	**1.18**	4.17	**0.301**
700	1.560	2.85	**0.112**	2.01	**0.047**			...	...	...	...	11.23	**3.43**	7.78	**1.35**	4.49	**0.343**
750	1.671	3.05	**0.127**	2.15	**0.054**	**14″**		...	...	...	...	12.03	**3.92**	8.33	**1.55**	4.81	**0.392**
800	1.782	3.25	**0.143**	2.29	**0.061**			...	...	...	...	12.83	**4.43**	8.88	**1.75**	5.13	**0.443**
850	1.894	3.46	**0.160**	2.44	**0.068**	2.02	**0.042**	...	...	...	...	13.64	**5.00**	9.44	**1.96**	5.45	**0.497**
900	2.005	3.66	**0.179**	2.58	**0.075**	2.13	**0.047**	...	...	...	...	14.44	**5.58**	9.99	**2.18**	5.77	**0.554**
950	2.117	3.86	**0.198**	2.72	**0.083**	2.25	**0.052**			...	...	15.24	**6.21**	10.55	**2.42**	6.09	**0.613**
1 000	2.228	4.07	**0.218**	2.87	**0.091**	2.37	**0.057**	**16″**		...	...	16.04	**6.84**	11.10	**2.68**	6.41	**0.675**
1 100	2.451	4.48	**0.260**	3.15	**0.110**	2.61	**0.068**			...	...	17.65	**8.23**	12.22	**3.22**	7.05	**0.807**
1 200	2.674	4.88	**0.306**	3.44	**0.128**	2.85	**0.080**	2.18	**0.042**	...	...	...	...	13.33	**3.81**	7.70	**0.948**
1 300	2.896	5.29	**0.355**	3.73	**0.150**	3.08	**0.093**	2.36	**0.048**	...	...	...	...	14.43	**4.45**	8.33	**1.11**
1 400	3.119	5.70	**0.409**	4.01	**0.171**	3.32	**0.107**	2.54	**0.055**					15.55	**5.13**	8.98	**1.28**
1 500	3.342	6.10	**0.466**	4.30	**0.195**	3.56	**0.122**	2.72	**0.063**	**18″**				16.66	**5.85**	9.62	**1.46**
1 600	3.565	6.51	**0.527**	4.59	**0.219**	3.79	**0.138**	2.90	**0.071**					17.77	**6.61**	10.26	**1.65**
1 800	4.010	7.32	**0.663**	5.16	**0.276**	4.27	**0.172**	3.27	**0.088**	2.58	**0.050**			19.99	**8.37**	11.54	**2.08**
2 000	4.456	8.14	**0.808**	5.73	**0.339**	4.74	**0.209**	3.63	**0.107**	2.87	**0.060**	**20″**		22.21	**10.3**	12.82	**2.55**
2 500	5.570	10.17	**1.24**	7.17	**0.515**	5.93	**0.321**	4.54	**0.163**	3.59	**0.091**					16.03	**3.94**
3 000	6.684	12.20	**1.76**	8.60	**0.731**	7.11	**0.451**	5.45	**0.232**	4.30	**0.129**	3.46	**0.075**	**24″**		19.24	**5.59**
3 500	7.798	14.24	**2.38**	10.03	**0.982**	8.30	**0.607**	6.35	**0.312**	5.02	**0.173**	4.04	**0.101**			22.44	**7.56**
4 000	8.912	16.27	**3.08**	11.47	**1.27**	9.48	**0.787**	7.26	**0.401**	5.74	**0.222**	4.62	**0.129**	3.19	**0.052**	25.65	**9.80**
4 500	10.03	18.31	**3.87**	12.90	**1.60**	10.67	**0.990**	8.17	**0.503**	6.46	**0.280**	5.20	**0.162**	3.59	**0.065**	28.87	**12.2**
5 000	11.14	20.35	**4.71**	14.33	**1.95**	11.85	**1.21**	9.08	**0.617**	7.17	**0.340**	5.77	**0.199**	3.99	**0.079**	...	...
6 000	13.37	24.41	**6.74**	17.20	**2.77**	14.23	**1.71**	10.89	**0.877**	8.61	**0.483**	6.93	**0.280**	4.79	**0.111**	...	...
7 000	15.60	28.49	**9.11**	20.07	**3.74**	16.60	**2.31**	12.71	**1.18**	10.04	**0.652**	8.08	**0.376**	5.59	**0.150**	...	...
8 000	17.82	...	...	22.93	**4.84**	18.96	**2.99**	14.52	**1.51**	11.47	**0.839**	9.23	**0.488**	6.38	**0.192**	...	...
9 000	20.05	...	...	25.79	**6.09**	21.34	**3.76**	16.34	**1.90**	12.91	**1.05**	10.39	**0.608**	7.18	**0.242**	...	...
10 000	22.28	...	...	28.66	**7.46**	23.71	**4.61**	18.15	**2.34**	14.34	**1.28**	11.54	**0.739**	7.98	**0.294**	...	...
12 000	26.74	...	...	34.40	**10.7**	28.45	**6.59**	21.79	**3.33**	17.21	**1.83**	13.85	**1.06**	9.58	**0.416**	...	...
14 000	31.19	...	...	...	...	33.19	**8.89**	25.42	**4.49**	20.08	**2.45**	16.16	**1.43**	11.17	**0.562**	...	...
16 000	35.65	...	...	...	...	...	...	29.05	**5.83**	22.95	**3.18**	18.47	**1.85**	12.77	**0.723**	...	...
18 000	40.10	...	...	...	...	...	...	32.68	**7.31**	25.82	**4.03**	20.77	**2.32**	14.36	**0.907**	...	...
20 000	44.56	...	...	...	...	...	...	36.31	**9.03**	28.69	**4.93**	23.08	**2.86**	15.96	**1.12**	...	...

For pipe lengths other than 100 feet, the pressure drop is proportional to the length. Thus, for 50 feet of pipe, the pressure drop is approximately one-half the value given in the table . . . for 300 feet, three times the given value, etc.

Velocity is a function of the cross sectional flow area; thus, it is constant for a given flow rate and is independent of pipe length.

Reproduced from Technical Paper No. 410, Courtesy Crane Co.

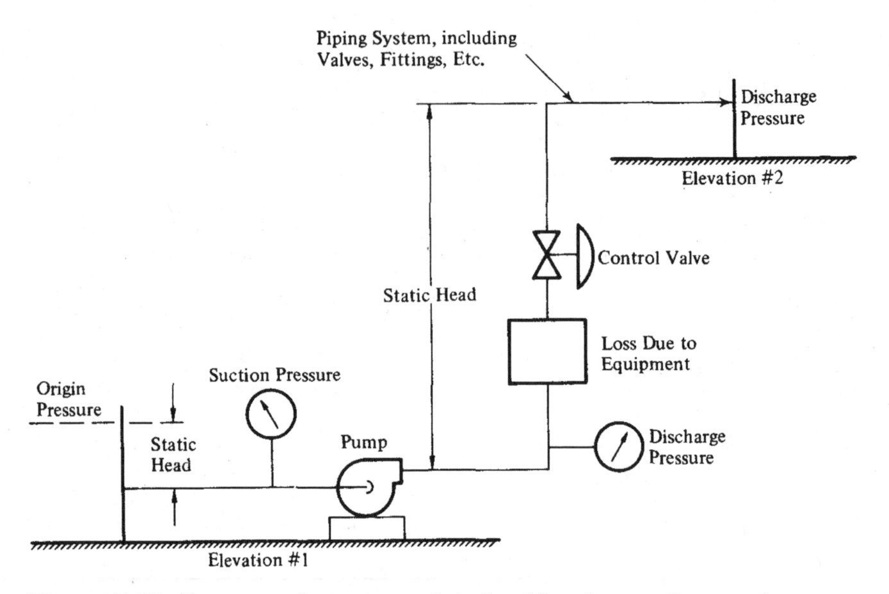

Figure 5-15. Pressure drops associated with pipe and pumping systems.

ANSWER

		8″	10″	12″
1.	Discharge pressure (psia)	80	80	80
2.	Static head (psi)	13.4	13.4	13.4
	Line Loss			
	Straight pipe (ft)	300	300	300
	6-90° EL - (ft)	87	111	132
	2-Gate valves (ft)	17	22	26
	Equivalent length (ft)	404	433	458
	Pi (psi/100 ft) from Table 5-3	2.55	0.808	0.339
3.	Line loss total (psi)	10.3	3.49	1.55
	Total discharge pressure (psi) (1 + 2 + 3)	103.7	96.89	94.95
	Origin pressure (psi)	15	15	15
	Static loss			
	Line loss (suction)	1	1	1
4.	Total suction pressure (psi)	14	14	14

Figure 5-16. Pump and piping worksheet.

No. of Fittings or Total Length		Le	Equivalent Length Total L1 Suction	Equivalent Length Total L1 Discharge
______	ft. of straight pipe		______	______
______	90° Ells ×	______	______	______
______	45° Ells ×	______	______	______
______	Tee Run ×	______	______	______
______	Tee Branch ×	______	______	______
______	Gate Valves ×	______	______	______
______	Globe Valves ×	______	______	______
______	Check Valves ×	______	______	______
	Total		______	______

Delivery Pressure	______	psia
Static Head	______	psi
Line Loss = ≤ × P_1 (Table 5-3)	______	psi
ΔP Control Valves (Allow 10 psi minimum per valve)	______	psi
ΔP Other Losses	______	psi
1. Total Discharge Pressure	______	psia
Suction Pressure		
Origin Pressure	______	psia
Static Head	______	psi
– Line Loss ≤ × P_1	______	psi
2. Total Suction	______	psia
Differential Pressure		
3. Discharge-Suction Pressure ΔP	______	psi

$$\text{Brake horsepower} = \frac{\text{GPM} \times \Delta P}{1715 \times \eta}$$

*Standard Radius Fittings—Butt Welded

Note: For energy comparisons only exit and entrance losses, reducers and expanders have been neglected.

	Pipe Size (inches) Le—Values													
Valve or Fitting Type	1	1-1/2	2	3	4	6	8	10	12	14	16	18	20	24
*90° El	1.5	3	3.5	6	7.5	11	14.5	18.5	22	22.5	29.5	33	37	44
*45° El	1	1.5	1.5	2.5	3.5	5	6.5	8.5	10	11.5	13.5	15	16.5	20
*Tee Run	1	1	1.5	2	3	4	5	7	8	9	11	12	13	16
*Tee Branch	5	7	10	14.5	19	29	39	48	58	68	78	87	97	116
Gate Valve (Full Open)	1	1.5	2	3	4.5	6.5	8.5	11	13	15	17	19.5	22	26
Globe Valve	28	42.5	57	85	113	170	227	283	340	397	453	510	567	686
Check Valve	11	17	22.5	34	45	67.5	90	112.5	135	157.5	180	202.5	225	270

5.	Differential pressure (psi)	89.7	82.89	80.95
6.	Brake horsepower	149.4	138	134.8
7.	Yearly energy cost $\left(0.045 \times 8000 \times \dfrac{hp \times .746}{n \times PF}\right)$	\$49,533	\$45,753	\$44,691
8.	Additional annual owning cost $P \times CR = P \times .162$		486	1,116
9.	Total yearly cost	\$49,533	\$46,239	\$45,807
			Choice	________

For six identical systems, a yearly savings before taxes of \$22,356 is possible by using the 12" piping system.

DISTILLATION COLUMNS

In chemical plants, a large percentage of the steam generated is used for distillation processes. A distillation column is, essentially, a column filled with trays. The liquid to be separated is heated in an exchanger called a reboiler, as illustrated in Figure 5-17. The reboiler is usually heated by steam or another heated fluid. The overhead vapors are condensed prior to discharge to the remainder of the process. A portion of the overhead is refluxed back to the column to improve separation efficiency. The traditional design uses a minimum number of trays inside the column in order to minimize first cost. The reflux back to the column is usually a set flow rate regardless of column feed rate.

Several ways to minimize steam usage are as follows:

1. The reflux to the column should be based on a fixed ratio overhead rate with overhead and bottom concentrations at minimum quality requirements. It should be noted that steam usage is directly dependent on the quantity of overhead vapors and bottoms liquids. A small increase in overhead and bottoms can increase steam usage significantly. Thus, by minimizing reflux, the energy usage will be decreased.

2. Distillation columns designers should lean to using more trays in the columns. More trays will reduce required reflux rate and, therefore, reduce steam usage.

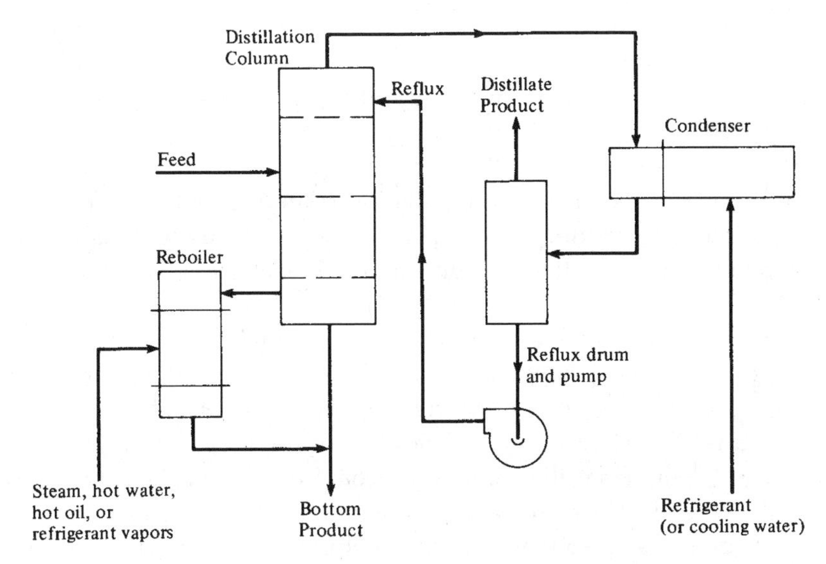

Figure 5-17. Typical flowsheet of distillation process.

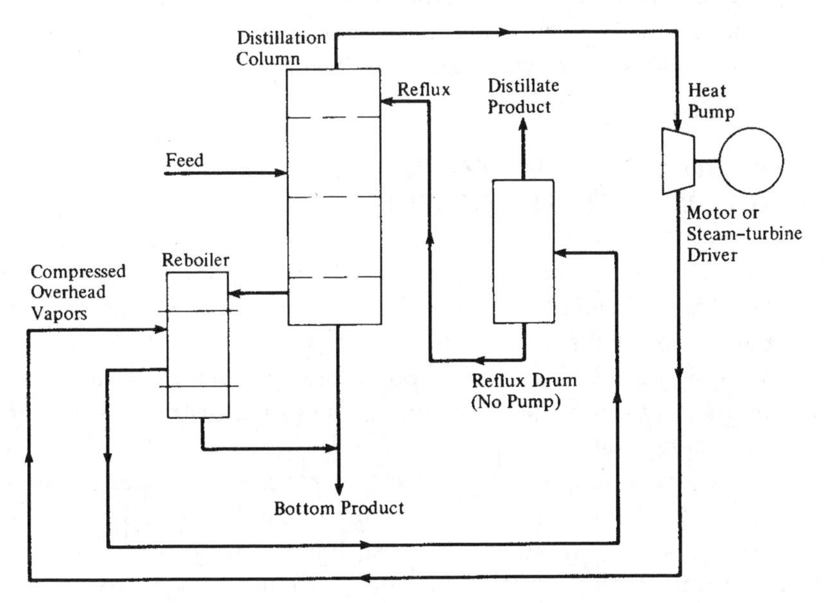

Figure 5-18. Typical flowsheet of vapor recompression distillation.

3. Distillation columns should be operated near flooding conditions. At reduced feed rates, separation efficiency decreases. To correct this condition, the normal top operating pressure should be reduced.

4. The feed location to the distillation column should be designed for the temperature and composition of the incoming liquid. If the process changes, the location should be changed.

5. For processes such as the separation of ethane from ethylene, or propane from propylene, or distillation processes run at lower pressures, vapor recompression distillation should be considered. Figure 5-17 shows a typical flow sheet for vapor recompression. The vapors from the column are compressed and then routed to the reboiler. The overhead vapor temperature is raised by the heat of compression so that it has the required ΔT to drive the reboiler. The condensed overhead vapors are then sent to the reflux drum and pressure controlled to provide reflux to the column without a reflux pump. The only additional energy requirements are for the compressor driver, which can be a steam turbine or heat pump.

INCORPORATION OF ENERGY UTILIZATION IN PROCUREMENT SPECIFICATIONS

The distillation column's first cost is greatly influenced by the number of trays inside the vessel. The more trays, the higher the first cost, but the lower the energy cost.

To design an energy-efficient pump-piping system requires more man hours. Optimum discharge pipe design as illustrated saves on lower operating costs.

These two examples emphasize that procurement specifications that encourage choices based on lowest lump sum bids greatly penalize an energy utilization program. The best time to save energy is during initial design. Procurement specifications should reflect "cost plus" bids where the client can ensure that energy utilization measures are considered during the design phase.

6

Heat Transfer

THE IMPORTANCE OF UNDERSTANDING THE PRINCIPLES OF HEAT TRANSFER

The principles of heat transfer are the fundamental building blocks needed to understand how heat losses occur and how they can be minimized. Heat transfer finds application in equipment sizing as well. For instance, a heat exchanger is used to transfer heat from one fluid to another. Thus, heat transfer applications are involved with energy transfer in equipment, piping systems, and building. In Chapter 7, heat transfer applications to building design will be presented. In this chapter, you will learn three modes of heat transfer, see how much energy is lost from uninsulated tanks and pipes, and discover how to use economic insulation thickness tables to reduce heat losses.

THREE WAYS HEAT IS TRANSFERRED

Heat transfer is determined by the effects of conduction, radiation and convection. The three modes can be thought of simply as follows:

Conduction—Heat transfer is based on one space surrendering heat while another one gains it by the ability of the dividing surface to conduct heat.

Radiation—Heat transfer is based on the properties of light, where no surface or fluid is needed to carry heat from one object to another.

Convection—Heat transfer is based on the exchange of heat between a fluid, gas, or liquid as it transverses a conducting surface.

Heat Transfer By Conduction

Various metals conduct heat differently. Metals are the best conductors of heat, while wood, asbestos, and felt are the poorer ones. The physical property which relates the ability of a material to transmit heat is referred to as the *conductivity* K of a material. The units of K are expressed as Btu • in/hr • ft^2 • °F or Btu • in per hr • ft^2 • degree F. Thus, K represents the amount of heat in Btu flowing through a one inch thick, one square foot area of homogeneous material in one hour with a temperature differential of one degree. Table 6-1 indicates typical conductivities of various metals at room temperature. Thermal conductivities of most pure metals decrease with an increase in temperature.

Table 6-1. Thermal conductivity at room temperature for various metals. Btu in/ft^2 • hr • °F.

Description	K
Aluminum (alloy 1100)	1536
Aluminum bronze (76% Cu, 22% Zn, 2% Al)	696
Brass:	
red (85% Cu, 15% Zn)	1044
yellow (65% Cu, 35% Zn)	828
Bronze	204
Copper (electrolytic)	2724
Gold	2064
Iron:	
Cast	331
Wrought	418.8
Nickel	412.8
Platinum	478.8
Silver	2940
Steel (mild)	314.4
Zinc:	
Cast	780
Hot rolled	744

For heat transfer calculations, it is satisfactory to assume a constant conductivity at the average temperature of the material. As an example, 0.23 carbon has a K of 350 at 100°F and varies linearly to 300 at 700°F.

Conduction Through A Flat Surface

When a flat plate is heated on one side and cooled on the other, as indicated in Figure 6-1, a flow of heat from the hot side to the cold side will occur. The flow of heat is defined as:

$$q = \frac{K}{d}A(t_2 - t_1) = \frac{KA}{d}\Delta T \tag{6-1}$$

where

q is the rate of heat flow Btuh
d is the thickness of the material in inches
A is the area of the plate in ft^2
$t_2 - t_1$ is the temperature difference ΔT, causing heat flow, °F.

Figure 6-2 illustrates a composite of several materials or a nonhomogenous material.

The flow of heat for this case is defined as:

$$q = UA\ (t_0 - t_5) = UA\ \Delta T \text{ Btuh} \tag{6-2}$$

where U is referred to as the conductance or coefficient of transmission of the material. Btu per hr • ft^2 • °F. In Figure 6-2, a surface film conductance is introduced. The surface or film conductance is

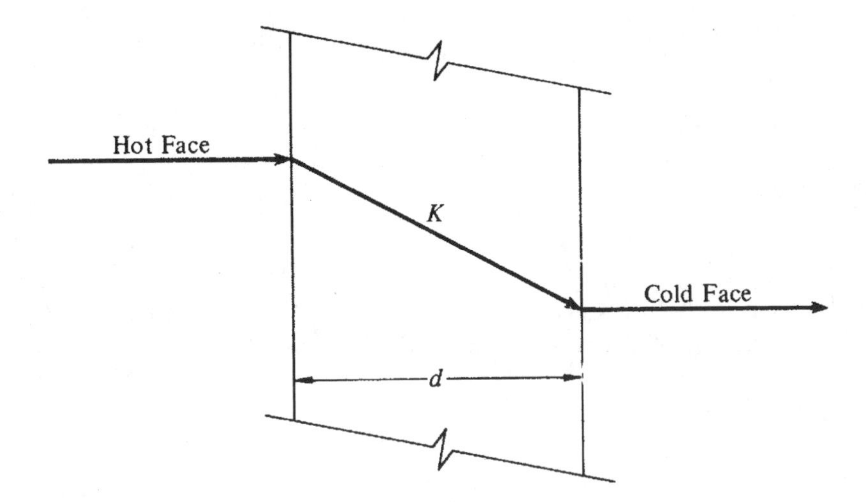

Figure 6-1. Temperature distribution for the single wall with conductivity *K*.

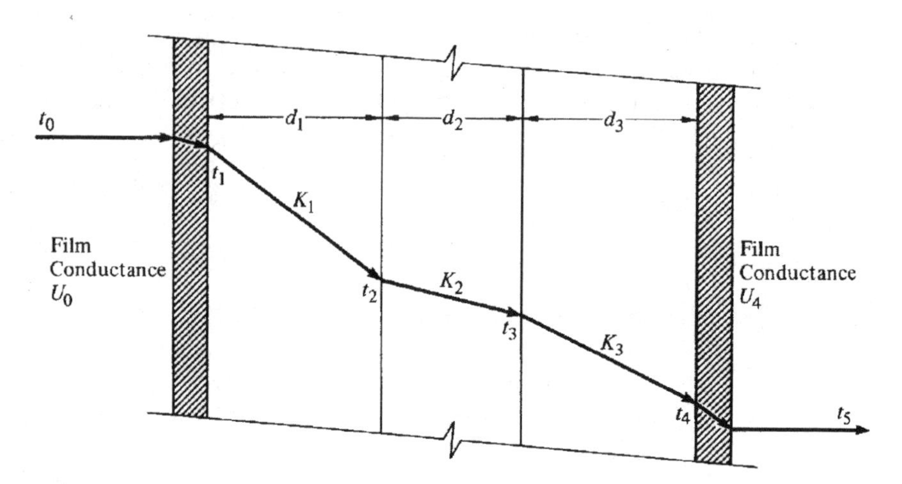

Figure 6-2. Temperature distribution for the composite wall.

the amount of heat transferred in Btu per hour from a surface to air or from air to a surface per square foot for one degree difference in temperature. The flow of heat for the composite material can also be specified in terms of the conductivity of the material and the conductance of the air film:

$$q = \frac{A(t_0 - t_5)}{1/U_0 + d_1/K_1 + d_2/K_2 + d_3/K_3 + 1/U_4}. \tag{6-3}$$

The conductance of a homogeneous material is expressed as

$$u = \frac{K}{D} \tag{6-4}$$

The reciprocal of conductance is referred to as the resistance (R). For a unit length and through a unit area, R becomes:

$$R = \frac{1}{U} = \frac{d}{K} \tag{6-5}$$

The resistance of a material is directly analogous to that of electrical circuits. Heat transfer problems are solved using electrical analogies.

Referring to Figure 6-2, each material is thought of as having a heat flow resistance of d/K or $1/U$.

In series circuits, resistances are added; thus

$$R = R_1 + R_2 + R_3 + R_4 + R_5 \qquad \text{(6-6}$$

$$R = \frac{1}{U_0} + \frac{d_1}{K_1} + \frac{d_2}{K_2} + \frac{d_3}{K_3} + \frac{1}{U_4} \qquad \text{(6-5)}$$

SIM 6-1

What is the heat loss through six inches of mild steel at room temperature, given the following:

$$\Delta T = 15°F$$
$$\text{Area} = 100 \text{ ft}^2$$

Exclude the effect of surface film conductance.

ANSWER

From Table 6-1, $K = 314.4$

$$q = \frac{314.4}{6} \times 100 \times 15 = 78{,}600 \text{ Btuh.}$$

SIM 6-2

The surface film resistance of galvanized steel for still air is 1.85. Calculate the heat flow when this factor is included in SIM 6-1.

ANSWR

$$q = \frac{A\Delta T}{R}$$

$$R = 2\ (1.85) + \frac{6}{314.4} = 3.70 + 0.019 = 3.719$$

$$q = \frac{100 \times 15}{3.719} = 403.3 \text{ Btuh.}$$

SIM 6-3

A 1 in. layer of felt fiberglass insulation having a conductivity of 0.3 is added to the plate of SIM 6-1. The outside air film resistance then becomes 0.68 while the inside air film resistance remains at 1.85. Comment on the heat loss of the plate with insulation.

ANSWER

$$R = 1.85 + 0.68 + 1 \frac{1}{0.3} + \frac{6}{314} = 5.88$$

$$q = \frac{100 \times 15}{5.88} = 255 \text{ Btuh.}$$

Conduction through a Cylindrical Surface

A common problem facing the engineer is the computation of heat flow through a cylindrical surface. The surface area of a cylinder is not constant as the distance from the center increases; thus, the basic heat flow equation must be modified.

Figure 6-3 indicates the heat flow through a composite cylinder. In this case, the heat flow is: (temperatures, conductivities, and film conductances are defined in the figure)

$$q = \frac{A(t_0 - t_4)}{(d_3/d_1)(1/U_0)+(d_3/2K_1)\log_e d_2/d_1+(d_3/2K_2)\log_e d_3/d_2+1/U_4} \tag{6-8}$$

where

$A = \pi L d_3$
L = length of pipe.

Heat Transfer By Radiation

Radiation is the transfer of radiant energy from a source to a receiver. Radiation from a source is partially absorbed by the receiver and partially reflected. The radiation emitted depends upon its surface emissivity, area, and temperature, as illustrated by the following equation:

$$q = \varepsilon\sigma A T^4 \tag{6-9}$$

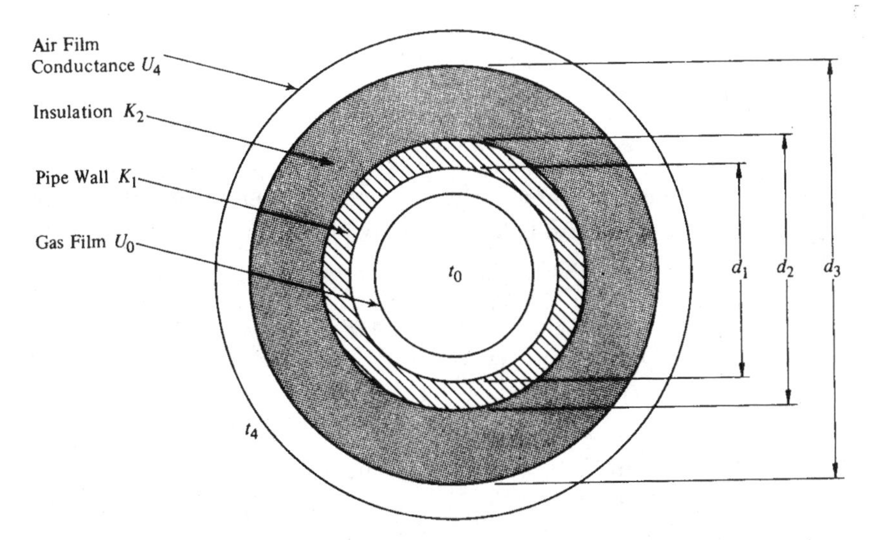

Figure 6-3. Heat flow through the composite cylindrical wall.

where

q = rate of heat, flow by radiation Btu/hr

ε = emissivity of a body, which is defined as the ratio of energy radiated by the actual body to that of a black body. $\varepsilon = 1$, for a black body

σ = Stefan Boltzmann Constant, 1.71×10^{-9} Btu/ft^2 • hr • T^4

A = surface area of body in square feet.

Heat Transfer By Convection

Convection is the transfer of heat between a fluid, gas, or liquid. Equation (6-2) is indicative of the basic form of convective heat transfer. U, in this case, represents the convection film conductance, Btu/ft^2 • hr • °F.

Heat transferred for heat exchanger applications is predominantly a combination of conduction and convection expressed as:

$$q = U_0 A \Delta T_m \tag{6-10}$$

where

q = rate of heat flow by convection, Btu/hr

U_0 = overall heat transfer coefficient Btu/ft^2 • hr • °F

A = area of the tubes in square feet

ΔT_m = logarithmic mean temperature difference and represents the situation where the temperatures of two fluids change as they transverse the surfce.

$$\Delta T_m = \frac{\Delta T_1 - \Delta T_2}{\text{Log}_e\ [\Delta T_1/\Delta T_2]} \qquad (6\text{-}11)$$

To understand the different logarithmic mean temperature relationships, Figure 6-4 should be used. Referring to Figure 6-4, the ΔT_m for the counterflow heat exchanger is:

$$\Delta T_m = \frac{(T_1 - T'_2) - (T_2 - T'_1)}{\text{Log}_e\ [T_1 - T'_2/T_2 - T'_1]} \qquad (6\text{-}12a)$$

The ΔT_m for the parallel flow heat exchanger is:

$$\Delta T_m = \frac{\Delta T_1 - \Delta T_2}{\text{Log}_e\ [\Delta T_1/\Delta T_2]} \qquad (6\text{-}12b)$$

Finned-Tube Heat Exchangers

When waste heat in exhaust gases is recovered for heating liquids for purposes such as providing domestic hot water, for heating the feedwater for steam boilers, or for hot water space heating, the finned-tube heat exchanger is generally used. Round tubes are connected together in bundles to contain the heated liquid and fins are welded or otherwise attached to the outside of the tubes to provide additional surface area for removing the waste heat in the gases. Figure 6-5 shows the usual arrangement for the finned-tube exchanger positioned in a duct and details of a typical finned-tube construction. This particular type of application is more commonly known as an economizer. The tubes are often connected all in series but can also be arranged in series-parallel bundles to control the liquid side pressure drop. The air side pressure drop is controlled by the spacing of the tubes and the number of rows of tubes within the duct. Finned-tube exchangers are available prepackaged in modular sizes or can be made up to custom specifications very rapidly from standard components. Temperature control of the heated liquid is usually provided by a bypass duct arrangement which varies the flow rate of hot gases over the heat exchanger. Materials for the tubes and the fins can be selected to withstand corrosive liquids and/or corrosive exhaust gases.

A. Counterflow

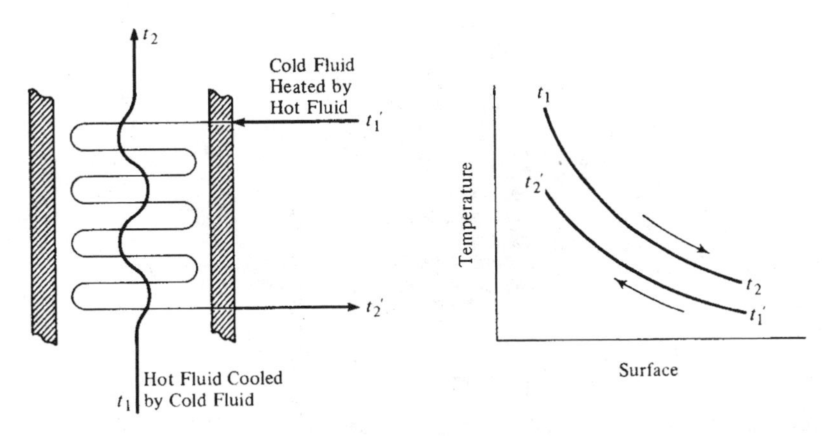

B. Parallel Flow

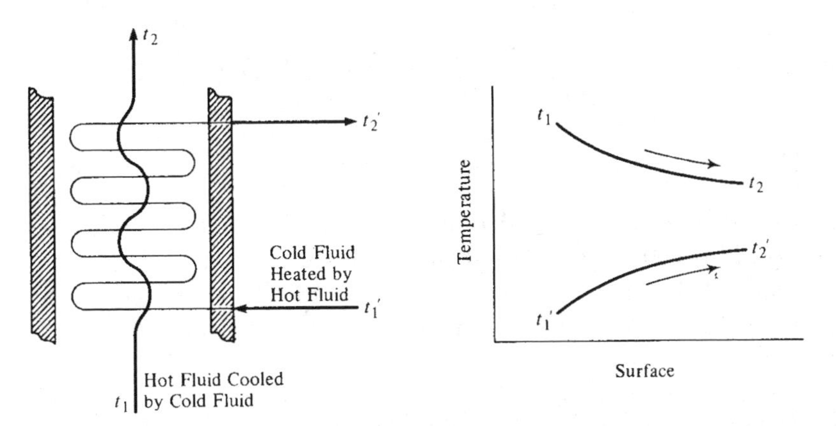

Figure 6-4. Temperature relationships for heat exchangers.

Finned-tube heat exchangers are used to recover waste heat in the low to medium temperature range from exhaust gases for heating liquids. Typical applications are domestic hot water heating, heating boiler feedwater, hot water space heating, absorption-type refrigeration or air conditioning, and heating process liquids.

Shell and Tube Heat Exchanger

When the medium containing waste heat is a liquid or a vapor which heats another liquid, then the shell and tube heat exchanger

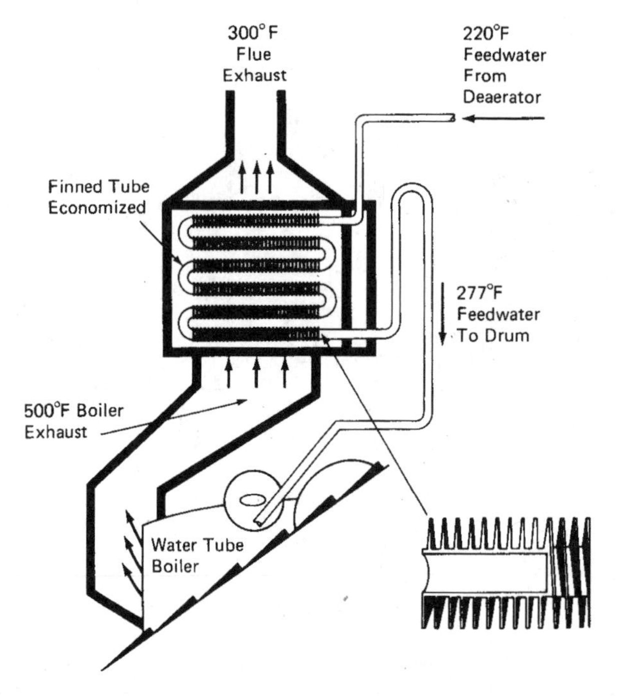

Figure 6-5. Finned tube gas to liquid regenerator (economizer).

must be used since both paths must be sealed to contain the pressures of their respective fluids. The shell contains the tube bundle, and usually internal baffles, to direct the fluid in the shell over the tubes in multiple passes. The shell is inherently weaker than the tubes so that the higher pressure fluid is circulated in the tubes while the lower pressure fluid flows through the shell. When a vapor contains the waste heat, it usually condenses, giving up its latent heat to the liquid being heated. In this application, the vapor is almost invariably contained within the shell. If the reverse is attempted, the condensation of vapors within small diameter parallel tubes causes flow instabilities. Tube and shell heat exchangers are available in a wide range of standard sizes with many combinations of materials for the tubes and shells.

Typical applications of shell and tube heat exchangers include heating liquids with the heat contained by condensates from refrigeration and air conditioning systems; condensate from process steam; coolants from furnace doors, grates, and pipe supports; coolants from engines, air compressors, bearings, and lubricants; and the condensates from distillation processes.

HOW TO ESTIMATE THE HEAT LOSS OF A VESSEL OR TANK

Heat loss calculations from a vessel or tank are complex, since conduction, convection, and radiation flows occur simultaneously. Figure 6-6 indicates the heat loss associated with a tank. As indicated, the total heat loss is the sum of the following:

$$q = q_1 + q_2 + q_3 + q_4 \tag{6-13}$$

where

- q_1 is the heat loss from liquid in the tank through the tank sidewalls to atmosphere (Btuh)
- q_2 is the heat loss from vapor in the tank through the sidewalls to atmosphere (Btuh)
- q_3 is the heat loss from the vapor in the tank through the roof to atmosphere
- q_4 is the heat loss from the bottom of the tank to the ground (Btuh).

To simplify calculations, the following are assumed:

1. Conductances of tank walls are neglected. (For metal tanks, as indicated in SIM 6-2, this assumption is valid. For other materials, the tank resistance should be added.)

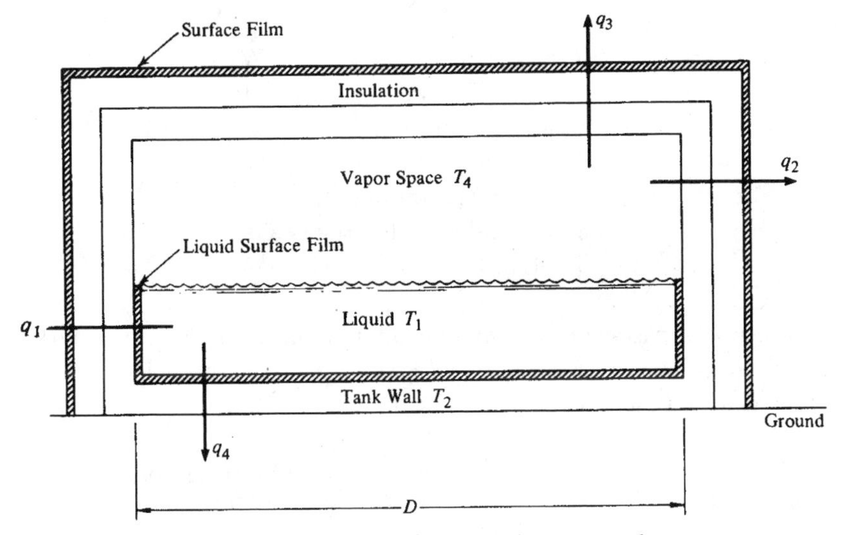

Figure 6-6. Losses from tank or vessel.

2. Liquid and vapor surface conductances are neglected for uninsulated tanks. Heat loss is based on air surface conductance.
3. Liquid and vapor conductances are neglected for insulated tanks. Heat loss is based on insulation conductance and surface air conductance.

A simplified procedure to determine approximate heat losses will be illustrated. This procedure simplifies computations of the various film coefficients. As an approximation, Figure 6-6 is used to compute the heat loss for uninsulated tanks based on the surface film conductances. This figure is used to compute q_1, q_2 and q_3 by equating q_t, to the q in question. The total heat loss is computed by multiplying the heat loss from Figure 6-6 by the surface area. Another term, ΔT_w, is introduced in this figure. To compute ΔT_w:

$$\Delta T_w = (T_w - T_A)\, W \qquad (6\text{-}14)$$

where

T_w is the fluid temperature, °F, T_A is the ambient temperature 70°F, and W is the correction factor which takes into account the process fluid.

W is defined as

Process Fluid	*W*
Agitated Liquid	1
Aqueous Solution	0.9
Fuel Oil	0.7
Condensing Vapor	1.0
Non-condensing Vapor	0.2
Fouling Liquids	0.4

Figure 6-7 is based on an ambient air temperature of 70°F, a surface emissivity of 0.9, and a wind velocity of zero. To use Figure 6-6 for other values, correct each term searately.

$$q_a \text{ (radiation)} = \frac{\text{New Emissivity}}{0.90} \times \begin{array}{c}\text{Value from graph} \\ \text{(use for other than 0.9)}\end{array} \qquad (6\text{-}15)$$

$$q_c \text{ (convection)} = \frac{\text{New Emissivity}}{0.90} \times \begin{matrix}\text{Value from graph} \\ \text{(use for other than 0.9)}\end{matrix} \quad (6\text{-}15)$$

$$q_t = q_a + q_c \quad (6\text{-}17)$$

Heat Loss To Ground

The last portion of Equation 6-18 deals with heat losses from the bottom of the tank when it rests on the ground. It is defined as:

$$q_4 = 2DK_4 (T_w - T_3) \quad (6\text{-}18)$$

where

D is the diameter of the tank in ft
K_4 is the conductance to ground
Use $K_4 = 0.8$ unless otherwise known
T_w is the liquid temperature, °F
T_3 is the ground temperature, °F

Assume ambient temperatures unless the ground temperature is specifically known.

Reducing Heat Loss of Vessels or Tanks

Insulation is the common method used to reduce the heat loss from a vessel or tank. Figure 6-8 shows the effect of various insulation thicknesses. This figure can be used as a guide to determine suggested economic thickness for various temperature differentials.

To use Figure 6-8, the temperature difference factor ΔT_1 should be corrected for wind velocities other than zero. To correct ΔT_1, multiply by the correction factors for various wind velocities given below:

Wind Velocity Correction Factors

Insulation Thickness	10 mph	30 mph
1	1.09	1.14
1-1/2	1.07	1.10
2	1.06	1.09
2-1/2	1.04	1.07
3	1.04	1.06
3-1/2	1.03	1.05

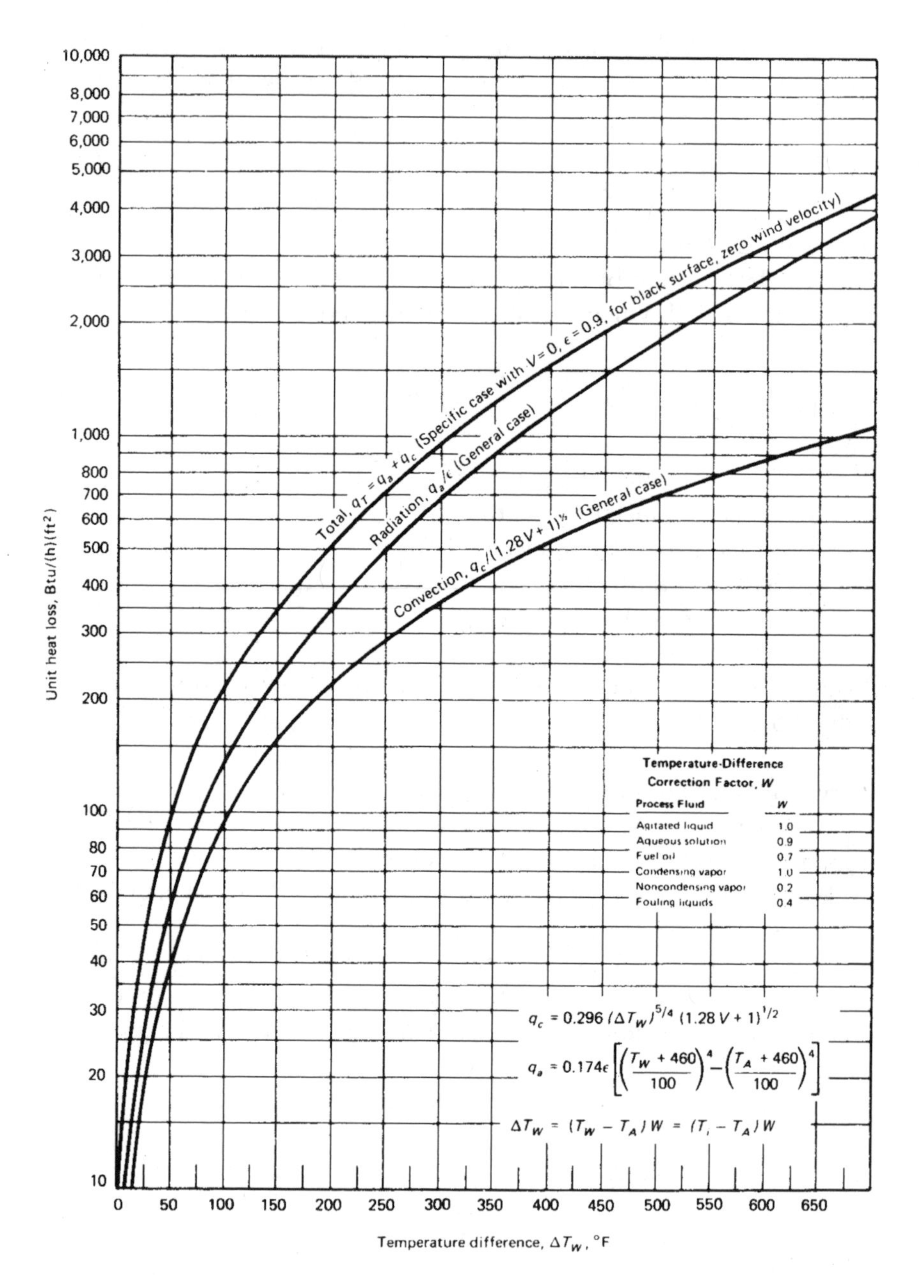

Figure 6-7. Heat loss for uninsulated tanks and vessels. [Reprinted by special permission from *Chemical Engineering* (May 27, 1974) Copyright © 1974 by McGraw-Hill, Inc., New York, NY)

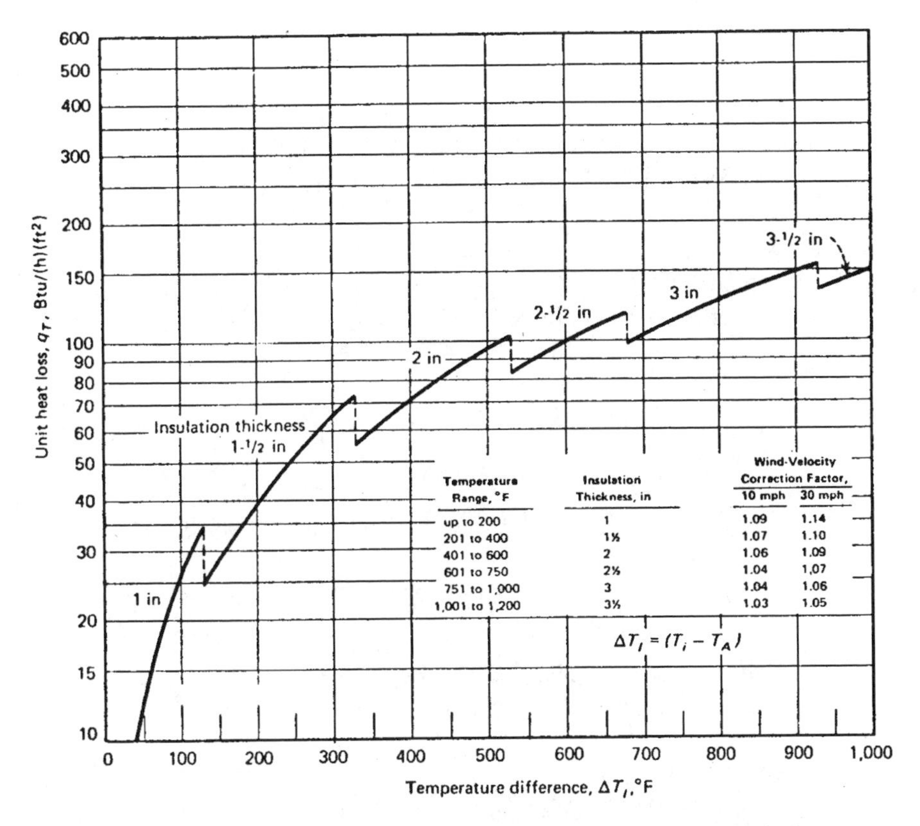

Temperature Range, °F	Insulation Thickness, in	Wind-Velocity Correction Factor, 10 mph	30 mph
up to 200	1	1.09	1.14
201 to 400	1½	1.07	1.10
401 to 600	2	1.06	1.09
601 to 750	2½	1.04	1,07
751 to 1,000	3	1.04	1.06
1,001 to 1,200	3½	1.03	1.05

INSULATED tanks, covered with calcium silicate, have heat losses based on negligible resistance to heat flow on process side. Values in chart are for wind velocity of zero, emissivity of 0.8, ambient air temperature of 70°F.

Figure 6-8. Heat loss for calcium silicate insulated tanks. [Reprinted by special permission from *Chemical Engineering* (May 27, 1974) Copyright © 1974 by McGraw-Hill, Inc., New York, NY]

SIM 6-4

A 15 ft diameter by 15 ft high carbon tank, supported at grade, is used to contain 250°F non-agitated process fluid. The fluid level is 10 ft high. Calculate the annual savings by adding 1-1/2 in. insulation, given the following:

Plant operation— 8760 hr/yr
Wind velocity— 10 mph

Ambient temperature—70°
Emissivity of tank—0.20
Vapor space has non-condensable vapors
Heat content of steam—926.1 Btu/lb
Steam cost—$6/1000 lb

ANSWER

Uninsulated Tank

Wetted Area of Tank

$A_1 = \pi DH = \pi \times 15 \times 10 = 471 \text{ ft}^2$

$\Delta T = (250 - 70) = 180°F$

$\Delta T_W = 0.9 \times 180 = 162$

From Figure 6-7

$q'_c = q_c \times (1.28 \times V + 1)^{1/2} = 180 \times 3.71 = 667.8 \text{ Bu/h} \bullet \text{ft}^2$

$$q'_c = q_a \times \frac{0.2}{0.9} = 250 \times 0.222 = 55.5 \text{ Btu/h} \bullet \text{ft}^2$$

$q_T = 667.8 + 55.5 = 723.3 \text{ Btu/h} \bullet \text{ft}^2$

$q_1 = 471 \times 723.3 = 340{,}674$ Btuh.

Heat Loss for Vapor Space

$A_2 = \pi \times 15\ (15 - 10) = 235.6 \text{ ft}^2$

$\Delta T_W = \Delta T \times 0.20 = 36$

$q'_c = 25 \times 3.71 = 92.75 \text{ Btu/h} \bullet \text{ft}^2$

$q'_a = 35 \times 0.222 = 7.77 \text{ Btu/h} \bullet \text{ft}^2$

$q_T = 100.52$ Btuh

$q_2 = 235.6 \times 100.52 = 23{,}682$ Btuh.

Heat Loss for Roof

$$A = \frac{\pi D^2}{4} = 176.7 \text{ ft}^2$$

$q_3 = 176.6 \times 100.52 = 17{,}761$ Btuh.

Heat Loss for Tank Bottom

$q_4 = 2 \times 15 \times 0.8\ (250 - 70) = 4{,}320$ Btuh.

Total Heat Loss = 386,437 Btuh

Insulated Tank

$\Delta T = (250 - 70) = 180°F$

From Figure 6-8,

$q_T = 35$ Btu/h • ft^2

$q'_T = 35 \times 1.07 = 37.45$ Btu/h • ft^2

Since the total tank is insulated,

$q = 37.45\ (A_1 + A_2 + A_3) = 37.45\ (471 + 235.6 + 176.7) = 33{,}079$ Btuh

Total heat loss = 33,079 + 4,320 = 37,339 Btuh

Energy savings with insulation =349,038 Btuh

$$\text{Yearly energy savings by reducing steam usage} = \frac{349{,}038 \times 8760}{926.1} \times \frac{\$6.00}{1000} = \$19{,}809/\text{year}$$

HOW TO ESTIMATE THE HEAT LOSS OF PIPING AND FLAT SURFACES

A simple method used to estimate heat losses for horizontal bare steel pipes and flat surfaces is to use Table 6-2. By knowing the temperature difference between the pipe surface and the ambient air, the heat loss is determined.

SIM 6-5

An existing process indicated in Scheme 1 of Figure 6-9 uses a heat exchanger to cool the discharge of the column before it enters the sewer. The traditional process supplies fluid directly to Column #1 at 104°F. The column is used as a stripper with 30 psig steam.

Scheme 2 illustrates a proposed modification. The proposed scheme replaces the bottom fluid exchanger with a larger exchanger and uses the bottom fluid to heat the process liquid; the bottom fluid in turn will be cooled. What recommendations should be made, given the following data?

The installed cost of heat exchanger #1 is $9000 with a 50 ft^2 area. The conductance U of the heat exchanger = 100

Table 6-2. Heat losses for horizontal bare steel pipe and flat surfaces.

In Btu per (Sq Ft of Pipe Surface) (Hour) (F Deg Temperature Difference Between Pipe and Air)

Pipe size inches	Linear Foot Factor	Temperature Difference F Deg Between Pipe Surface and Surrounding Air at 80 F																			
		50	100	150	200	250	300	350	400	450	500	550	600	650	700	750	800	850	900	950	1000
1/2	0.220	2.12	2.48	2.80	3.10	3.42	3.74	4.07	4.47	4.86	5.28	5.72	6.19	6.69	7.22	7.79	8.39	9.03	9.70	10.42	11.18
3/4	0.275	2.08	2.43	2.74	3.04	3.35	3.67	4.00	4.40	4.79	5.21	5.65	6.12	6.61	7.15	7.71	8.31	8.95	9.62	10.34	11.09
1	0.344	2.04	2.38	2.69	2.99	3.30	3.61	3.94	4.33	4.72	5.14	5.58	6.05	6.54	7.07	7.64	8.23	8.87	9.55	10.26	11.02
1-1/4	0.435	2.00	2.34	2.64	2.93	3.24	3.55	3.88	4.27	4.66	5.07	5.51	5.97	6.47	7.00	7.56	8.16	8.79	9.47	10.18	10.94
1-1/2	0.497	1.98	2.31	2.61	2.90	3.20	3.52	3.84	4.23	4.62	5.03	5.47	5.93	6.43	6.96	7.52	8.12	8.75	9.43	10.14	10.89
2	0.622	1.95	2.27	2.56	2.85	3.15	3.46	3.78	4.17	4.56	4.97	5.41	5.87	6.37	6.89	7.45	8.05	8.68	9.36	10.07	10.82
2-1/2	0.753	1.92	2.23	2.52	2.81	3.11	3.42	3.74	4.12	4.51	4.92	5.36	5.82	6.31	6.84	7.40	7.99	8.63	9.30	10.01	10.77
3	0.916	1.89	2.20	2.49	2.77	3.07	3.37	3.69	4.08	4.46	4.87	5.31	5.77	6.26	6.79	7.35	7.94	8.57	9.25	9.96	10.71
3-1/2	1.047	1.87	2.18	2.46	2.74	3.04	3.34	3.66	4.05	4.43	4.84	5.27	5.73	6.23	6.75	7.31	7.91	8.54	9.21	9.92	10.67
4	1.178	1.85	2.16	2.44	2.72	3.01	3.32	3.64	4.02	4.40	4.81	5.25	5.71	6.20	6.72	7.28	7.87	8.51	9.18	9.89	10.64
4-1/2	1.309	1.84	2.14	2.42	2.70	2.99	3.30	3.61	4.00	4.38	4.79	5.22	5.68	6.17	6.69	7.25	7.85	8.48	9.15	9.86	10.61
5	1.456	1.83	2.13	2.40	2.68	2.97	3.28	3.59	3.97	4.35	4.76	5.20	5.65	6.15	6.68	7.23	7.82	8.45	9.12	9.83	10.58
6	1.734	1.80	2.10	2.37	2.65	2.94	3.24	3.55	3.94	4.32	4.72	5.16	5.61	6.10	6.63	7.19	7.78	8.41	9.08	9.79	10.54
7	1.996	1.79	2.08	2.35	2.63	2.91	3.21	3.53	3.91	4.29	4.69	5.13	5.58	6.07	6.60	7.15	7.75	8.38	9.05	9.76	10.51
8	2.258	1.77	2.06	2.33	2.60	2.89	3.19	3.50	3.88	4.26	4.67	5.10	5.56	6.05	6.57	7.12	7.72	8.35	9.02	9.73	10.48
9	2.520	1.76	2.05	2.31	2.59	2.87	3.17	3.48	3.86	4.24	4.65	5.08	5.53	6.02	6.54	7.10	7.69	8.32	8.99	9.70	10.45
10	2.814	1.75	2.03	2.30	2.57	2.85	3.15	3.46	3.84	4.22	4.62	5.05	5.51	6.00	6.52	7.08	7.67	8.30	8.97	9.68	10.43
12	3.338	1.73	2.01	2.27	2.54	2.83	3.12	3.43	3.81	4.19	4.59	5.02	5.48	5.96	6.48	7.04	7.63	8.26	8.93	9.64	10.39
14	3.655	1.72	2.00	2.26	2.53	2.81	3.11	3.41	3.79	4.17	4.57	5.00	5.46	5.94	6.47	7.02	7.61	8.24	8.91	9.62	10.37
16	4.189	1.70	1.98	2.24	2.51	2.79	3.08	3.39	3.77	4.14	4.55	4.98	5.43	5.92	6.44	6.99	7.59	8.21	8.88	9.59	10.34
18	4.717	1.69	1.96	2.22	2.49	2.77	3.07	3.37	3.75	4.12	4.53	4.96	5.41	5.90	6.42	6.97	7.56	8.19	8.86	9.57	10.32
20	5.236	1.68	1.95	2.21	2.47	2.75	3.05	3.36	3.73	4.11	4.51	4.94	5.39	5.88	6.40	6.95	7.54	8.17	8.84	9.55	10.29
24	6.283	1.66	1.93	2.19	2.45	2.73	3.02	3.33	3.70	4.07	4.48	4.90	5.36	5.84	6.36	6.92	7.51	8.14	8.80	9.51	10.26
Vertical Surface																					
		1.84	2.14	2.42	2.70	3.00	3.30	3.62	4.00	4.38	4.79	5.22	5.68	6.17	6.70	7.26	7.85	8.48	9.15	9.86	10.62
Horizontal Surface Facing Upward																					
		2.03	2.37	2.67	2.97	3.28	3.59	3.92	4.31	4.70	5.12	5.56	6.02	6.52	7.05	7.61	8.21	8.85	9.52	10.24	10.99
Horizontal Surface Facing Downward																					
		1.61	1.86	2.11	2.36	2.64	2.93	3.23	3.60	3.97	4.37	4.80	5.25	5.73	6.25	6.80	7.39	8.02	8.69	9.39	10.14

Notes:

1. To find losses per linear foot, multiply square foot losses by factors in column 2.
2. Areas of flat surfaces are four square feet or more.
3. For pipe sizes larger than 24 inches, use the losses for 24 inch pipe.

Reprinted by permission from *ASHRAE Handbook of Fundamentals* 1972.

Scheme 1 Existing Process

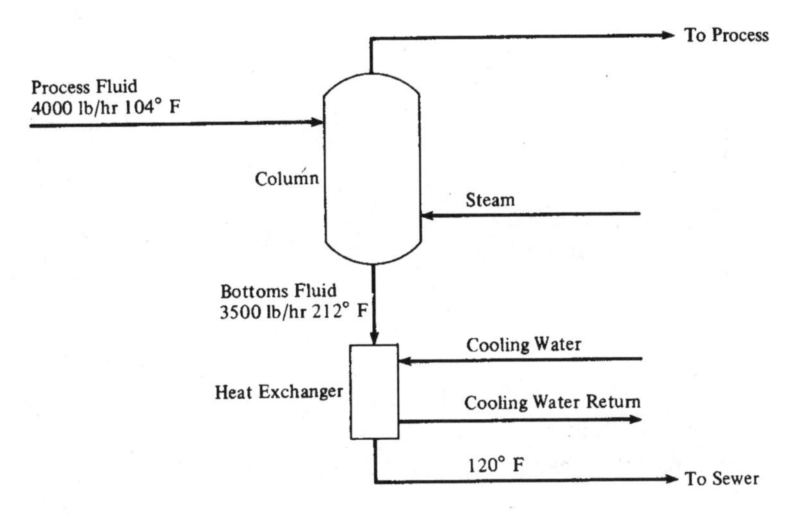

Scheme 2 Proposed Modification to Process

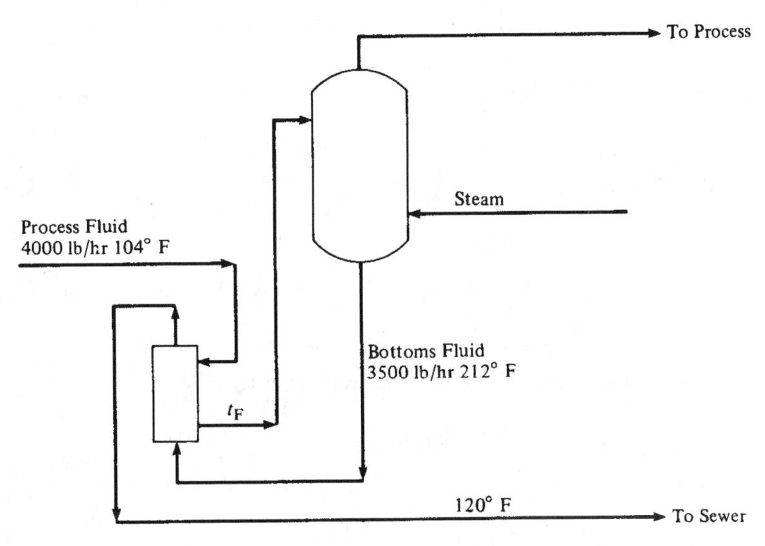

Figure 6-9. Process schemes for Ajax Plant.

30 psig steam is used
Plant operates 8760 hours/yr
Steam cost $6.00/1000#

Note: To compute the cost for an exchanger of increased area, use the following aproximation:

$$\text{New cost} = \left[\frac{\text{new are}}{\text{original area}}\right]^{0.6} \times \text{original price} \tag{6-19}$$

Analysis

The first analysis indicates that the hot bottom fluid can be used to heat the process fluid.

The amount of heat supplied to the process fluid is

$$q = MCp\ \Delta T = 3500\ \#/\text{hr} \times 1(212 - 120) = 322{,}000\ \text{Btuh}$$

30 psig = 30 + 14.7 = 44.7 psia steam
From Steam Table 12-20, h_{fg} = 928.6 Btu/#

Energy Savings of Stripper

$$\#/\text{hr of 30 psig steam saved} = \frac{322{,}000\ \text{Btuh}}{928.6\ \text{Btu}/\#} = 347\ \#/\text{hr}$$

$$\text{Steam savings} = \frac{\$6.00}{1000} \text{ x } 347 \text{ x } 8760 = \$18{,}237$$

The temperature of the process fluid will rise to:

$$MCp\ \Delta T\ (\text{process}) = 322{,}000\ \text{Btuh}$$

$$\Delta t = \frac{322{,}000}{4000 \times 1} = 80°\text{F}$$

$$t_f = 104 + 80 = 184°\text{F}.$$

The required area of heat exchanger will increase from Scheme 1

$$q = U_0 A \,\Delta T_m = 100\, A\Delta T_m = 32{,}00 \tag{6-20}$$

$$\Delta T_m = \frac{(212 - 184) - (120 - 104)}{\log_e\left(\frac{212 - 184}{120 - 104}\right)}$$

$$\Delta T_m = \frac{82 - 16}{\log_e\left(\frac{28}{16}\right)} = \frac{12}{\log_e(1.75)} = 21.4$$

$$A = \frac{3220}{21.4} = 150.4 \text{ ft}^2$$

$$\text{New installed cost} = \left(\frac{\text{area new}}{\text{area old}}\right)^{0.6} \times \text{old price}$$

$$= \left(\frac{150.4}{50}\right)^{0.6} \times \$9000 = \$17{,}400.$$

With an annual energy savings of $18,237, the investment in additional heat exchanger capacity is desirable.

7

Reducing Building Energy Losses

ENERGY LOSSES DUE TO HEAT LOSS AND HEAT GAIN

Depending on the time of year, a heat loss or a heat gain wastes energy. For example, a heat loss during the winter means wasted energy in heating the building. Similarly, during the summer months, a heat gain means wasted energy in cooling the building. The building construction affects the heat loss and heat gain. Figure 7-1 illustrates the total heat gain of the building. The flow of heat is always from one temperature to a colder temperature. The heat loss of a building is illustrated by Figure 7-2. In this case, the building is considered the "hot body."

In the context of this book, heat loss refers to heating loads, while heat gain refers to cooling loads. By considering building materials and constructions, the associated heat loss and heat gains can be reduced.

In this chapter, you will see: how to apply handy Building Construction Tables to solve most heat transfer problems, how substitutions of building materials saves energy, and how to apply different types of glass to save energy.

CONDUCTIVITY THROUGH BUILDING MATERIALS

The best conductors of heat are metals. Insulations such as wood, asbestos, cork, and felt are poor conductors. Conductance is widely used because many materials used in the construction of buildings are non-homogeneous.

Table 7-1 should be used when the surface conductance, which is the transfer of heat from air to a surface or from a surface to air, is

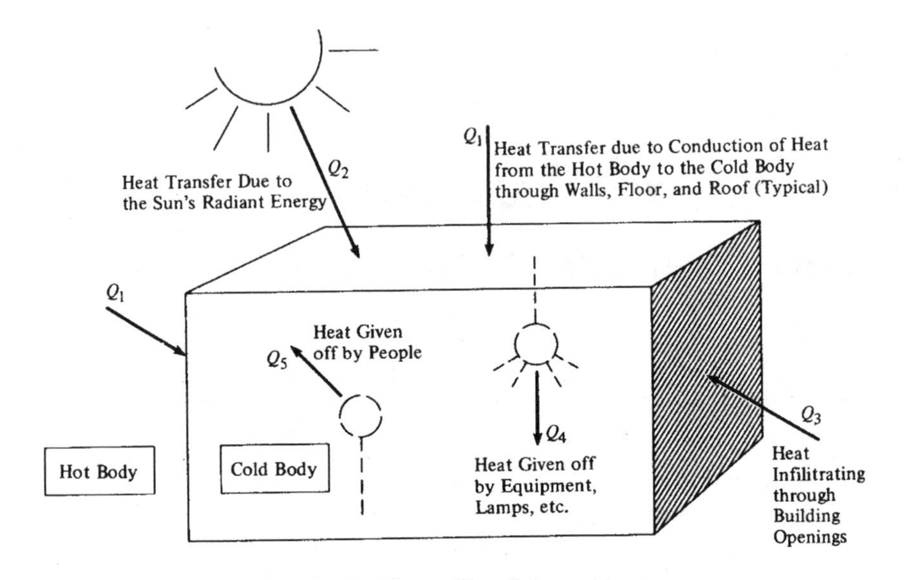

Figure 7-1. Heat gain of a building.

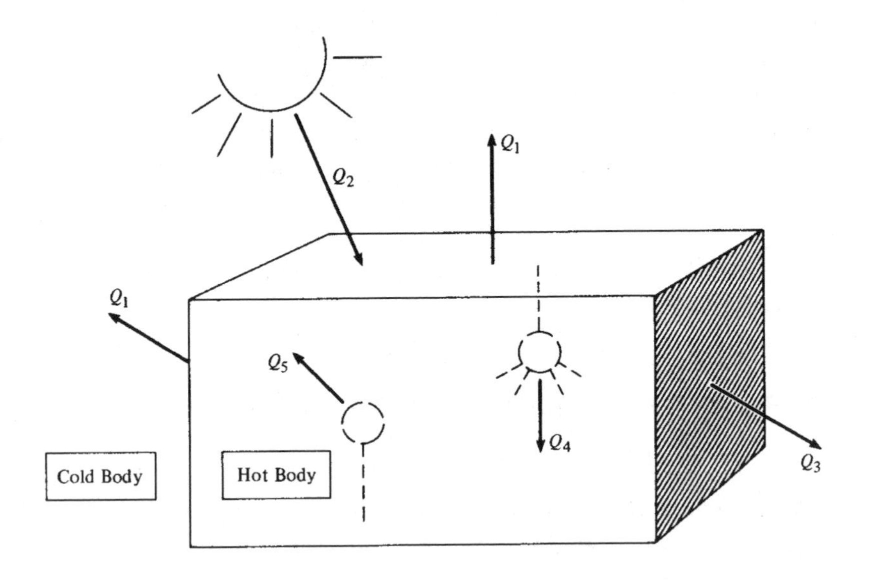

Figure 7-2. Heat loss of a building.

required. Table 7-1 also lists air space conductances for various positions and heat directions. Note that the conductance will change from the heat gain case to the heat loss case. The data for a 3/4-in. air space shows the insulating properties of air. When one surface of the air space is covered with aluminum foil, the resistance increases; this, in turn, will reduce the heat loss.

The heat transfer coefficients for various building materials are given in Table 7-2.

Table 7-1. Thermal resistances for surface films and air spaces.

Medium & Position	Direction of Heat Flow	Building Materials: wood, paper, glass, masonry $\varepsilon = 0.90$	Galvanized Steel $\varepsilon = 0.20$	Aluminum Foil $\varepsilon = 0.05$
A. Surface Films		*R*	*R*	*R*
1. Still Air				
a) Horizontal	Up	0.61	1.10	1.32
b) Horizontal	Down	0.92	2.70	4.55
c) Vertical	Horizontal	0.68	1.85	1.70
2. Moving Air				
a) 15 mph Wind (Winter)	Any	0.17	—	—
b) 7.5 mph Wind (Summer)	Any	0.25	—	—
B. Air Space 3/4"		$\varepsilon = 0.82$	$\varepsilon = 0.20$	$\varepsilon = 0.05$
1. Air Mean Temp. 90°F/0°F.**		*R*	*R*	*R*
a) Horizontal	Up	0.76/1.02*	1.63/1.78	2.26/2.16
b) Horizontal	Down	0.84/1.31	2.08/2.88	3.25/4.04
c) Slope 45°	Up	0.81/1.13	1.90/2.13	2.81/2.71
d) Slope 45°	Down	0.84/1.31	2.09/2.88	3.24/4.04
e) Vertical	Horizontal	0.84/1.28	2.10/2.73	3.28/3.76

*Assume an average ΔT of 10°F.

**For resistance at temperatures other than 90°F or 0°F, interpolate between the two values—typical.

Reprinted by permission from *ASHRAE Handbook of Fundamentals* 1972.

Table 7-2. Heat transfer coefficients of building materials.*

Material	Description	Conductivity k#	Conductance c +
building boards	asbestos-cement board	4.0	
	gypsum or plaster board...1/2 in.		2.25
	plywood	0.80	
	plywood...3/4 in.		1.07
	sheathing (impregnated or coated)	0.38	
	sheathing (impregnated or coated) 25/32 in.		0.49
	wood fiber—hardboard type	1.40	
insulating materials	blanket and batt:		
	mineral wool fibers (rock, slag, or glass)	0.27	
	wood fiber	0.25	
	boards and slabs:		
	cellular glass	0.39	
	corkboard	0.27	
	glass fiber	0.25	
	insulating roof deck ...2 in.		0.18
masonry materials	loose fill:		
	mineral wool (glass, slag, or rock)	0.27	
	vermiculite (expanded)	0.46	
	concrete:		
	cement mortar	5.0	
	lightweight aggregates, expanded shale, clay, slate, slags; cinder; pumice; perlite; vermiculite	1.7	
	sand and gravel or stone aggregate	12.0	
	stucco	5.0	

Table 7-2. (*Continued*)

masonry materials	brick, tile, block, and stone:		
	brick, common	5.0	
	brick, face	9.0	
	tile, hollow clay, 1 cell deep, 4 in		0.90
	tile, hollow clay, 2 cells, 8 in.		0.54
	block, concrete, 3 oval core:		
	sand & gravel aggregate 4 in.		1.40
	sand & gravel aggregate 8 in.		0.90
	cinder aggregate...4 in.		0.90
	cinder aggregate...8 in.		0.58
	stone, lime or sand	12.50	
plastering materials	cement plaster, sand aggregate	5.0	
	gypsum plaster:		
	lightweight aggregate...1/2 in.		3.12
	lt. wt. agg. on metal lath ...3/4 in.		2.13
	perlite aggregate	1.5	
	sand aggregate	5.6	
	sand aggregate on metal lath 3/4 in.		7.70
	vermiculite aggregate	1.7	
roofing	asphalt roll roofing		6.50
	built-up roofing ...3/8 in.		3.00
siding materials	asbestos-cement, 1/4 in. lapped		4.76
	asphalt insulating (1/2 in. board)		0.69
	wood, bevel, 1/2 × 8, lapped		1.23
woods	maple, oak, and similar hardwoods	1.10	
	fir, pine, and similar softwoods	0.80	
	fir, pine & sim. softwoods 25/32 in.		1.02

*Extracted with permission from *ASHRAE Guide and Data Book*, 1965.
#Conductivity given in Btu in. per hr sq ft F
+Conductance given in Btu per hr sq ft F
Courtesy of the Trane Company.

SIM 7-1

Calculate the heat loss through 10,000 ft^2 of building wall, as indicated by Figure 7-3.

Assume a temperature differential of 17°F. When using Table 7-2, remember: $R = d/K$.

ANSWER

Item	*Description*	*Resistance*	*Reference*
1	Outside air film at 15 mph	0.17	Table 7-1
2	4" brick	0.44	Table 7-2: $R = 4/9$
3	Mortar	0.10	Table 7-2: $R = 1/2/5$
4	Block	1.11	Table 7-2: $R = 1/.9$
5	Gypsum	0.44	Table 7-2: $R = 1/2/.25$
6	Inside film	0.68	Table 7-1
	Total resistance	2.94	

$$U = 1/R = 0.34$$
$$q = U\ A\ \Delta\ T$$
$$= 0.34 \times 10{,}000 \times 17 = 57{,}800 \text{ Btu/h.}$$

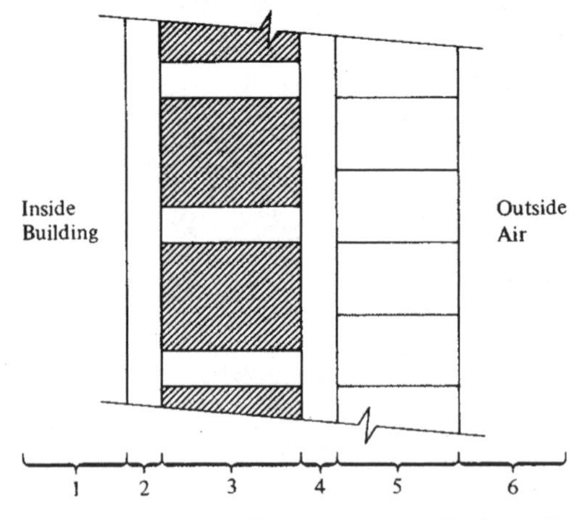

1 Inside Air Film
2 1/2" Plaster Board Interior Finish
3 8" Concrete Block, Sand & Gravel Aggregate
4 1/2" Cement Mortar
5 4" Brick Exterior
6 Outside Air Film @ 15 mph Wind

Figure 7-3. Typical wall construction.

SIM 7-2

In order to reduce the heat loss through the building wall, the engineer is recommending substituting cinder aggregate for the 8″ concrete block wall. (Assume the structural considerations are the same.) Comment on the recommendation.

ANSWER

The conductance of the cinder aggregate concrete block wall is 0.58, as compared to 0.90, thus the resistance will increase from 1.11 to 1.72. The effect is to increase the overall resistance of the wall to 3.54.

$U = 1/R - 0.28$

$q = 0.28 \times 10{,}000 \times 17 = 47{,}600$ Btuh or 10,200 Btuh are saved with the change in wall construction. The total system should be evaluated from an energy conservation viewpoint.

SIM 7-3

Calculate the heat gain during the summer for the construction of SIM 7-1.

ANSWER

The only item which changes is the outside air film resistance:

Item	*Description*	*Resistance*
1	Outside air film at 7-1/2 mps	0.25
Thus, total resistance =		3.02

$$U = \frac{1}{3.02} = 0.3$$

$q = 56{,}100$ Btuh.

Handy Tables to Save Time

Energy analysis of composite building constructions can be simplified using Tables 7-3 through 7-13. These tables have been reprinted with the permission of ASHRAE and the Trane Company. To use them simply find the column and row which describes the type of construction. The corresponding coefficient of transmission (conductance) is then found.

Table 7-14 is another useful table which illustrates design temperatures for various cities in the United States.

Table 7-3. Coefficients of transmission (U) of solid masonry walls.

Coefficients are expressed in Btu per (hour) (square foot) (Fahrenheit degree difference in temperature between the air on the two sides), and are based on an outside wind velocity of 15 mph.

Exterior Construction[3]		None	Interior Finish										No.
			Plas. 5/8 in on Wall		Metal Lath and 3/4 in Plas. on Furring		Gypsum Lath (3/8 in.) and 1/2 in Plas. on Furring			Insul. Bd. Lath (1/2 in.) and 1/2 in. Plas on Furring		Wood Lath and 1/2 in. Plas.	
Resistance →			(Sand Agg.) 0.11	(Lt. Wt. Agg.) 0.39	(Sand Agg.) 0.13	(Lt. Wt. Agg.) 0.47	No. Plas 0.32	(Sand Agg.) 0.41	(Lt. Wt. Agg.) 0.64	No. Plas. 1.43	(Sand Agg.) 1.52	(Sand Agg.) 0.40	
		U	U	U	U	U	U	U	U	U	U	U	
Material	R	A	B	C	D	E	F	G	H	I	J	K	
Brick (face and common)[4]													
(6 in.)	0.61	0.68	0.64	0.54	0.39	0.34	0.36	0.35	0.33	0.26	0.25	0.35	1
(8 in.)	1.24	0.48	0.45	0.41	0.31	0.28	0.30	0.29	0.27	0.22	0.22	0.29	2
(12 in.)	2.04	0.35	0.33	0.30	0.25	0.23	0.24	0.23	0.22	0.19	0.19	0.23	3
(16 in.)	2.84	0.27	0.26	0.25	0.21	0.19	0.20	0.20	0.19	0.16	0.16	0.20	4
Brick (common only)													
(8 in.)	1.60	0.41	0.39	0.35	0.28	0.26	0.27	0.26	0.25	0.21	0.20	0.26	5
(12 in.)	2.40	0.31	0.30	0.27	0.23	0.21	0.22	0.22	0.21	0.18	0.17	0.22	6
(16 in.)	3.20	0.25	0.24	0.23	0.19	0.18	0.19	0.18	0.18	0.16	0.15	0.18	7
Stone (lime and sand)													
(8 in.)	0.64	0.67	0.63	0.53	0.39	0.34	0.36	0.35	0.32	0.26	0.25	0.35	8
(12 in.)	0.96	0.55	0.52	0.45	0.34	0.31	0.32	0.31	0.29	0.24	0.23	0.31	9
(16 in.)	1.28	0.47	0.45	0.40	0.31	0.28	0.29	0.28	0.27	0.22	0.22	0.29	10
(24 in.)	1.92	0.36	0.35	0.32	0.26	0.24	0.25	0.24	0.23	0.19	0.19	0.24	11
Hollow clay tile													
(8 in.)	1.85	0.36	0.36	0.32	0.26	0.24	0.25	0.25	0.23	0.20	0.19	0.25	12
(10 in.)	2.22	0.33	0.31	0.29	0.24	0.22	0.23	0.22	0.21	0.18	0.18	0.23	13
(12 in.)	2.50	0.30	0.29	0.27	0.22	0.21	0.22	0.21	0.20	0.17	0.17	0.21	14

Poured concrete												
30 lb. per cu ft												
(4 in.)	4.44	0.19	0.19	0.18	0.16	0.15	0.15	0.15	0.14	0.13	0.13	0.15	15
(6 in.)	6.66	0.13	0.13	0.13	0.12	0.11	0.11	0.11	0.11	0.10	0.10	0.11	16
(8 in.)	8.88	0.10	0.10	0.10	0.09	0.09	0.09	0.09	0.09	0.08	0.08	0.09	17
(10 in.)	11.10	0.08	0.08	0.08	0.08	0.07	0.08	0.08	0.07	0.07	0.07	0.08	18
80 lb. per cu ft													
(6 in.)	2.40	0.31	0.30	0.27	0.23	0.21	0.22	0.22	0.21	0.18	0.17	0.22	19
(8 in.)	3.20	0.25	0.24	0.23	0.19	0.18	0.19	0.18	0.18	0.16	0.15	0.18	20
(10 in.)	4.00	0.21	0.20	0.19	0.17	0.16	0.16	0.16	0.15	0.14	0.14	0.16	21
(12 in.)	4.80	0.18	0.17	0.17	0.15	0.14	0.14	0.14	0.14	0.12	0.12	0.14	22
140 lb. per cu ft													
(6 in.)	0.48	0.75	0.69	0.58	0.41	0.36	0.38	0.37	0.34	0.27	0.26	0.37	23
(8 in.)	0.64	0.67	0.63	0.53	0.39	0.34	0.36	0.35	0.32	0.26	0.25	0.35	24
(10 in.)	0.80	0.61	0.57	0.49	0.36	0.32	0.34	0.33	0.31	0.25	0.24	0.33	25
(12 in.)	0.96	0.55	0.52	0.45	0.34	0.31	0.32	0.31	0.29	0.24	0.23	0.31	26
Concrete block													
(Gravel Agg.) (8 in.)	1.11	0.52	0.48	0.43	0.33	0.29	0.31	0.30	0.28	0.23	0.22	0.30	27
(12 in.)	1.28	0.47	0.45	0.40	0.31	0.28	0.29	0.28	0.27	0.22	0.22	0.29	28
(Cinder Agg.) (8 in.)	1.72	0.39	0.37	0.34	0.27	0.25	0.26	0.25	0.24	0.20	0.20	0.25	29
(12 in.)	1.89	0.36	0.35	0.32	0.26	0.24	0.25	0.24	0.23	0.19	0.19	0.24	30
(Lt. Wt. Agg.) (8 in.)	2.00	0.35	0.34	0.31	0.26	0.23	0.24	0.24	0.22	0.19	0.19	0.24	31
(12 in.)	2.27	0.32	0.31	0.28	0.24	0.22	0.23	0.22	0.21	0.18	0.18	0.22	32

[3]If stucco or structural glass is applied to the exterior, the additional resistance value of 0.10 would have a negligible effect on the U value.
[4]Brick, 6 in. (5-1/2 in. actual) is assumed to have no backing. Walls 8, 12 and 16 in. have 4 in. of face brick and balance of common brick.
Extracted with permission from 1965 *ASHRAE Guide and Data Book*.
Courtesy of the Trane Company.

Table 7-4. Coefficients of transmission (U) of masonry partitions.

Coefficients are expressed in Btu per (hour) (square foot) (Fahrenheit degree difference in temperature between the air on the two sides), and are based on still air (no wind) conditions on both sides.

Type of Partition		Surface Finish					Number
		None	Plas. (lt. wt. agg.) 5/8 in.		Plas. (sand agg.) 5/8 in.		
			One Side	Two Sides	One Side	Two Sides	
Resistance →			0.39	0.78	0.11	0.22	
Material	R	U	U	U	U	U	
		A	B	C	D	E	
Hollow concrete block							
(Cinder agg.)							
(3 in.)	0.86	0.45	0.38	0.33	0.43	0.41	1
(4 in.)	1.11	0.40	0.35	0.31	0.39	0.37	2
(8 in.)	1.72	0.32	0.29	0.26	0.31	0.30	3
(12 in.)	1.89	0.31	0.27	0.25	0.30	0.29	4
(Lt. Wt. agg.)							
(3 in.)	1.27	0.38	0.33	0.30	0.36	0.35	5
(4 in.)	1.50	0.35	0.31	0.27	0.34	0.32	6
(8 in.)	2.00	0.30	0.27	0.24	0.29	0.28	7
(12 in.)	2.27	0.28	0.25	0.23	0.27	0.26	8
(Gravel agg.)							
(8 in.)	1.11	0.40	0.35	0.31	0.39	0.37	9
(12 in.)	1.28	0.38	0.33	0.29	0.36	0.38	10

Hollow clay tile							
(3 in.)	0.80	0.46	0.39	0.34	0.44	0.42	11
(4 in.)	1.11	0.40	0.35	0.31	0.39	0.37	12
(6 in.)	1.52	0.35	0.31	0.27	0.33	0.32	13
(8 in.)	1.85	0.31	0.28	0.25	0.30	0.29	14
Hollow gypsum tile							
(3 in.)	1.35	0.37	0.32	0.29	0.35	0.34	15
(4 in.)	1.67	0.33	0.29	0.26	0.32	0.31	16
Solid plaster walls							
Gypsum Lath. (1/2 in.) and Plas.							
3/4 in. each side							
(lt. wt. agg.)	1.39	0.36	—	—	—	—	17
(Sand agg.)	0.71	0.48	—	—	—	—	18
1 in. each side							
(lt. wt. agg.)	1.73	0.32	—	—	—	—	19
(Sand agg.)	0.81	0.46	—	—	—	—	20
Metal lath and plas.[1]							
2 in. total thickness							
(lt. wt. agg.)	1.28	0.38	—	—	—	—	21
(Sand agg.)	0.36	0.58	—	—	—	—	22
2-1/2 in. total thickness							
(lt. wt. agg.)	1.60	0.34	—	—	—	—	23
(Sand agg.)	0.45	0.55	—	—	—	—	24
Glass and glass blocks	See Table 3-2						

[1]Metal Core and Supports disregarded. Plaster troweled smooth both sides.
Extracted with permission from 1965 *ASHRAE Guide and Date Book.*
Courtesy of the Trane Company.

Table 7-5. Coefficients of transmission (U) of frame walls.

These coefficients are expressed in Btu per (hour) (square foot) (Fahrenheit degree difference in temperature between the air on the two sides), and are based on an outside wind velocity of 15 mph.

Exterior[1]			Interior finish		Type of Sheathing[3]						Number
					none, building paper	gypsum board 1/2 in.	plywood 5/16 in.	wood 25/32 in. and building paper	insulation board sheathing 1/2 in.	insulation board sheathing 25/32 in.	
			Resistance	→	0.06	0.45	0.39	1.04	1.32	2.06	
Material	R	Av. R	Material	R	U	U	U	U	U	U	
					A	B	C	D	E	F	
			none	—	0.43	0.37	0.38	0.30	0.28	0.23	23
			gypsum bd. (3/8 in.)	0.32	0.28	0.25	0.25	0.22	0.20	0.18	24
			gypsum lath (3/8 in.) and 1/2 in. plas. (lt. wt. agg.)	0.64	0.25	0.23	0.23	0.20	0.19	0.17	25
wood shingles over insul.: backer bd. (5/16 in.)	1.40	1.42[2]	gypsum lath (3/8 in.) and 1/2 in. plas. (sand agg.)	0.41	0.27	0.24	0.25	0.21	0.20	0.18	26
asphalt insul. siding	1.45		metal lath and 3/4 in. plas. (lt. wt. agg.)	0.47	0.27	0.24	0.24	0.21	0.20	0.17	27
			metal lath and 3/4 in. plas. (sand agg.)	0.13	0.29	0.26	0.27	0.23	0.21	0.18	28
			insul. bd. (1/2 in.)	1.43	0.21	0.20	0.20	0.18	0.17	0.15	29
			insul. bd. lath (1/2 in.) and 1/2 in. plas. (sand agg.)	1.52	0.21	0.19	0.19	0.17	0.16	0.15	30
			plywood (1/4 in.)	0.31	0.28	0.25	0.25	0.22	0.20	0.18	31
			wood panels (3/4 in.)	0.94	0.24	0.22	0.22	0.19	0.18	0.16	32
			wood lath and 1/2 in. plas. (sand agg.)	0.40	0.27	0.24	0.25	0.21	0.20	0.18	33

			none	—	0.91	0.67	0.70	0.48	0.42	0.32	34
			gypsum bd. (3/8 in.)	0.32	0.42	0.36	0.37	0.30	0.27	0.23	35
			gypsum. lath (3/8 in.) and 1/2 in. plas. (lt. wt. agg.)	0.64	0.37	0.32	0.33	0.27	0.25	0.21	36
			gypsum lath (3/8 in.) and 1/2 in. plas. (sand agg.)	0.41	0.40	0.35	0.36	0.29	0.27	0.22	37
asbestos-cement siding	0.21										
stucco[5] 1 in	0.20	0.19[2]	metal lath and 3/4 in. plas. (lt. wt. agg.)	0.47	0.39	0.34	0.35	0.28	0.26	0.22	38
			metal lath and 3/4 in. plas. (sand agg.)	0.13	0.45	0.39	0.40	0.31	0.29	0.24	39
asphalt roll siding	0.15		insul. bd. (1/2 in.)	1.43	0.29	0.26	0.26	0.22	0.21	0.18	40
			insul. bd. lath (1/2 in.) and 1/2 in. plas. (sand agg.)	1.52	0.28	0.25	0.26	0.22	0.21	0.18	41
			plywood (1/4 in.)	0.31	0.42	0.36	0.37	0.30	0.27	0.23	42
			wood panels (3/4 in.)	0.94	0.33	0.29	0.30	0.25	0.23	0.20	43
			wood lath and 1/2 in. plas. (sand agg.)	0.40	0.40	0.35	0.36	0.29	0.27	0.22	44

[1]Note that although several types of exterior finish may be grouped because they have approximately the same thermal resistance value, it is not implied that all types may be suitable for application over all types of sheathing listed.
[2]Average resistance of items listed. This average was used in computation of U values shown.
[3]Building paper is not included except where noted.
[4]Small air space between building paper and brick veneer neglected.
[5]Where stucco is applied over insulating board or gypsum sheathing, building paper is generally required, but the change in U value is negligible.
Extracted with permission from 1965 *ASHRAE Guide and Data Book*.
Courtesy of the Trane Company.

Table 7-6. Coefficients of transmission (U) of masonry walls.

Coefficients are expressed in Btu per (hour) (square foot) (Fahrenheit degree difference in temperature between the air on the two sides), and are based on an outside wind velocity of 15 mph.

Exterior Facing			Backing		None	Interior Finish										Number
						Plas. 5/8 in. on Wall		Metal Lath and 3/4 in. Plas. on Furring		Gypsum Lath (3/8 in.) and 1/2 in. Plas. on Furring			Insul. Bd. Lath (1/2 in.) and 1/2 in. Plas. on Furring		Wood Lath and 1/2 in. Plas.	
			Resistance	↳		(Sand Agg.) 0.11	(Lt. Wt. Agg.) 0.39	(Sand Agg.) 0.13	(Lt. Wt. Agg.) 0.47	No. Plas 0.32	(Sand Agg.) 0.41	(Lt. Wt. Agg.) 0.64	No. Plas. 1.43	(Sand Agg.) 1.52	(Sand Agg.) 0.40	
Material	R	Av. R	Material		U	U	U	U	U	U	U	U	U	U	U	
					R	A	B	C	D	E	F	G	H	I	J K	
			Concrete block													
			(Cinder agg.)													
			(4 in.)	1.11	0.41	0.39	0.35	0.28	0.26	0.27	0.26	0.25	0.21	0.20	0.26	1
			(8 in.)	1.72	0.33	0.32	0.29	0.24	0.22	0.23	0.23	0.21	0.18	0.18	0.23	2
			(12 in.)	1.89	0.31	0.30	0.28	0.23	0.21	0.22	0.22	0.21	0.18	0.17	0.22	3
			(lt. wt. agg.)													
Face brick			(4 in.)	1.50	0.35	0.34	0.31	0.25	0.23	0.24	0.24	0.22	0.19	0.19	0.24	4
4 in.	0.44		(8 in.)	2.00	0.30	0.29	0.27	0.23	0.21	0.22	0.21	0.20	0.17	0.17	0.21	5
Stone			(12 in.)	2.27	0.28	0.27	0.25	0.21	0.20	0.20	0.20	0.19	0.17	0.16	0.20	6
4 in.	0.32		(sand agg.)													
Precast			(4 in.)	0.71	0.49	0.46	0.41	0.32	0.29	0.30	0.29	0.27	0.22	0.22	0.29	7
Concrete		.39	(8 in.)	1.11	0.41	0.39	0.35	0.28	0.26	0.27	0.26	0.25	0.21	0.20	0.26	8
(sand agg.)			(12 in.)	1.28	0.38	0.37	0.33	0.27	0.25	0.26	0.25	0.24	0.20	0.20	0.25	9
4 in.	0.32															
6 in.	0.48		Hollow clay tile													
			(4 in.)	1.11	0.41	0.39	0.35	0.28	0.26	0.27	0.26	0.25	0.21	0.20	0.26	10
			(8 in.)	1.85	0.31	0.30	0.28	0.23	0.22	0.22	0.22	0.21	0.18	0.18	0.22	11
			(12 in.)	2.50	0.26	0.25	0.24	0.20	0.19	0.19	0.19	0.18	0.16	0.16	0.19	12

			Concrete													
			(sand agg.)													
			(4 in.)	0.32	0.60	0.56	0.49	0.36	0.32	0.34	0.33	0.31	0.25	0.24	0.33	13
			(6 in.)	0.48	0.55	0.52	0.45	0.34	0.31	0.32	0.31	0.29	0.24	0.23	0.31	14
			(8 in.)	0.64	0.51	0.48	0.42	0.32	0.29	0.31	0.30	0.28	0.23	0.22	0.30	15
			Concrete block													
			(cinder agg.)													
			(4 in.)	1.11	0.36	0.35	0.32	0.26	0.24	0.25	0.24	0.23	0.19	0.19	0.24	16
			(8 in.)	1.72	0.29	0.29	0.26	0.22	0.21	0.21	0.21	0.20	0.17	0.17	0.21	17
			(12 in.)	1.89	0.28	0.27	0.25	0.21	0.20	0.21	0.20	0.19	0.17	0.17	0.20	18
			(lt. wt. agg.)													
			(4 in.)	1.50	0.32	0.30	0.28	0.23	0.22	0.22	0.22	0.21	0.18	0.18	0.22	19
			(8 in.)	2.00	0.27	0.26	0.25	0.21	0.20	0.20	0.20	0.19	0.16	0.16	0.20	20
Common brick			(12 in.)	2.27	0.25	0.25	0.23	0.20	0.19	0.19	0.19	0.18	0.16	0.16	0.19	21
4 in.	0.80		(sand agg.)													
Precast			(4 in.)	0.71	0.42	0.40	0.36	0.29	0.26	0.27	0.27	0.25	0.21	0.21	0.27	22
Concrete		.72	(8 in.)	1.11	0.36	0.35	0.32	0.26	0.24	0.25	0.24	0.23	0.19	0.19	0.24	23
(sand			(12 in.)	1.28	0.34	0.33	0.30	0.25	0.23	0.24	0.23	0.22	0.19	0.18	0.23	24
agg.)																
8 in.	0.64															
			Hollow clay tile													
			(4 in.)	1.11	0.36	0.35	0.32	0.26	0.24	0.25	0.24	0.23	0.19	0.19	0.24	25
			(6 in.)	1.85	0.28	0.28	0.26	0.22	0.20	0.21	0.20	0.19	0.17	0.17	0.20	26
			(12 in.)	2.50	0.24	0.23	0.22	0.19	0.18	0.18	0.18	0.17	0.15	0.15	0.18	27
			Concrete													
			(sand agg.)													
			(4 in.)	0.32	0.50	0.48	0.42	0.32	0.29	0.30	0.30	0.28	0.23	0.22	0.30	28
			(6 in.)	0.48	0.47	0.44	0.39	0.31	0.28	0.29	0.28	0.27	0.22	0.22	0.28	29
			(8 in.)	0.64	0.43	0.41	0.37	0.29	0.27	0.28	0.27	0.26	0.21	0.21	0.27	30

Extracted with permission from 1965 ASHRAE Guide and Data Book.
Courtesy of the Trane Company.

Table 7-7. Coefficients of transmission (U) of masonry cavity walls.

Coefficients are expressed in Btu per (hour) (square foot) (Fahrenheit degree difference in temperature between the air on the two sides), and are based on an outside wind velocity of 15 mph.

Exterior Construction			Inner Section		None	Interior Finish: Plas. 5/8 in. on Wall		Metal Lath and 3/4 in. Plas. on Furring		Gypsum Lath (3/8 in.) and 1/2 in. Plas. on Furring			Insul. Bd. Lath (1/2 in.) and 1/2 in. Plas. on Furring		Wood Lath and 1/2 in. Plas.	Number
			Resistance ↳			(Sand Agg.) 0.11	(Lt. Wt. Agg.) 0.39	(Sand Agg.) 0.13	(Lt. Wt. Agg.) 0.47	No Plas. 0.32	(Sand Agg.) 0.41	(Lt. Wt. Agg.) 0.64	No Plas. 1.43	(Sand Agg.) 1.52	(Sand Agg.) 0.40	
Material	R	Av. R	Material		U	U	U	U	U	U	U	U	U	U	U	
					R	A	B	C	D	E	F	G	H	I	J K	
Face Brick (4 in.)		.44	Concrete block (4 in.) (gravel agg.)	0.71	0.34	0.32	0.30	0.25	0.23	0.23	0.23	0.22	0.19	0.18	0.23	1
			(cinder agg.)	1.11	0.30	0.29	0.27	0.22	0.21	0.21	0.21	0.20	0.17	0.17	0.21	2
			(lt. wt. agg.)	1.50	0.27	0.26	0.24	0.21	0.19	0.20	0.19	0.19	0.16	0.16	0.19	3
			Common brick (4 in.)	0.80	0.33	0.32	0.29	0.24	0.22	0.23	0.23	0.21	0.18	0.18	0.23	4
			Clay tile (4 in.)	1.11	0.30	0.29	0.27	0.22	0.21	0.21	0.21	0.20	0.17	0.17	0.21	5

			Concrete block (4 in.)													
Common brick			(gravel agg.)	0.71	0.30	0.29	0.27	0.23	0.21	0.22	0.21	0.20	0.18	0.17	0.21	6
(4 in.)	0.80		(cinder agg.)	1.11	0.27	0.26	0.25	0.21	0.19	0.20	0.20	0.19	0.16	0.16	0.20	7
Concrete block		0.76	(lt. wt. agg.)	1.50	0.25	0.24	0.22	0.19	0.18	0.19	0.18	0.18	0.15	0.15	0.18	8
(gravel agg.)																
(4 in.)	0.71		Common brick													
			(4 in.)	0.80	0.30	0.29	0.27	0.22	0.21	0.21	0.21	0.20	0.17	0.17	0.21	9
			Clay tile (4 in.)	1.11	0.27	0.26	0.25	0.21	0.19	0.20	0.20	0.19	0.16	0.16	0.20	10
			Concrete block (4 in.)													
			(gravel agg.)	0.71	0.27	0.27	0.25	0.21	0.20	0.20	0.20	0.19	0.17	0.16	0.20	11
Concrete			(cinder agg.)1.11	0.25	0.24	0.23	0.19	0.18	0.19	0.18	0.18	0.16	0.15	0.18	12	
block			(lt. wt. agg.)	1.50	0.23	0.22	0.21	0.18	0.17	0.17	0.17	0.17	0.15	0.14	0.17	13
(cinder agg.)			1.11													
(4 in.)			Common brick													
			(4 in.)	0.80	0.27	0.26	0.24	0.21	0.19	0.20	0.20	0.19	0.16	0.16	0.20	14
			Clay tile (4 in.)	1.11	0.25	0.24	0.23	0.19	0.18	0.19	0.18	0.18	0.16	0.15	0.18	15

Extracted with permission from 1965 *ASHRAE Guide and Data Book*.
Courtesy of the Trane Company.

Table 7-8. Coefficients of transmission (U) of frame partitions or interior walls.

Coefficients are expressed in Btu per (hour) (square foot) (Fahrenheit degree difference in temperature between the air on the two sides), and are based on still air (no wind) conditions on both sides.

Type of Interior Finish		Single Partition (Finish on only one side of studs)	Double Partition (Finish on both sides of studs)	Number
Material	R	U	U	
		A	B	
Gypsum bd. (3/8 in.)	0.32	0.60	0.34	1
Gypsum lath (3/8 in.) and 1/2 in. Plas. (lt. wt. agg.)	0.64	0.50	0.28	2
Gypsum lath (3/8 in.) and 1/2 in. Plas. (sand agg.)	0.41	0.56	0.32	3
Metal lath and 3/4 in. Plas. (lt. wt. agg.)	0.47	0.55	0.31	4
Metal lath and 3/4 in. Plas. (sand agg.)	0.13	0.67	0.39	5
Insul. bd. (1/2 in.)	1.43	0.36	0.19	6
Insul. bd. lath (1/2 in.) and 1/2 in. Plas. (sand agg.)	1.52	0.35	0.19	7
Plywood: (1/4 in.)	0.31	0.60	0.34	8
(3/8 in.)	0.47	0.55	0.31	9
(1/2 in.)	0.63	0.50	0.28	10
Wood Panels (3/4 in.)	0.94	0.43	0.24	11
Wood-lath and 1/2 in. Plas. (sand agg.)	0.40	0.57	0.32	12
Sheet-Metal Panels adhered to Wood (Framing)	0	0.74	0.43	13
Glass and Glass Blocks		See Table 3-2		

Extracted with permission from 1965 *ASHRAE Guide and Data Book.*
Courtesy of the Trane Company.

Table 7-9. Coefficients of transmission (U) of frame walls.

These coefficients are expressed in Btu per (hour) (square foot) (Fahrenheit degree difference in temperature between the air on the two sides), and are based on an outside wind velocity of 15 mph.

Exterior[1]			Interior finish		Type of Sheathing[3]						
					none, building paper	gupsum board 1/2 in.	plywood 5/16 in.	wood 25/32 in. and building paper	insulation board sheathing 1/2 in.	insulation board sheathing 25/32 in.	
			Resistance →		0.06	0.45	0.39	1.04	1.32	2.06	
					U	U	U	U	U	U	
Material	R	Av R	Material	R	A	B	C	D	E	F	Number
			None	—	0.57	0.47	0.48	0.36	0.33	0.27	1
			Gypsum bd. (3/8 in.)	0.32	0.33	0.29	0.30	0.25	0.23	0.20	2
			Gypsum lath (3/8 in.) & 1/2 in. plas. (lt. wt. agg.)	0.64	0.30	0.27	0.27	0.23	0.22	0.19	3
			Gypsum lath (3/8 in.) & 1/2 in. plas. (sand agg.)	0.41	0.32	0.28	0.29	0.24	0.23	0.19	4
Wood Siding Drop-(1 in. × 8 in.)	0.79		Metal lath and 3/4 in. plas. (lt. wt. agg.)	0.47	0.31	0.28	0.28	0.24	0.22	0.19	5
Bevel (1/2 in. × 8 in.)	0.81	0.85[2]	Metal lath and 3/4 in. plas. (sand. agg.)	0.13	0.35	0.31	0.31	0.26	0.24	0.21	6
Wood Shingles 7-1/2 in. exposure	0.87		Insul. bd. (1/2 in.)	1.43	0.24	0.22	0.22	0.19	0.18	0.16	7
Wood Panels (3/4 in.)	0.94		Insul. bd. lath (1/2 in.) & 1/2 in. plas. (sand agg.)	1.52	0.24	0.22	0.22	0.19	0.18	0.16	8
			Plywood (1/4 in.)	0.31	0.33	0.29	0.30	0.25	0.23	0.20	9
			Wood panels (3/4 in.)	0.94	0.27	0.25	0.25	0.22	0.20	0.18	10
			Wood lath and 1/2 in. plas. (sand agg.)	0.40	0.32	0.28	0.29	0.24	0.23	0.19	11
			None	—	0.73	0.56	0.58	0.42	0.38	0.30	12
			Gypsum bd. (3/8 in.)	0.32	0.37	0.33	0.33	0.27	0.25	0.21	13
			Gypsum lath (3/8 in.) & 1/2 in. plas. (lt. wt. agg)	0.64	0.33	0.30	0.30	0.25	0.24	0.20	14
			Gypsum lath (3/8 in.) & 1/2 in. plas. (sand agg.)	0.41	0.36	0.32	0.32	0.27	0.25	0.21	15
Face-brick veneer[4]	0.44	0.45[2]	Metal lath and 3/4 in. plas. (lt. wt. agg.)	0.47	0.35	0.31	0.32	0.26	0.25	0.21	16
Plywood (3/8 in.)	0.47		Metal lath and 3/4 in. plas. (sand agg.)	0.13	0.40	0.35	0.36	0.29	0.27	0.22	17
			Insul. bd. (1/2 in.)	1.43	0.26	0.24	0.24	0.21	0.20	0.17	18
			Insul. bd. lath (1/2 in.) and 1/2 in. plas. (sand agg.)	1.52	0.26	0.23	0.24	0.21	0.19	0.17	19
			Plywood (1/4 in.)	0.31	0.38	0.33	0.33	0.27	0.26	0.21	20
			Wood panels (3/4 in.)	0.94	0.30	0.27	0.28	0.23	0.22	0.19	21
			Wood lath and 1/2 in. plas. (sand agg.)	0.40	0.36	0.32	0.32	0.27	0.25	0.21	22

[1]Note that although several types of exterior finish may be grouped because they have approximately the same thermal resistance value, it is not implied that all types may be suitable for application over all types of sheathing listed.
[2]Average resistance of items listed. This average was used in computation of U values shown.
[3]Building paper is not included except where noted.
[4]Small air space between building paper and brick veneer neglected.
[5]Where stucco is applied over insulating board or gypsum sheathing, building paper is generally required, but the change in U value is negligible.
Extracted with permission from 1965 *ASHRAE Guide and Data Book*.
Courtesy of the Trane Company.

Table 7-10. Coefficients of transmission (U) of frame construction ceilings and floors.

Coefficients are expressed in Btu per (hour) (square foot) (Fahrenheit degree difference between the air on the two sides) and are based on still air (no wind) conditions on both sides

Direction of Heat	Heat Flow Upward (winter conditions)						Heat Flow Downward (summer conditions)						
	Type of Floor												
	Wood Subfloor (25/32 in.), felt, and—						Wood Subfloor (25/32 in.), felt, and—						
Type of Ceiling	None	Wood subfloor (25/32 in.)	Cement (1-1/2 in.) and ceramic tile (1/2 in.)	Hardwood floor (3/4 in.)	Plywood (5/8 in.) and floor tile or linoleum (1/8 in.)	Insul. bd. (3/8 in.) and hard bd. (1/4 in.) and floor tile or linoleum (1/8 in.)	None	Wood subfloor (25/32 in.)	Cement (1-1/2 in.) and ceramic tile (1/2 in.)	Hardwood floor (3/4 in.)	Plywood (5/8 in.) and floor tile[1] or linoleum (1/8 in.)	Insul. bd. (3/8 in.) and hard bd. (1/4 in.) and floor tile or linoleum (1/8 in.)	Number
Resistance ↳	—	0.98	1.38	1.72	1.87	2.26	—	0.98	1.38	1.72	1.87	2.26	

Material	R	U	U	U	U	U	U	U	U	U	U	U	U	
		A	B	C	D	E	F	G	H	I	J	K	L	
None	—	—	0.45	0.38	0.34	0.32	0.29	—	0.35	0.31	0.28	0.26	0.24	1
Gypsum bd. (3/8 in.)	0.32	0.65	0.30	0.27	0.24	0.23	0.22	0.46	0.24	0.22	0.21	0.20	0.19	2
Gypsum lath (3/8 in.) and 1/2 in. Plas. (lt. wt. agg.)	0.64	0.54	0.27	0.24	0.23	0.22	0.20	0.40	0.22	0.21	0.19	0.19	0.17	3
Gypsum lath (3/8 in.) and 1/2 in. Plas. (sand agg.)	0.41	0.61	0.29	0.26	0.24	0.23	0.21	0.44	0.24	0.22	0.20	0.20	0.18	4
Metal lath and 3/8 in. Plas. (lt. wt. agg.)	0.47	0.59	0.28	0.26	0.23	0.23	0.21	0.43	0.23	0.21	0.20	0.19	0.18	5
Metal lath and 3/8 in. Plas. (sand agg.)	0.13	0.74	0.31	0.28	0.26	0.25	0.22	0.51	0.25	0.23	0.21	0.21	0.19	6
Insul bd. (1/2 in.)	1.43	0.38	0.22	0.20	0.19	0.19	0.17	0.31	0.19	0.18	0.17	0.16	0.15	7
Insul. bd. lath (1/2 in.) and 1/2 in. Plas. (sand agg.)	1.52	0.36	0.22	0.20	0.19	0.18	0.17	0.30	0.19	0.17	0.17	0.16	0.15	8
Acoustical Tile (1/2 in.) on gypsum bd. (3/8 in.)	1.51[2]	0.37	0.22	0.20	0.19	0.18	0.17	0.30	0.19	0.17	0.17	0.16	0.15	9
(1/2 in.) on furring	1.19	0.41	0.24	0.22	0.20	0.19	0.18	0.33	0.20	0.19	0.17	0.17	0.16	10
(3/4 in.) on gypsum bd. (3/8 in.)	2.10[2]	0.30	0.19	0.18	0.17	0.17	0.15	0.25	0.17	0.16	0.15	0.15	0.14	11
(3/4 in.) on furring	1.78	0.33	0.21	0.19	0.18	0.17	0.16	0.28	0.18	0.17	0.16	0.15	0.15	12
Wood lath and 1/2 in. Plas. (sand agg.)	0.40	0.62	0.29	0.26	0.24	0.23	0.21	0.45	0.24	0.22	0.20	0.20	0.18	13

[1]Includes asphalt, rubber, and plastic tile (1/2 in.), ceramic tile, or terrazzo (1 in.).
[2]Includes thermal resistance of 3/8 in. gypsum wall board.
Extracted with permission from 1965 *ASHRAE Guide and Data Book*.
Courtesy of the Trane Company.

Table 7-11. Coefficients of transmission (U) of concrete floor-ceiling constructions (Summer Conditions, Downward Flow).

Coefficients are expressed in Btu per (hour) (square foot) (Fahrenheit degree difference in temperature between the air on the two sides), and are based on still air (no wind) conditions on both sides.

Type of deck			Type of finish floor		Type of ceiling														Number
					Ceiling applied directly to slab					Suspended ceiling									
										Gypsum bd. 3/8 in. and plas.			Metal lath and plas.		Acoustical tile				
					None	Plas.		Acoustical tile-glued							On furring or channels		On gypsum bd. (3/8 in.)		
						(lt. wt. agg.) 1/8 in.	(sand agg.) 1/8 in.	1/2 in.	3/4 in.	No plas.	(lt. wt. agg.) 1/2 in.	(sand agg.) 1/2 in.	(lt. wt. agg.) 3/4 in.	(sand agg.) 3/4 in.	1/2 in.	3/4 in.	1/2 in.	3/4 in.	
			Resistance	↳	—	0.08	0.02	1.19	1.78	0.32	0.64	0.41	0.47	0.13	1.19	1.78	1.51	2.10	
Material	R	AV. R	Material	AV. R	U O	U P	U Q	U R	U S	U T	U U	U V	U W	U X	U Y	U Z	U Z′	U Z″	
Concrete[4] (sand agg.) (4 in.) 0.32 (6 in.) 0.48		0.40	None	—	0.45	0.43	0.44	0.29	0.25	0.28	0.26	0.27	0.27	0.30	0.23	0.20	0.21	0.19	1
			Floor tile[5] or linoleum (1/8 in.).	0.05	0.44	0.42	0.43	0.29	0.25	0.28	0.26	0.27	0.27	0.29	0.22	0.20	0.21	0.19	2
			Wood block (13/16 in.) on slab	0.74	0.34	0.33	0.33	0.24	0.21	0.23	0.22	0.23	0.23	0.24	0.19	0.17	0.18	0.17	3
			Floor on sleepers Plywood subfloor 5/8 in.), felt and floor tile[5] or linoleum (1/8 in.)	0.89	0.23	0.23	0.23	0.18	0.17	0.19	0.18	0.19	0.18	0.20	0.16	0.15	0.15	0.14	4

		Wood subfloor (25/32 in.), felt and hardwood (3/4 in.)	1.72	0.20	0. 19	0.20	0.16	0.15	0.16	0.15	0.16	0.16	0.17	0.14	0.13	0.14	0.13	5
		None	—	0.39	0.38	0.39	0.27	0.23	0.26	0.24	0.25	0.25	0.27	0.21	0.19	0.20	0.18	6
		Floor tile[5] or linoleum (1/8 in.)	0.05	0.38	0.37	0.38	0.26	0.23	0.26	0.24	0.25	0.25	0.27	0.21	0.19	0.20	0.18	7
Concrete[4] (sand agg.) (8 in.) 0.64 (10 in.) 0.80		Wood block (13/16 in.) on slab	0.74	0.30	0.30	0.30	0.22	0.20	0.22	0.20	0.21	0.21	0.23	0.18	0.16	0.17	0.16	8
	0.72	Floor on sleepers Plywood subfloor (5/8 in.), felt and floor tile[5] or linoleom (1/8 in.)	0.89	0.22	0.21	0.22	0.17	0.16	0.18	0.17	0.18	0.17	0.18	0.15	0.14	0.15	0.14	9
		Wood subfloor (25/32 in.), felt and hardwood (3/4 in.)	1.72	0.19	0. 18	0. 18	0. 15	0.14	0.16	0.15	0.15	0.15	0.16	0.14	0.13	0.13	0.12	10

[4]Concrete is assumed to have a thermal conductivity k of 12.0.
[5]Includes asphalt, rubber, and plastic tile (1/8 in.), ceramic tile on terrazzo (1 in.).
Extracted with permission from 1965 *ASHRAE Guide and Data Book.*
Courtesy of the Trane Company.

Table 7-12. Coefficients of transmission (U) of flat masonry roofs with built-up roofing with and without Suspended Ceiling (Summer Conditions, Downward Flow).
These coefficients are expressed in Btu per (hour) (square foot) (Fahrenheit degree difference in temperature between the air on the two sides), and are based on an outside wind velocity of 7.5 mph.

Type of Deck	Type of Form		Type of Ceiling: Roof Insulation—No Ceiling: C Value of Roof Insulation							Type of Ceiling: Suspended Ceiling: Gypsum bd. (3/8 in.) and plas.			Metal lath and plas.		Acoustical Tile				Number
			NONE	0.72	0.36	0.24	0.19	0.15	0.12	No plas.	lt. wt. agg. (1/2″)	sand agg. (1/2″)	lt. wt. agg. (3/4″)	sand agg. (3/4″)	on furring or channels 1/2 in.	on furring or channels 3/4 in.	on gypsum bd. (3/8 in.) 1/2 in.	on gypsum bd. (3/8 in.) 3/4 in.	
Resistance →			—	1.39	2.78	4.17	5.26	6.67	8.33	0.32	0.64	0.41	0.47	0.13	1.19	1.78	1.51	2.10	
			U	U	U	U	U	U	U	U	U	U	U	U	U	U	U	U	
Material R	Material	R	A′	B′	C′	D′	E′	F′	G′	H′	I′	J′	K′	L′	M′	N′	O′	P′	
Concrete slab[4] (gravel agg.)																			
(4 in.)...0.32	Temporary	—	0.55	0.31	0.22	0.17	0.14	0.12	0.10	0.32	0.29	0.31	0.30	0.34	0.25	0.22	0.23	0.20	1
(6 in.)...0.48	Temporary	—	0.51	0.30	0.21	0.16	0.14	0.12	0.10	0.30	0.28	0.30	0.29	0.32	0.24	0.21	0.22	0.20	2
(8 in.)...0.64	Temporary	—	0.47	0.28	0.20	0.16	0.14	0.11	0.10	0.29	0.27	0.28	0.28	0.31	0.23	0.20	0.22	0.19	3
Lt. wt. agg.[5]																			
(2 in.)...2.22	Corrugated Metal[3]	0	0.27	0.20	0.15	0.13	0.11	0.10	0.08	0.20	0.19	0.20	0.19	0.21	0.17	0.15	0.16	0.15	4
	Insul. bd. (1 in.)	2.78	0.15	0.13	0.11	0.09	0.09	0.08	0.07	0.13	0.12	0.13	0.13	0.13	0.12	0.11	0.11	0.10	5
	Insul. bd. (1-1/2 in.)	4.17	0.13	0.11	0.09	0.08	0.08	0.07	0.06	0.11	0.11	0.11	0.11	0.11	0.10	0.09	0.10	0.09	6
	Glass fib. bd. (1 in.)	4.00	0.13	0.11	0.10	0.08	0.08	0.07	0.06	0.11	0.11	0.11	0.11	0.11	0.10	0.10	0.10	0.09	7
(3 in.)...3.33	Corrugated Metal[3]	0	0.21	0.16	0.13	0.11	0.10	0.09	0.08	0.16	0.16	0.16	0.16	0.17	0.14	0.13	0.14	0.13	8
	Insul. bd. (1 in.)	2.78	0.13	0.11	0.10	0.08	0.08	0.07	0.06	0.11	0.11	0.11	0.11	0.11	0.10	0.10	0.10	0.09	9

	Insul. bd. (1-1/2 in.)	4.17	0.11	0.10	0.08	0.08	0.07	0.06	0.06	0.10	0.09	0.10	0.10	0.10	0.09	0.09	0.09	0.08	10
	Glass fib. bd. (1 in.)	4.00	0.11	0.10	0.09	0.08	0.07	0.06	0.06	0.10	0.10	0.10	0.10	0.10	0.09	0.09	0.09	0.08	11
(4 in.)...4.44	Corrugated Metal[3]	0	0.17	0.14	0.11	0.10	0.09	0.08	0.07	0.14	0.13	0.14	0.14	0.14	0.12	0.12	0.12	0.11	12
	Insul. bd. (1 in.)	2.78	0.11	0.10	0.09	0.08	0.07	0.06	0.06	0.10	0.10	0.10	0.10	0.10	0.09	0.09	0.09	0.09	13
	Insul. bd. (1-1/2 in.)	4.17	0.10	0.09	0.08	0.07	0.07	0.06	0.05	0.09	0.09	0.09	0.09	0.09	0.08	0.08	0.08	0.08	14
	Glass fib. bd. (1 in.)	4.00	0.10	0.09	0.08	0.07	0.07	0.06	0.05	0.09	0.09	0.09	0.09	0.09	0.08	0.08	0.08	0.08	15
Gypsum slab[7]																			
(2 in.)...1.20	Gypsum bd. (1/2 in.)	0.45	0.32	0.22	0.17	0.14	0.12	0.10	0.09	0.22	0.21	0.22	0.22	0.23	0.19	0.17	0.18	0.16	16
	Insul. bd. (1 in.)	2.78	0.18	0.15	0.12	0.10	0.09	0.08	0.07	0.15	0.14	0.15	0.14	0.15	0.13	0.12	0.13	0.12	17
	Insul. bd. (1-1/2 in.)	4.17	0.15	0.12	0.10	0.09	0.08	0.07	0.07	0.12	0.12	0.12	0.12	0.13	0.11	0.10	0.11	0.10	18
	Asbestos-cement bd.[6] (1/4 in.)	0.06	0.34	0.23	0.18	0.14	0.12	0.10	0.09	0.25	0.23	0.24	0.24	0.26	0.20	0.18	0.19	0.17	19
	Glass fib. bd. (1 in.)	4.00	0.15	0.12	0.11	0.09	0.08	0.07	0.07	0.13	0.12	0.12	0.12	0.13	0.11	0.11	0.11	0.10	20
(3 in.)...1.80	Gypsum bd. (1/2 in.)	0.45	0.27	0.19	0.15	0.13	0.11	0.10	0.08	0.20	0.19	0.19	0.19	0.21	0.17	0.15	0.16	0.15	21
	Insul. bd. (1 in.)	2.78	0.16	0.13	0.11	0.10	0.09	0.08	0.07	0.14	0.13	0.13	0.13	0.14	0.12	0.11	0.12	0.11	22
	Insul. bd. (1-1/2 in.)	4.77	0.13	0.11	0.10	0.09	0.08	0.07	0.06	0.11	0.11	0.11	0.11	0.12	0.10	0.11	0.10	0.10	23
	Asbestos-cement bd. (1-1/4 in.)	0.06	0.30	0.21	0.16	0.13	0.12	0.10	0.09	0.21	0.20	0.21	0.21	0.22	0.18	0.16	0.17	0.16	24
	Glass fib. bd. (1 in.)	4.00	0.14	0.12	0.10	0.09	0.08	0.07	0.06	0.12	0.11	0.12	0.11	0.12	0.11	0.10	0.10	0.10	25
(4 in.)...2.40	Gypsum bd. (1/2 in.)	0.45	0.23	0.17	0.14	0.12	0.10	0.09	0.08	0.18	0.17	0.17	0.17	0.18	0.15	0.14	0.15	0.13	26
	Insul. bd. (1 in.)	2.78	0.15	0.12	0.11	0.09	0.08	0.07	0.07	0.13	0.12	0.12	0.12	0.13	0.11	0.11	0.11	0.10	27
	Insul. bd. (1-1/2 in.)	4.17	0.12	0.11	0.09	0.08	0.08	0.07	0.06	0.11	0.10	0.11	0.11	0.11	0.10	0.09	0.10	0.09	28
	Asbestos-cement bd.[3] (1/4 in.)	0.06	0.25	0.19	0.15	0.12	0.11	0.09	0.08	0.19	0.18	0.19	0.19	0.20	0.16	0.15	0.16	0.14	29
	Glass fib. bd. (1 in.)	4.00	0.13	0.11	0.09	0.08	0.08	0.07	0.06	0.11	0.11	0.11	0.11	0.11	0.10	0.09	0.10	0.09	30

[3]U values would also apply if slab were poured on metal lath, paper-backed wire, fabric, or asbestos-cement board (1/4 in.).
[4]Concrete assumed to have a thermal conductivity k of 12.0 and a density of 140 lb per cu ft.
[5]Concrete assumed to have a thermal conductivity k of 0.90 and a density of 30 lb per cu ft.
[6]Gypsum slab 2-1/4 in. thick since this is recommended practice.
[7]Gypsum fiber concrete with 12-1/2 percent wood chips (thermal conductivity k = 1.66).
Extracted with permission from *1965 ASHRAE Guide and Data Book.*

Table 7-13. Coefficients of transmission (U) of wood or metal construction, flat roofs and ceilings (summer conditions, downward flow). Coefficients are expressed in Btu per (hour) (square foot) (Fahrenheit degree difference in temperature between the air on the two sides), and are based upon an outside wind velocity of 7.5 mph.

		Insulation added on top of deck[4]		Type of Ceiling											
					Gypsum bd. (3/8 in. and plas.			Metal lath and plas.		Insul bd. (1/2 in.)	Acoustical tile				
Type of deck (built-up roof in all cases)		Con. duct- ance insul.	Re- sist- ance	None	None	Lt. wt. agg. 1/2 in.	Sand agg. 1/2 in.	Lt. wt. agg. 3/4 in.	Sand agg. 3/4 in.	plain (1.43) or 1/2 in. plas. sand agg. 1.52	On furring		On gypsum bd. (1/2 in.)		Number
											1/2 in.	3/4 in.	1/2 in.	3/4 in.	
Resistance → ↓					0.32	0.64	0.41	0.47	0.13	1.47	1.19	1.78	1.51	2.10	
Material	R	C	R	U A′	U B′	U C′	U D′	U E′	U F′	U G′	U H′	U I′	U J′	U K′	
		None	—	0.40	0.26	0.24	0.26	0.25	0.28	0.20	0.22	0.19	0.20	0.18	1
		0.72	1.39	0.26	0.19	0.18	0.19	0.19	0.20	0.16	0.17	0.15	0.16	0.14	2
		0.36	2.78	0.19	0.15	0.15	0.15	0.15	0.16	0.13	0.14	0.13	0.13	0.12	3
Wood[3] 1 in.	0.98	0.24	4.17	0.15	0.13	0.12	0.12	0.12	0.13	0.11	0.11	0.11	0.11	0.10	4
		0.19	5.26	0.13	0.11	0.11	0.11	0.11	0.11	0.10	0.10	0.10	0.10	0.09	5
		0.15	6.67	0.11	0.10	0.09	0.10	0.09	0.10	0.09	0.09	0.08	0.09	0.08	6
		0.12	8.33	0.09	0.08	0.08	0.08	0.08	0.08	0.08	0.08	0.07	0.08	0.07	7
		None	—	0.28	0.21	0.19	0.20	0.20	0.22	0.17	0.18	0.16	0.17	0.15	8
		0.72	1.39	0.20	0.16	0.15	0.16	0.16	0.17	0.14	0.14	0.13	0.14	0.13	9
		0.36	2.78	0.16	0.13	0.13	0.13	0.13	0.13	0.11	0.12	0.11	0.11	0.11	10
Wood[3] 2 in.	2.03	0.24	4.17	0.13	0.11	0.11	0.11	0.11	0.11	0.10	0.10	0.10	0.10	0.09	11
		0.19	5.26	0.11	0.10	0.10	0.10	0.10	0.10	0.09	0.09	0.09	0.09	0.08	12
		0.15	6.67	0.10	0.09	0.09	0.09	0.09	0.09	0.08	0.08	0.08	0.08	0.08	13
		0.12	8.33	0.08	0.08	0.07	0.07	0.08	0.08	0.07	0.07	0.07	0.07	0.07	14

		None	—	0.21	0.17	0.16	0.16	0.16	0.17	0.14	0.15	0.13	0.14	0.13	15
		0.72	1.39	0.16	0.13	0.13	0.13	0.13	0.14	0.12	0.12	0.11	0.12	0.11	16
		0.36	2.78	0.13	0.11	0.11	0.11	0.11	0.12	0.10	0.10	0.10	0.10	0.09	17
Wood[3] 3 in.	3.23	0.24	4.17	0.11	0.10	0.10	0.10	0.10	0.10	0.09	0.09	0.09	0.09	0.08	18
		0.19	5.26	0.10	0.09	0.09	0.09	0.09	0.09	0.08	0.08	0.09	0.08	0.08	19
		0.15	6.67	0.09	0.08	0.08	0.08	0.08	0.08	0.07	0.07	0.08	0.07	0.07	20
		0.12	8.33	0.08	0.07	0.07	0.07	0.07	0.07	0.07	0.07	0.07	0.06	0.06	21
Preformed slabs-wood fiber and cement binder															
2 in.	3.60	None	—	0.20	0.16	0.15	0.15	0.15	0.16	0.13	0.14	0.15	0.13	0.12	22
3 in.	5.40	None	—	0.14	0.12	0.12	0.12	0.12	0.13	0.11	0.11	0.12	0.11	0.10	23
		None	—	0.67	0.36	0.32	0.34	0.34	0.38	0.25	0.27	0.23	0.25	0.22	24
		0.72	1.39	0.35	0.24	0.22	0.23	0.23	0.25	0.19	0.20	0.18	0.19	0.17	25
Flat metal		0.36	2.78	0.23	0.18	0.17	0.18	0.17	0.19	0.15	0.16	0.14	0.15	0.14	26
roof deck	0	0.24	4.17	0.18	0.14	0.14	0.14	0.14	0.15	0.12	0.13	0.12	0.12	0.11	27
		0.19	5.26	0.15	0.12	0.12	0.12	0.12	0.13	0.11	0.11	0.11	0.11	0.10	28
		0.15	6.67	0.12	0.11	0.10	0.10	0.10	0.11	0.09	0.10	0.09	0.09	0.09	29
		0.12	8.33	0.10	0.09	0.09	0.09	0.09	0.09	0.08	0.08	0.08	0.08	0.08	30

[3]Wood deck 1, 2, and 3 in. is assumed to be 25/32, 1-5/8, and 2-5/8, in. thick, respectively. The thermal conductivity k is assumed to be 0.80.
[4]If a vapor barrier is used beneath roof insulation it will have a negligible effect on the U value.
Extracted with permission from *1965 ASHRAE Guide and Data Book.*
Courtesy of *the Trane Company.*

Table 7-14. Outside design temperatures and latitudes for various cities in the United States.

STATE	CITY	DESIGN TEMP DB	DESIGN TEMP WB	NORTH LATITUDE, DEGREES
ALABAMA	ANNISTON	95	75	33
	BIRMINGHAM	95	78	33
	MOBILE	95	80	31
ALASKA	JUNEAU	65	52	58
ARIZONA	PHOENIX	105	76	33
	TUCSON	105	72	32
ARKANSAS	LITTLE ROCK	95	78	35
CALIFORNIA	FRESNO	105	74	37
	LOS ANGELES	90	70	34
	SACRAMENTO	100	72	38
	SAN FRANCISCO	85	65	38
COLORADO	DENVER	95	64	40
	PUEBLO	95	65	38
CONNECTICUT	HARTFORD	93	75	42
DELAWARE	WILMINGTON	95	78	40
DISTRICT OF COLUMBIA	WASHINGTON	95	78	39
FLORIDA	JACKSONVILLE	95	78	30
	MIAMI	91	79	26
GEORGIA	ATLANTA	95	76	34
	SAVANNAH	95	78	32
HAWAII	HONOLULU	83	73	21
IDAHO	BOISE	95	65	44
	POCATELLO	95	65	43
ILLINOIS	CHICAGO	95	75	43
	PEORIA	96	76	43
	SPRINGFIELD	98	77	42
INDIANA	EVANSVILLE	95	78	38
	FORT WAYNE	95	75	41
	INDIANAPOLIS	95	76	40

IOWA	DES MOINES	95	78	42
	DUBUQUE	95	78	42
	SIOUX CITY	95	78	42
KANSAS	KANSAS CITY	100	76	39
	WICHITA	100	75	38
KENTUCKY	ASHLAND	95	76	38
	LOUISVILLE	95	78	38
LOUISIANA	NEW ORLEANS	95	80	30
	SHREVEPORT	100	78	32
MAINE	PORTLAND	90	73	44
MARYLAND	BALTIMORE	95	78	39
MASSACHUSETTS	BOSTON	92	75	42
	HOLYOKE	93	75	42
MICHIGAN	DETROIT	95	75	42
	GRAND RAPIDS	95	75	43
MINNESOTA	DULUTH	93	73	47
	MINNEAPOLIS	95	75	45
MISSISSIPPI	JACKSON	95	78	32
MISSOURI	KANSAS CITY	100	76	39
	SPRINGFIELD	100	75	37
	ST. LOUIS	95	78	39
MONTANA	BILLINGS	90	66	46
	HELENA	95	67	46
NEBRASKA	LINCOLN	95	78	41
	OMAHA	95	78	41
NEVADA	RENO	95	65	39
NEW HAMPSHIRE	CONCORD	90	73	43
NEW JERSEY	TRENTON	95	78	40
NEW MEXICO	ALBUQUERQUE	95	70	35
	SANTA FE	95	65	36
NEW YORK	ALBANY	93	75	43
	BUFFALO	93	73	43
	NEW YORK	95	75	41

Table 7-14. (*Continued*)

NORTH CAROLINA	ASHEVILLE	93	75	36
	GREENSBORO	95	78	36
NORTH DAKOTA	BISMARCK	95	73	47
OHIO	CINCINNATI	95	78	39
	CLEVELAND	95	75	42
OKLAHOMA	OKLAHOMA CITY	101	77	35
	TULSA	101	77	36
OREGON	PORTLAND	90	68	45
PENNSYLVANIA	PHILADELPHIA	95	78	40
	PITTSBURGH	95	75	40
RHODE ISLAND	PROVIDENCE	93	75	42
SOUTH CAROLINA	CHARLESTON	95	78	33
	GREENVILLE	95	75	35
SOUTH DAKOTA	RAPID CITY	95	70	44
TENNESSEE	CHATTANOOGA	95	76	35
	MEMPHIS	95	78	35
TEXAS	DALLAS	100	78	33
	EL PASO	100	69	32
	GALVESTON	95	80	29
	HOUSTON	95	80	30
	SAN ANTONIO	100	78	29
UTAH	SALT LAKE CITY	95	64	41
VIRGINIA	NORFOLK	95	78	37
	RICHMOND	95	78	38
	ROANOKE	95	78	37
WASHINGTON	SEATTLE	85	65	48
	SPOKANE	95	65	48
WEST VIRGINIA	CHARLESTON	95	75	42
WISCONSIN	EAU CLAIRE	95	75	45
	MADISON	95	75	43
	MILWAUKEE	95	75	43
WYOMING	CHEYENNE	95	65	41

Courtesy of the Trane Company.

SIM 7-4

The engineer is evaluating a 10,000 ft^2, 4 in. common brick exterior wall with 4 in. cinder aggregate concrete block backing with a 5/8 in. sand aggregate plaster on interior walls.

a. Comment on adding a 1-inch glass fiber insulating board prior to the interior finish.

b. Instead of insulating board, comment on providing a 3/4 inch air space with aluminum foil on one side. (Use the data for 0°F.)

ANSWER

From Table 7-6,

a. U=0.35 (Coordinate B-16)

or

$R = 1/0.35 = 2.85$

The effect of the insulating material can be seen by using Table 7-2 with

$$R = \frac{1}{0.25} = 4$$

Thus, Total $R = 2.85 + 4 = 6.85$

$$U = 1/6.85 = 0.14.$$

Energy Savings

$$q = (0.35 - 0.14) \times 10{,}000 \times 17 = 35{,}700 \text{ Btuh}$$

b. From Table 7-1,

The effect of the air space at 0°F is 3.76

Thus, R Total = 2.85 + 3.76 = 6.61

Energy Savings

$$q = \left(0.35 - \frac{1}{6.61}\right) \times 10{,}000 \times 17 = 33{,}781 \text{ Btuh.}$$

THE EFFECT OF SUNLIGHT

Heat from the sun's rays greatly increases heat gain of a building. If the building energy requirements were mainly due to cooling, then this gain should be minimized. Solar energy affects a building in the following ways:

1. *Raises the surface temperature:* Thus a greater temperature differential will exist at roofs than at walls.

2. A large percentage of direct solar radiation and diffuse sky radiation *passes through* transparent materials, such as glass.

Surface Temperatures

The temperature of a wall or roof depends upon:

(a) the angle of the sun's rays
(b) the color and roughness of the surface
(c) the reflectivity of the surface
(d) the type of construction.

When an engineer is specifying building materials, he should consider the above factors. A simple example is color. The darker the surface, the more solar radiation will be absorbed. Obviously, white surfaces have a lower temperature than black surfaces after the same period of solar heating. Another factor is that smooth surfaces reflect more radiant heat than do rough ones.

In order to properly take solar energy into account, the angle of the sun's rays must be known. If the latitude of the plant is known, the angle can be determined.

Tables 7-15 and 7-16 illustrate typical temperature differentials for the roof and walls, based on different locations in the United States and various material. Table 7-16 is simplified to indicate Dark (D) and Light (L) color materials only. A full set of tables can be found in the *ASHRAE Guide and Data Book.*

SIM 7-5

An engineer is evaluating three designs for a roof. In the first case the construction is as follows: a 2 in. lightweight aggregate deck with no insulation. The second case is the same as the first, except a 1 in. glass fiberboard form with roof insulation of $R = 4.17$ is used. The third case is the same as case two, except a 4 in. light weight aggregate is used instead of the 2 inch used in case one. The roof area is 5000 ft^2. What is the maximum heat gain for August 24, if the plant were located in Indianapolis, Indiana?

ANSWER

From Table 7-12, the conductance for each case is as follows:

Case 1: $U = 0.27$ (A′ – 4)
Case 2: $U = 0.13$ (D′ – 4)
Case 3: $U = 0.10$ (D′ – 12)

To find the latitude of the city, Table 7-14 is used. From Table 7-14, Indianapolis is at 40 degrees N. Latitude. Thus, Table 7-15 can be used to find the temperature differentials. For other latitudes, refer to the *ASHRAE Guide and Data Book.*

From Table 7-15 at 2:00 P.M. Case 1 and 2: $\Delta T = 60$
From Table 7-15 at 4:00 P.M. Case 3: $\Delta T = 54$

Thus,

Case 1: q max $= 0.27 \times 5000 \times 60 = 81{,}000$ Btuh
Case 2: q max $= 0.13 \times 5000 \times 60 = 39{,}000$ Btuh
Case 3: q max $= 0.10 \times 5000 \times 52 = 27{,}000$ Btuh
Case 3 reduces the heat gain through the roof by 54,000 Btuh.

Sunlight and Glass Considerations

A danger in the energy conservation movement is to take steps backward. A simple example would be to exclude glass from building designs because of the poor conductance and solar heat gain factors of clear glass. The engineer needs to evaluate various alternate glass constructions and coatings in order to maintain and improve the aesthetic qualities of good design while minimizing energy inefficiencies. Table 7-17 illustrates several categories of glass which are presently available. It should be noted that the method to reduce heat gain of glass due to conductance is to provide an insulating air space.

To reduce the solar radiation that passes through glass, several techniques are available. Heat absorbing glass (tinted glass) is very popular. Reflective glass is gaining popularity, as it greatly reduces solar heat gains. Figure 7-4 illustrates a building utilizing reflective glass.

To calculate the relative heat gain through glass, a simple method is illustrated below:

$$Q = UA\,(t_0 - t_1) + A \times S_1 \times S_2$$

where

Q is the total heat gain for each glass orientation (Btuh).
U is the conductance of the glass (Btu/h-ft^2 -°F)
A is the area of glass; The area used should include framing,

since it will generally have a poor conductance as compared with the surrounding material. (ft^2)

$t_0 - t_1$ is the temperature difference between the inside temperature and outside ambient. (°F)

S_1 is the shading coefficient; S_1 takes into account external shades, such as venetian blinds and draperies, and the qualities of the glass, such as tinting and reflective coatings.

S_2 is the solar heat gain factor and is determined from Table 7-18; This factor takes into account direct and diffused radiation from the sun. Diffused radiation is basically caused by reflections from dust particles and moisture in the air.

When using Table 7-18, determine the data and hour which give the maximum heat gain when the contributions of all walls are considered. To use the table for directions corresponding to N, NW, W, SW, and S use the hours at the bottom of the table.

CAUTION: When using reflective glass always insure that the reflections from the building do not increase the solar radiation on adjacent buildings.

Figure 7-4. Reflective glass installation Mountain Bell Plaza, Phoenix, Arizona. ***(Courtesy of PPG Industries.)***

Table 7-15. Total equivalent temperature differentials for roofs for April 20 and August 24 #.

Description of roof construction*	Sun Time								
	A.M.					P.M.			
	8	10	12	2	4	6	8	10	12
LIGHT CONSTRUCTION ROOFS—EXPOSED TO SUN									
1" Wood** or 1" Wood + 1" or 2" Insulation	14.0	40.0	56.0	64.0	52.0	21.0	12.0	6.0	2.0
MEDIUM CONSTRUCTION ROOFS—EXPOSED TO SUN									
2" Concrete or 2" Concrete + 1" or 2" insulation or 2" Wood	8.0	32.0	50.0	60.0	52.0	34.0	16.0	8.0	4.0
2" Gypsum or 2" Gypsum + 1" insulation 1" Wood or 2" Wood or 2" Concrete or 2" Gypsum (last four: + 4" rock wool in furred ceiling)	2.0	22.0	42.0	54.0	56.0	44.0	22.0	12.0	8.0
4" Concrete or 4" Concrete with 2" insulation	2.0	22.0	40.0	52.0	54.0	42.0	24.0	14.0	8.0
HEAVY CONSTRUCTION ROOFS—EXPOSED TO SUN									
6" Concrete	6.0	8.0	26.0	40.0	48.0	46.0	34.0	20.0	14.0
6" Concrete + 2" Insulation	8.0	8.0	22.0	36.0	44.0	46.0	36.0	22.0	16.0
ROOFS COVERED WITH WATER—EXPOSED TO SUN									
Light construction roof with 1" water	2.0	6.0	18.0	24.0	20.0	16.0	12.0	4.0	2.0
Heavy construction roof with 1" water	0	0	–2.0	12.0	16.0	18.0	16.0	12.0	8.0
Any roof with 6" water	0	2.0	2.0	8.0	12.0	12.0	10.0	6.0	2.0
ROOFS WITH ROOF SPRAYS—EXPOSED TO SUN									
Light construction	2.0	6.0	14.0	20.0	18.0	16.0	12.0	4.0	2.0
Heavy construction	0	0	4.0	10.0	14.0	16.0	14.0	12.0	8.0
ROOFS IN SHADE									
Light construction	–2.0	2.0	8.0	14.0	16.0	14.0	10.0	4.0	2.0
Medium construction	–2.0	0	4.0	10.0	14.0	14.0	12.0	8.0	4.0
Heavy construction	0	0	2.0	6.0	10.0	12.0	12.0	10.0	6.0

*Includes 3/8" felt roofing with or without slag. May also be used for shingle roof.
**Nominal thickness of wood.
#Table 7-15 is for 40 degrees North latitude. It may also be used for 40 degrees South latitude for February 20 and October 23.
Courtesy of the Trane Company.

Table 7-16. Total equivalent temperature differentials for walls for April 20 and August 24*.

North Latitude	SUN TIME A.M. 8 D	8 L	10 D	10 L	12 D	12 L	P.M. 2 D	2 L	4 D	4 L	6 D	6 L	8 D	8 L	10 D	10 L	12 D	12 L	South Latitude
	Exterior Color D = Dark L = Light																		
Frame																			
NE	19	9	21	11	14	11	14	12	16	16	16	16	12	12	8	6	4	4	SE
E	32	16	38	20	34	18	14	14	16	16	16	16	12	12	8	8	4	4	E
SE	18	10	33	21	34	22	28	19	18	16	16	16	12	12	8	6	4	4	NE
S	-2	-2	9	3	32	18	41	26	33	25	20	17	13	13	9	9	5	5	N
SW	-2	-2	2	0	8	6	31	26	46	32	49	33	29	24	8	6	4	4	NW
W	-2	-2	2	2	8	8	22	14	42	30	50	36	24	24	10	10	4	4	W
NW	-2	-2	2	0	8	6	14	12	24	21	37	26	31	23	8	6	4	4	SW
(shade) N	-2	-2	0	0	6	6	12	12	16	16	14	14	10	10	6	6	2	2	S (shade)
4″ Brick or stone veneer + frame																			
NE	0	-2	21	11	18	10	11	8	14	12	16	16	14	14	12	12	8	6	SE
E	4	2	32	16	33	19	16	16	14	14	16	16	14	14	12	10	8	8	E
SE	5	0	26	14	35	21	31	20	21	17	16	16	14	14	12	10	8	8	NE
S	-2	-2	0	0	19	11	34	22	35	24	26	20	14	14	10	10	6	6	N
SW	3	0	2	0	4	4	15	10	38	26	42	30	40	28	12	10	8	8	NW
W	2	0	2	2	6	4	12	10	28	20	42	30	44	30	18	16	8	8	W
NW	-2	-2	0	0	4	4	10	8	14	14	29	22	32	24	13	12	8	8	SW
(shade) N	-2	-2	0	0	2	2	8	8	12	12	14	14	14	14	10	10	6	6	S (shade)
8″ hollow tile or 8″ cinder block																			
NE	2	2	2	2	18	10	15	10	11	8	14	12	15	14	14	12	10	10	SE
E	6	4	14	6	26	14	28	16	22	14	14	12	16	14	16	12	12	10	E
SE	5	2	5	2	21	12	25	16	24	17	17	14	17	14	14	12	10	8	NE
S	3	3	3	3	6	3	19	11	34	20	35	21	26	18	15	12	11	8	N
SW	5	2	5	2	5	2	9	7	15	13	31	21	35	24	31	21	10	8	NW
W	6	4	6	4	6	4	8	6	12	10	20	16	32	24	34	24	20	16	W
NW	2	2	2	2	3	2	5	4	10	8	14	12	22	19	28	22	11	10	SW
(shade) N	0	0	0	0	0	0	2	2	8	8	12	12	12	12	12	12	8	8	S (shade)

8″ Brick or 12″ hollow tile or 12″ cinder block																				
	NE	4	4	4	4	10	4	15	9	14	9	11	8	12	10	12	12	11	10	SE
	E	10	8	1	8	16	10	20	12	20	12	16	10	16	12	16	12	14	12	E
	SE	11	7	9	7	9	7	18	14	23	16	20	15	15	12	15	12	15	13	NE
	S	8	5	8	5	8	5	8	5	16	10	22	14	22	16	16	13	14	11	N
	SW	11	7	9	7	9	7	11	7	13	9	15	10	24	15	29	19	24	17	NW
	W	10	6	8	6	8	8	10	8	12	8	16	10	22	18	26	18	26	18	W
	NW	4	4	4	4	4	4	5	4	7	6	10	8	12	10	17	15	18	15	SW
(shade)	N	2	2	2	2	2	2	2	2	4	4	8	8	10	10	10	10	8	8	S (shade)
12″ Brick																				
	NE	9	8	9	7	9	6	9	6	11	6	12	7	12	7	11	8	11	8	SE
	E	14	10	14	10	14	10	12	8	14	10	16	12	16	12	16	10	16	10	E
	SE	13	8	13	9	13	9	13	9	13	9	16	11	18	13	18	13	15	10	N
	S	12	9	13	10	10	7	10	7	10	7	13	7	16	10	18	12	17	11	NE
	SW	13	8	13	9	13	9	13	9	13	9	13	11	13	11	15	11	17	13	NW
	W	14	10	14	10	14	10	12	8	12	8	12	8	12	8	14	10	18	12	W
	NW	9	8	9	7	9	6	9	6	9	6	9	6	9	7	11	8	11	8	SW
(shade)	N	6	6	4	4	4	4	4	4	4	4	4	4	4	4	6	6	8	8	S (shade)
8″ Concrete or Stone or 6″ or 8″ concrete block																				
	NE	7	5	5	4	5	4	6	4	11	8	15	11	17	13	15	11	11	9	SE
	E	8	6	16	10	26	14	26	14	20	12	16	12	16	12	14	12	12	10	E
	SE	9	4	9	7	21	14	23	16	22	15	17	15	15	12	15	13	13	11	N
	S	5	3	5	3	8	3	18	10	23	18	25	17	19	16	14	11	12	9	NE
	SW	9	4	7	4	9	4	11	6	18	13	27	20	29	19	27	20	13	11	NW
	W	8	6	8	6	8	6	10	8	14	10	22	16	30	20	28	20	16	12	W
	NW	5	4	5	2	5	4	6	6	8	8	13	11	20	15	21	16	9	8	SW
(shade)	N	2	2	2	2	2	2	4	4	6	6	8	8	10	10	8	8	6	6	S (shade)
12″ Concrete or Stone																				
	NE	7	5	7	4	7	4	13	9	14	9	11	9	11	10	13	12	11	10	SE
	E	12	8	10	8	12	8	20	12	20	14	18	12	14	12	16	12	16	12	E
	SE	11	7	11	7	9	7	18	11	20	13	20	13	17	13	15	12	15	13	N
	S	11	8	8	5	8	5	8	5	16	10	20	15	22	17	19	13	14	11	NE
	SW	11	7	11	7	9	7	9	7	11	9	13	11	22	17	24	17	22	15	NW
	W	12	8	10	8	10	8	12	8	12	8	14	10	18	12	26	16	24	16	W
	NW	7	5	7	4	7	4	7	5	7	6	9	8	11	10	18	13	20	15	SW
(shade)	N	2	2	2	2	2	2	2	2	4	4	6	6	9	8	10	10	8	8	S (shade)

*Table 7-16 is for 40 degrees latitude. It may be used for South latitude for February 20 and October 23.

Courtesy of the Trane Company.

Table 7-17. Typical glass characteristics.

Monolithic Glass Clear and Tinted

Glass	Thickness		Transmittance		Reflectance	Relative Heat Gain		U Value		Shading Coefficients					
											Venetian Blinds		Draperies		
	in	mm	Average Daylight %	Total Solar %	Average Daylight %	Btu/hr-sq ft	W/m²	Btu/hr-sq ft/°F	W/m²°k	No Shade	Light	Med	Light	Med	Dark
Sheet	SS	2.5	91	87											
	DS	3	90	86		215	678			1.00					
	3/16	5	90	83		211	665			.98	.55	.64	.56	.61	.70
Clear	1/8	3	90	83		215	678			1.00					
	3/16	5	89	79		205	646			.95					
	1/4	6	88	77		201	634			.93	.55	.64	.56	.61	.70
	5/16	8	88	77	8	201	634			.93					
Clear Heavy duty	3/8	10	87	75		199	627			.92					
	1/2	12	86	71		191	602			.88					
	5/8	15	85	67		185	583			.85	.54	.62	.52	.56	.66
	3/4	19	83	63		179	564	1.10	6.3	.82					
	7/8	22	81	59		171	539			.78	.54	.60	.50	.53	.62
Blue-green	1/8	3	83	63		179	564			.82	.54	.60	.50	.53	.62
	3/16	5	79	55	7	161	508			.73					
	1/4	6	75	47		155	489			.70	.53	.57	.45	.47	.54
Grey	3/16	5	51	53	6	163	514			.74	.53	.58	.47	.49	.51
	1/4	6	44	46		151	476			.68	.52	.56	.44	.47	.52
	5/16	8	35	38	5	137	432			.61	.47	.50	.42	.44	.49
	3/8	10	28	31		125	394			.55	.43	.45	.40	.42	.45
	1/2	12	19	22	4	111	350			.48	.38	.40	.30	.38	.40

Bronze	1/8	3	68	65		183	577					.84	.54	.61	.51	.54	.64
	3/16	5	59	55	6	165	520					.75	.53	.58	.47	.49	.57
	1/4	6	52	49		155	489					.70	.53	.57	.45	.47	.54
	5/16	8	44	40		141	444					.63	.48	.51	.43	.45	.50
	3/8	10	38	34	5	131	413					.58	.45	.48	.41	.43	.47
	1/2	12	28	24		113	356					.49	.38	.40	.30	.38	.40
Thermopane Insulating Glass (inboard light clear)						**Air space** 1/2″	12 mm	**Air space** 1/4″	6 mm	1/2″	12 mm						
Clear	1/8	3	80	69	15	182	574					.87	.51	.57	.50	.53	.58
	3/16	5	79	62	14	172	542					.82	.51	.56	.48	.52	.57
	1/4	6	77	59	14	166	523					.79	.49	.54	.41	.50	.56
Blue-green	1/8	3	75	52	13	148	466					.70	.44	.48	.44	.47	.52
	3/16	5	70	43	12	132	416					.62	.39	.43	.41	.42	.43
	1/4	6	66	36	12	120	378	.60	3.4	.55	3.1	.56	.36	.39	.38	.41	.44
Grey	3/16	5	45	42	8	128	403					.60	.38	.42	.40	.43	.47
	1/4	6	39	35	7	120	370					.54	.35	.38	.37	.40	.43
Bronze	1/8	3	61	54	11	150	473					.71	.44	.48	.44	.47	.52
	3/16	5	53	43	9	122	385					.62	.34	.43	.41	.43	.48
	1/4	6	46	38	8	120	378					.56	.36	.39	.38	.41	.44
Thermopane Xi								Airspace 3/16		5mm							
Sheet	SS&DS		81	75	15	188	593	.58		3.4		.90	.51	.57	.50	.53	.58

Courtesy of Libbey-Owens-Ford Company.

Table 7-17. Typical glass characteristics (*Continued*).

Laminated Vari-Tran

GLASS	THICKNESS in	THICKNESS mm	VARI-TRAN COATING Color	VARI-TRAN COATING Number	TRANSMITTANCE Average Daylight %	TRANSMITTANCE Average Daylight Tol	TRANSMITTANCE Total Solar %	REFLECTANCE Average Daylight %	REFLECTANCE Average Daylight Tol	RELATIVE[1] HEAT GAIN Btu/hr-sq ft	RELATIVE[1] HEAT GAIN W/m²	U VALUE Btu/hr-sq ft/°F	U VALUE W/m²°k	SHADING COEFFICIENTS No Shade	Venetian Blinds Light	Venetian Blinds Med	Draperies Light	Draperies Med	Draperies Dark
				1-108	8	±1.5	9	43		71	224			.28	.21	.23	.21	.22	.23
			Silver	1-114	14	±2.0	15	33	±3.0	89	281			.37	.27	.31	.28	.29	.31
Clear	1/4	6		1-120	20	±2.5	19	27		95	299	1.05	6.0	.40	.29	.33	.30	.32	.34
				1-208	8	±1.5	8	27		73	230			.29	.22	.24	.22	.22	.24
			Golden	1-214	14	±2.0	13	24	±3.0	85	268			.35	.26	.29	.26	.27	.29
				1-220	20	±2.5	20	21		99	312			.42	.31	.35	.32	.33	.36

Monolithic Vari-Tran

GLASS	THICKNESS in	THICKNESS mm	VARI-TRAN COATING Color	VARI-TRAN COATING Number	TRANSMITTANCE Average Daylight %	TRANSMITTANCE Average Daylight Tol	TRANSMITTANCE Total Solar %	REFLECTANCE Average Daylight %	REFLECTANCE Average Daylight Tol	RELATIVE[1] HEAT GAIN Btu/hr-sq ft	RELATIVE[1] HEAT GAIN W/m²	U VALUE Btu/hr-sq ft/°F	U VALUE W/m²°k	SHADING COEFFICIENTS No Shade	Venetian Blinds Light	Venetian Blinds Med	Draperies Light	Draperies Med	Draperies Dark
				1-108	8	±1.5	9	44		57	180	.80	4.5	.23	.18	.19	.18	.18	.19
			Silver	1-114	14	±2.0	16	33	±3.0	82	258	.85	4.8	.35	.26	.29	.26	.28	.29
Clear				1-120	20	±2.5	20	27		91	287	.90	5.1	.39	.28	.32	.29	.31	.33
				1-208	8	±1.5	9	28		63	199	.80	4.5	.26	.20	.22	.20	.20	.21
			Golden	1-214	14	±2.0	14	26	±3.0	76	240	.85	4.8	.32	.24	.27	.24	.25	.26
				1-220	20	±2.5	21	24		93	293	.90	5.1	.40	.29	.33	.30	.32	.34
B/Green			Blue	2-350	50	±5.0	35	18	±3.0	129	407	1.05	6.0	.57	.42	.48	.42	.44	.48
	1/4	6		3-108[2]	8	±1.5	11	11		78	246	.85	4.8	.33	.25	.28	.25	.26	.27
Grey				3-114[2]	14	±2.0	17	9	±2.0	93	293	.90	5.1	.40	.29	.33	.30	.32	.34
Tuf-Flex			Grey	3-120[2]	20	±2.5	24	7		106	334	1.00	5.7	.46	.34	.38	.35	.36	.39
Grey				3-134	34	±4.0	36	6		135	426	1.05	6.0	.60	.44	.50	.43	.46	.50
Bronze				4-108[2]	8	±1.5	9	14		74	233	.85	4.8	.31	.24	.26	.24	.24	.25
Tuf-Flex			Bronze	4-114[2]	14	±2.0	14	11	±2.0	89	281	.90	5.1	.38	.28	.32	.29	.30	.32
				4-120[2]	20	±2.5	20	9		99	312	.95	5.4	.43	.32	.35	.32	.34	.37
Bronze				4-134	34	±4.0	31	6		127	400	1.05	6.0	.56	.42	.47	.41	.43	.47

Thermopane Vari-Tran (inboard light clear except Vari-Tran 2-350-2)

				1-108	7	±1.5	7	44		41	129			.17	.16	.16	.15	.15	.15
			Silver	1-114	13	±2.0	14	33	±3.0	59	186			.26	.23	.23	.22	.22	.23
Clear				1-120	18	±2.5	16	27		67	211			.30	.26	.27	.26	.26	.27
				1-208	7	±1.5	7	28		43	136	.50	2.8	.18	.17	.17	.16	.16	.16
			Golden	1-214	13	±2.0	12	26	±3.0	55	173			.24	.21	.22	.21	.21	.22
				1-220	18	±2.5	17	24		69	218			.31	.27	.28	.27	.27	.28
Blue-				2-350	45	±5.0	28	20		98	309	.55	3.1	.45	.38	.39	.36	.38	.41
Green	1	25	Blue	2-350-2[3]	38	±5.0	20	20	±3.0	96	303			.44	.37	.38	.35	.37	.40
Grey				3-108[2]	7	±1.5	9	11		53	167	.50	2.8	.23	.20	.21	.20	.20	.21
Tuf-Flex				3-114[2]	13	±2.0	14	9	±2.0	65	205			.29	.25	.26	.25	.25	.26
			Grey	3-120[2]	18	±2.5	20	7		76	240	.55	3.1	.34	.29	.30	.28	.29	.31
Grey				3-134	30	±4.0	29	7		102	322			.47	.39	.41	.38	.40	.43
				4-108[2]	7	±1.5	7	14		49	154			.21	.19	.20	.18	.18	.19
Bronze				4-114[2]	13	±2.0	11	11	±2.0	61	192	.50	2.8	.27	.24	.24	.23	.23	.24
Tuf-Flex			Bronze	4-120[2]	18	±2.5	15	9		70	221	.55	3.1	.31	.27	.28	.27	.27	.28
Bronze				4-134	30	±4.0	25	7		94	296			.43	.36	.37	.35	.37	.40

[1]When ASHRAE Solar Heat Gain Factor is 200 Btu/hr-sq. ft. and outdoor air is 14°F. warmer than indoor air, with no indoor shading.
[2]Tempered only may be furnished in 5/16″ (8 mm) thickness in some sizes.
[3]May require both lights be tempered.

ADDITIONAL TECHNICAL DATA UPON APPLICATION

The satisfactory performance of all LOF products requires selection of the appropriate type, size and thickness of glass, and proper glazing in adequate sash. The information contained in this catalog is intended as a general guide in planning your glass requirements. Libbey-Owens-Ford assumes no responsibility for its use or application.

Table 7-18. Solar heat gain factors for 40° north latitude.

Solar Heat Gain Factors

June 21

	AM					PM			
	8	**9**	**10**	**11**	**12**	**1**	**2**	**3**	**4**	
N	29	33	35	37	38	37	35	33	29	N
NE	156	113	62	40	38	37	35	31	26	NW
E	215	192	145	80	41	37	35	31	26	W
SE	152	161	148	116	71	41	36	31	26	SW
S	29	45	69	88	95	88	69	45	29	S
	4	**3**	**1**	**2**	**12**	**11**	**10**	**9**	**8**	
	PM					AM				

September 21

	AM						PM		
	8	**9**	**10**	**11**	**12**	**1**	**2**	**3**	**4**	
N	16	22	26	29	30	29	26	22	16	N
NE	87	47	28	29	30	29	26	22	16	NW
E	205	195	148	77	32	29	26	22	16	W
SE	199	226	221	192	141	77	30	23	16	SW
S	71	124	165	191	200	191	165	124	71	S
	4	**3**	**2**	**1**	**12**	**11**	**10**	**9**	**8**	
	PM						AM			

Courtesy of Libbey-Owens-Ford Company.

Applying Insulation

As illustrated in this chapter, insulation offers one of the best methods to improve the efficiency of building designs. Table 7-19 offers a comparison of various insulation materials. In practice, the thermal conductivities of insulation should be based on the manufacturer's data.

Table 7-19. Thermal conductivity (k) of industrial insulation (design values)[a] (for mean temperatures indicated).

Expressed in Btu per (hour) (square foot) (Fahrenheit degree temperature difference per in.)

Form	Material (Composition)	Accepted Max Temp for Use,* °F	Typical Density (lb/cu ft)	Typical Conductivity k at Mean Temp °F													
				-100	-75	-50	-25	0	25	50	75	100	200	300	500	700	900
BLANKETS and FELTS	MINERAL FIBER																
	(Rock, Slag, or Glass)																
	Blanket, Metal Reinforced	1200	6-12									0.26	0.32	0.39	0.54		
		1000	2.5-6									0.24	0.32	0.40	0.61		
	Mineral Fiber, Glass																
	Blanket, Flexible, Fine-fiber	350	0.65				0.25	0.26	0.28	0.30	0.33	0.36	0.53				
	Organic Bonded		0.75				0.24	0.25	0.27	0.29	0.32	0.34	0.48				
			1.0				0.23	0.24	0.25	0.27	0.29	0.32	0.43				
			1.5				0.21	0.22	0.23	0.25	0.27	0.28	0.37				
			2.0				0.20	0.21	0.22	0.23	0.25	0.26	0.33				
			3.0				0.19	0.20	0.21	0.22	0.23	0.24	0.31				
	Blanket, Flexible,																
	Textile-Fiber	350	0.65				0.27	0.28	0.29	0.30	0.31	0.32	0.50	0.68			
	Organic Bonded		0.75				0.26	0.27	0.28	0.29	0.31	0.32	0.48	0.66			
			1.0				0.24	0.25	0.26	0.27	0.29	0.31	0.45	0.60			
			1.5				0.22	0.23	0.24	0.25	0.27	0.29	0.39	0.51			
			3.0				0.20	0.21	0.22	0.23	0.24	0.25	0.32	0.41			
	Felt, Semi-Rigid Organic	400	3-8						0.24	0.25	0.26	0.27	0.35	0.44			
	Bonded	850	3	0.16	0.17	0.18	0.19	0.20	0.21	0.22	0.23	0.24	0.35	0.55			
	Laminated & Felted	1200	7.5											0.35	0.45	0.60	
	Without Binder																
	VEGETABLE and																
	ANIMAL FIBER																
	Hair Felt or Hair																
	Felt plus Jute	180	10						0.26	0.28	0.29	0.30					

Table 7-19 (*Continued*)

Form	Material (Composition)	Accepted Max Temp for Use,* °F	Typical Density (lb/cu ft)	Typical Conductivity k at Mean Temp °F													
				-100	-75	-50	-25	0	25	50	75	100	200	300	500	700	900
BLOCKS, BOARDS, and PIPE INSULATION	ASBESTOS																
	Laminated Asbestos Paper	700	30									0.40	0.45	0.50	0.60		
	Corrugated & Laminated Asbestos Paper																
	4-ply	300	11-13								0.54	0.57	0.68				
	6-ply	300	15-17								0.49	0.51	0.59				
	8-ply	300	18-20								0.47	0.49	0.57				
	MOLDED AMOSITE and BINDER	1500	15-18									0.32	0.37	0.42	0.52	0.62	0.72
	85% MAGNESIA	600	11-12									0.35	0.38	0.42			
	CALCIUM SILICATE	1200	11-13									0.35	0.41	0.44	0.52	0.62	0.72
		1800	12-15												0.63	0.74	0.95
	CELLULAR GLASS	800	9			0.32	0.33	0.35	0.36	0.38	0.40	0.42	0.48	0.55			
	DIATOMACEOUS SILICA	1600	21-22												0.64	0.68	0.72
		1900	23-25												0.70	0.75	0.80
	MINERAL FIBER																
	Glass, Organic Bonded, Block and Boards	400	3-10	0.16	0.17	0.18	0.19	0.20	0.22	0.24	0.25	0.26	0.33	0.40			
	Non-Punking Binder	1000	3-10									0.26	0.31	0.35	0.52		
	Pipe Insulation, slag or glass	350	3-4					0.20	0.21	0.22	0.23	0.24	0.29				
		500	3-10				0.20	0.22	0.24	0.25	0.26	0.33	0.40				
	Inorganic Bonded Block	1000	10-15										0.33	0.38	0.45	0.55	
		1800	15-24									0.32	0.37	0.42	0.52	0.62	0.74
	Pipe Insulation, slag or glass	1000	10-15									0.33	0.35	0.45	0.55		
	MINERAL FIBER																
	Resin Binder		15			0.23	0.24	0.25	0.26	0.28	0.29						
	Rigid Polystyrene																
	Extruded, R-12 exp	170	3.5	0.16	0.16	0.15	0.16	0.16	0.17	0.18	0.19	0.20					
	Extruded, R-12 exp	170	2.2	0.16	0.16	0.17	0.16	0.17	0.18	0.19	0.20						

	Extruded	170	1.8	0.17	0.18	0.19	0.20	0.21	0.23	0.24	0.25	0.27				
	Molded Beads	170	1	0.18	0.20	0.21	0.23	0.24	0.25	0.26	0.28					
	Polyurethane**															
	R-11 exp	210	1.5-2.5	0.16	0.17	0.18	0.18	0.18	0.17	0.16	0.16	0.17				
	RUBBER, Rigid Foamed	150	4.5						0.20	0.21	0.22	0.23				
	VEGETABLE and ANIMAL FIBER															
	Wool Felt (Pipe Insulation)	180	20						0.28	0.30	0.31	0.33				
INSULATING CEMENTS	MINERAL FIBER															
	(Rock, Slag, or Glass)															
	With Colloidal Clay Binder	1800	24-30									0.49	0.55	0.61	0.73	0.85
	With Hydraulic Setting Binder	1200	30-40									0.75	0.80	0.85	0.95	
LOOSE FILL	Cellulose insulation (Milled pulverized paper or wood pulp)		2.5-3							0.26	0.27	0.29				
	Mineral fiber, slag, rock or glass		2-5			0.19	0.21	0.23	0.25	0.26	0.28	0.31				
	Perlite (expanded)		5-8	0.25	0.27	0.29	0.30	0.32	0.34	0.35	0.37	0.39				
	Silica aerogel		7.6			0.13	0.14	0.15	0.15	0.16	0.17	0.18				
	Vermiculite (expanded)		7-8.2			0.39	0.40	0.42	0.44	0.45	0.47	0.49				
			4-6			0.34	0.35	0.38	0.40	0.42	0.44	0.46				

[a]Representative values for dry materials as selected by the ASHRAE Technical Committee 2.4 on Insulation. They are intended as design (not specification) values for materials of building construction for normal use. For the thermal resistance of a particular product, the user may obtain the value supplied by the manufacturer or secure the results of unbiased tests.

*These temperatures are generally accepted as maximum. When operating temperature approaches these limits the manufacturer's recommendations should be followed.

**These are values for aged board stock.

Reprinted by permission from *ASHRAE Handbook of Fundamentals*, 1972.

WINDOW TREATMENTS

Several types of window treatments to reduce losses have become available. This section describes some of the products on the market based on information supplied by manufacturers. No claims are made concerning the validity or completeness described. The summary is based on "Windows For Energy Efficient Buildings" as prepared by the Lawrence Berkeley Laboratory for U.S. DOE under contract W-7405-ENG-48.

Solar Control

Solar Control Films

A range of tinted and reflective polyester films is available to adhere to inner window surfaces to provide solar control for existing clear glazing. Films are typically two- or three-layer laminates composed of metalized, transparent and/or tinted layers. Films are available with a wide range of solar and visible light transmittance values, resulting in shading coefficients as low as 0.24. Most films are adhered with precoated pressure sensitive adhesives. Reflective films will reduce winter U values by about 20%. (Note that a new solar control film, which provides a U value of 0.68, is described in the Thermal Barriers section below). Films adhered to glass improve the shatter resistance of glazing and reduce transmission, thus reducing fading of furnishings.

Fiberglass Solar Control Screens

Solar control screens provide sun and glare control as well as some reduction in winter heat loss. Screens are woven from vinyl-coated glass strands and are available in a variety of colors. Depending on color and weave, shading coefficients of 0.3 to 0.5 are achieved. Screens are durable, maintenance-free, and provide impact resistance. They are usually applied on the exterior of windows and may (1) be attached to mounting rails and stretched over windows, (2) mounted in rigid frames and installed over windows, or (3) made into roller shades which can be retracted and stored as desired.

Motorized Window Shading System

A variety of plastic and fabric shades is available for use with a motorized window shading system. A reversible motor is located within the shade tube roller and contains a brake mechanism to stop and hold in any position. Motor controls may be gauged and operated locally or from a master station. Automatic photoelectric controls are available that (1) monitor sun intensity and angle and adjust shade position to provide solar control and (2) employ an internal light sensor and provide a preset level of internal ambient light.

Exterior Sun Control Louvers

Operable external horizontal and vertical louver systems are offered for a variety of building sun control applications. Louvers are hinged together and can be rotated in unison to provide the desired degree of shading for any sun position. Operation may be manual or electric; electrical operation may be initiated by manual switches, time clock, or sun sensors. Louvers may be closed to reduce night thermal losses. Sun control elements are available in several basic shapes and in a wide range of sizes.

External Venetian Blinds

Externally mounted, all-weather venetian blinds may be manually operated from within a building or electrically operated and controlled by means of automatic sun sensors, time controls, manual switches, etc. Aluminum slats are held in position with side guides and controlled by weatherproof lifting tapes. Slats can be tilted to modulate solar gain, closed completely, or restricted to admit full light and heat. Blinds have been in use in Europe for many years and have been tested for resistance to storms and high winds.

Adjustable Louvered Windows

Windows incorporating adjustable external louvered shading devices are available. Louvers are extruded aluminum or redwood, 3 to 5 inches wide, and are manually controlled. Louvers may be specified on double-hung, hinged, or louvered-glass windows. When open, the louvers provide control of solar gain and glare; when closed, they provide privacy and security.

Solar Shutters

The shutter is composed of an array of aluminum slats set at 45° or 22-1/2° from the vertical to block direct sunlight. Shutters are designed for external application and may be mounted vertically in front of the window or projected outward from the bottom of the window. Other rolling and hinged shutters are stored beside the window and roll or swing into place for sun control, privacy, or security.

Thermal Barriers

Multilayer, Roll-up Insulating Window Shade

A multilayer window shade stores in a compact roll and utilizes spacers to separate the aluminized plastic layers in the deployed position, thereby creating a series of dead air spaces. A five-layer shade combined with insulated glass provides R8 thermal resistance.

Thermal Barrier, Roll-up Insulating Window Shade

A ThermoShade thermal barrier is a roll-up shade composed of hollow, lens-shaped, rigid, white PVC slats with virtually no air leakage through connecting joints. The side tracking system reduces window infiltration. Designed for interior installation and manual or automatic operation.

Using the Insulating Window Shade

When added to a window, the roll-up insulating shade provides R4.5 for a single-glazed window or R5.5 for a double-glazed window. Quilt is composed of fabric outer surfaces and two polyester fiberfill layers sandwiched around a reflective vapor barrier. Quilt layers are ultrasonically welded. Shade edges are enclosed in a side track to reduce infiltration.

Reflective, Perforated Solar Control Laminate

Laminate of metalized weatherable polyester film and black vinyl is then perforated with 225 holes/in.^{2}, providing 36 percent open area. Available in a variety of metallized and nonmetallized colors, the shading coefficients vary from 0.30 to 0.35 for externally mounted screens and 0.37 to 0.45 for the material adhered to the inner glass surface. The laminate is typically mounted in aluminum screen frames which are hung externally, several inches from the window; it can also be utilized in a roll-up form. Some reduction in winter U value can be expected with external applications.

Semi-transparent Window Shades

Roll-up window shades are made from a variety of tinted or reflective solar control film laminates. These shades provide most of the benefits of solar control film applied directly to glass but provide additional flexibility and may be retracted on overcast days or when solar heat gain is desired. Shades are available with spring operated and gravity (cord and reel) operated rollers as well as motorized options. Shading coefficients as low as 0.13 are achieved and a tight fitting shade provides an additional air space and thus reduced U-value.

Louvered Metal Solar Screens

The solar screen consists of an array of tiny louvers which are formed from a sheet of thin aluminum. The louvered aluminum sheet is then installed in conventional screen frames and may be mounted against a window in place of a regular insect screen or mounted away from the building to provide free air circulation around the window. View to the outside is maintained while substantially reducing solar gain. Available in a light green or black finish with shading coefficients of 0.21 or 0.15, respectively.

Operable External Louver Blinds

Solar control louver blinds, mounted on the building exterior, can be controlled manually or automatically by sun and wind sensors. Slats can be tilted to modulate light, closed completely, or retracted to admit full light and heat. Developed and used extensively in Europe, they provide summer sun control, control of natural light, and reduction of winter heat loss.

Louvered Metal Solar Screens

A solar screen consists of an array of tiny fixed horizontal louvers which are woven in place. Louvers are tilted at 17° to provide sun control. Screen material is set in metal frames which may be permanently installed in a variety of configurations or designed for removal. Installed screens have considerable wind and impact resistance. The standard product (17 louvers/inch) has a shading coefficient of 0.23; low sun angle variant (23 louvers/inch) has a shading coefficient of 0. 15. Modest reductions in winter U value have been measured.

Insulating Solar Control Film

A modified solar control film designed to be adhered to the interior of windows provides conventional solar control function and has

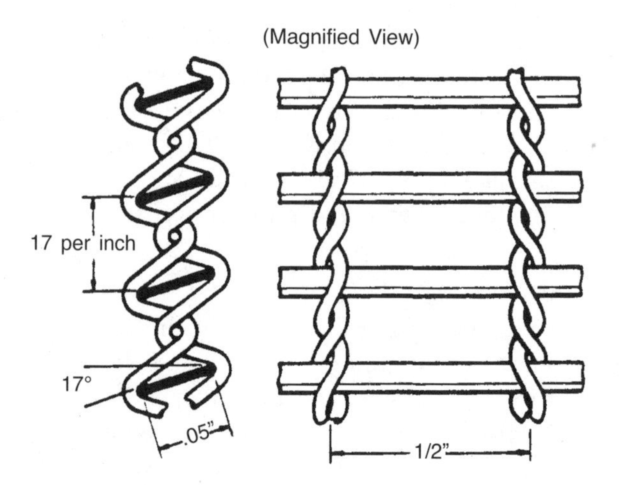

greatly improved insulating properties. Film emissivity is 0.23 to 0.25, resulting in a U value of 0.68 Btu/ft^2 hr-°F under winter conditions, compared to 0.87 for conventional solar control films and 1.1 for typical single-glazed windows.

Interior Storm Windows

These low-cost, do-it-yourself interior storm windows have a rigid plastic glazing panel. The glazing panel may be removed for cleaning or summer storage. It reduces infiltration losses as well as conductive/convective heat transfer.

Retrofit Insulating Glass System

Single glazing is converted to double glazing by attaching an extra pane of glass with neoprene sealant. A desiccant-filled aluminum spacer absorbs moisture between the panes. An electric resistance wire embedded in the neoprene is heated with a special power source. This hermetically seals the window. New molding can then be applied if desired.

Infiltration

Weather-strip Tape

This polypropylene film is scored along its centerline so that it can be easily formed into a "V" shape. It has a pressure-sensitive adhesive on one leg of the "V" for application to seal cracks around doors and windows. On an average fitting, double-hung window, infiltration is reduced by over 70%. It can be applied to rough or smooth surfaces.

ADVANCED OPTICAL TECHNOLOGIES

Opportunities For High Performance Window Products

The product mix of the window industry of today reflects a number of innovations designed to improve the energy performance, optical clarity, and aesthetics of residential and commercial windows. Low-E coatings, tints, reflective films and coatings, gas fills, and new frame materials and designs are features that offer substantial energy savings to consumers, and opportunities for window manufacturers to add value to their product line at a time of increased energy and environmental awareness.

While a good window today is almost twice as energy efficient as the standard double glazed units of the early '80s, most of the energy saving potential for advanced windows has yet to be captured. Researchers are working to further improve on current technologies, expand high performance window applications to retrofit markets, and ultimately turn windows into dynamic energy supply systems that continuously respond to the energy and daylighting needs of buildings and their occupants.

"Smart Windows" with Switchable Glazing

The next technological step beyond selective glazing is the capability to dynamically alter the solar-optical properties of a window. A switchable glazing can change its properties upon some external stimulus. Some types of switchable glazings respond directly to an environmental variable such as heat or light. A more controllable, and thus more desirable type is a glazing whose properties are altered via a low voltage electrical signal. Building occupants could control heat and light transmitted through the windows by simply adjusting a switch.

Some switchable glazings have been around for years. For example, photochromic glass that darkens in bright light is common in sunglasses but is not yet available in large sheets for windows. Several new switchable products are becoming available. One is a combination photochromic and heat-sensitive thermochromic material intended largely for skylights. When its temperature rises, this material changes from clear to a reflective milky white to block heat gain from the sun. Another newly introduced product uses a thin layer of liquid crystals sandwiched between glass.

"Cool Windows" Using Selective Glazings

Window glass and plastic films can be coated with microscopically thin metallic layers or incorporate tints that reflect or absorb some or all of the wavelengths of the solar energy spectrum. A desirable "selective glazing" would allow solar energy wavelengths in the visible spectrum (daylight) to pass through a window while reflecting near infrared (heat) and ultraviolet (fabric damaging) wavelengths. Reflective coatings do a better job of reflecting the sun's energy than absorbing glazings. These "cool windows" reduce cooling loads by up to 50% but still look clear to the human eye.

First generation selective coatings have already been used successfully by several glazing manufacturers in products aimed at commercial and sunbelt residential applications. These high performance silver-based coatings require protection within a sealed glazing assembly to avoid degradation.

Superwindows

A superwindow is a highly insulating window with a heat loss so low that it performs better than an insulated wall in winter, since the sunlight that it admits is greater than its heat loss over a 24-hour period. Superwindow research focuses on the design, development and testing of new windows and window components such as frames and edge seals, and on new materials and techniques to reduce thermal losses across window systems. An objective of the research is marketable window products with insulating values up to five times as high as conventional double glazed windows. Another objective is for the vast majority of window manufacturers to incorporate several of these technologies into their windows. The increased availability of very high-R windows plus incremental improvements in the mainstream windows would effect a shift in average performance from less than R-2 to over R-5, and result in significant energy savings.

8

Heating, Ventilation and Air-Conditioning System Optimization

EFFICIENT USE OF HEATING AND COOLING EQUIPMENT SAVES DOLLARS

Most people have probably experienced improper air conditioning or heating controls which waste energy. Energy is saved when efficient heating, ventilation, and air conditioning (HVAC) systems are used. In this chapter, you will learn how to compare the efficiency of various systems and apply the heat pump to save energy, see how various refrigeration systems can be used to save energy, learn the basics of air conditioning design from an energy conservation viewpoint, and begin to apply the computer approach for energy conservation.

Measuring System Efficiency by Using the Coefficient of Performance

The coefficient of performance (COP) is the basic parameter used to compare the performance of refrigeration and heating systems. COP for cooling and heating applications is defined as follows:

$$\text{COP (Cooling)} = \frac{\text{Rate of Net Heat Removal}}{\text{Total Energy Input}} \tag{8-1}$$

$$\text{CPOP (Heating, Heat Pump*)} = \frac{\text{Rate of Useful Heat Delivered*}}{\text{Total Energy Input}} \tag{8-2}$$

*For heat pump applications, exclude supplemental heating.

APPLYING THE HEAT PUMP TO SAVE ENERGY

The heat pump has gained wide attention due to its high potential COP. The heat pump in its simplest form can be thought of as a window air conditioner. During the summer, the air on the room side is cooled while air is heated on the outside air side. If the window air conditioner is turned around in the winter, some heat will be pumped into the room. Instead of switching the air conditioner around, a cycle reversing valve is used to switch functions. This valve switches the function of the evaporator and condenser, and refrigeration flow is reversed through the device. *Thus, the heat pump is heat recovery through a refrigeration cycle.* Heat is removed from one space and placed in another. In Chapter 7, it was seen that the direction of heat flow is from hot to cold. Basically, energy or pumping power is needed to make heat flow "up hill." The mechanical refrigeration compressor "pumps" absorbed heat to a higher level for heat rejection. The refrigerant gas is compressed to a higher temperature level so that the heat absorbed by it, during the evaporation or cooling process, is rejected in the condensing or heating process. Thus, the heat pump provides cooling in the summer and heating in the winter. The source of heat for the heat pump can be from one of three elements: air, water or the ground.

Air-to-air Heat Pumps

Heat exists in air down to 460°F below zero. Using outside air as a heat source has its limitations, since the efficiency of a heat pump drops off as the outside air level drops below 55°F. This is because the heat is more dispersed at lower temperatures, or more difficult to capture. Thus, heat pumps are generally sized on cooling load capacities. Supplemental heat is added to compensate for declining capacity of the heat pump. This approach allows for a realistic first cost and an economical operating cost.

Heat Pumps Do Save Energy

An average of 2 to 3 times as much heat can be moved for each kW input compared to that produced by use of straight resistance heating. Heat pumps can have a COP of greater than 3 in industrial processes, depending on temperatures. Commercially available heat pumps range in size from two to three tons for residences to up to 40 tons for commercial and industrial users. Figure 8-1 illustrates a simple scheme for determining the supplemental heat required when using an air-air heat pump.

Hydronic Heat Pump

The hydronic heat pump is similar to the air-to-air unit, except the heat exchange is between water and refrigerant instead of air to refrigerant, as illustrated in Figure 8-2. Depending on the position of the reversing valve, the air heat exchanger either cools or heats room air. In the case of cooling, heat is rejected through the water cooled condenser to the building water. In the case of heating, the reversing valve causes the water to refrigerant heat exchanger to become an evaporator. Heat is then absorbed from the water and discharged to the room air.

Imagine several hydronic heat pumps connected to the same building water supply. In this arrangement, it is conceivable that while one unit is providing cool air to one zone, another is providing hot air to another zone; the first heat pump is providing the heat source for the second unit, which is heating the room. This illustrates the principle of energy conservation. In practice, the heat rejected by the cooling units does not equal the heat absorbed. An additional evaporative cooler is added to the system to help balance the loads. A better heat source would be the water from wells, lakes, or rivers, which is thought of as

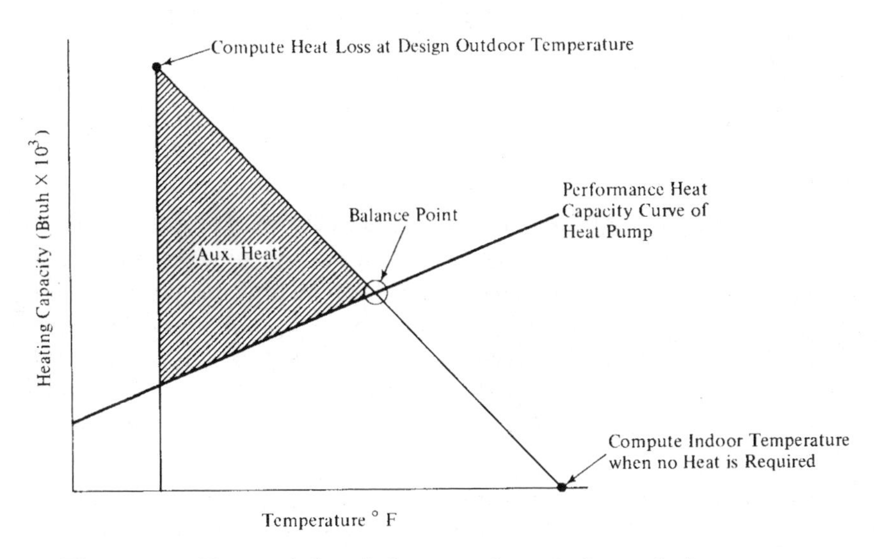

Figure 8-1. Determining balance point of air to air heat pump.

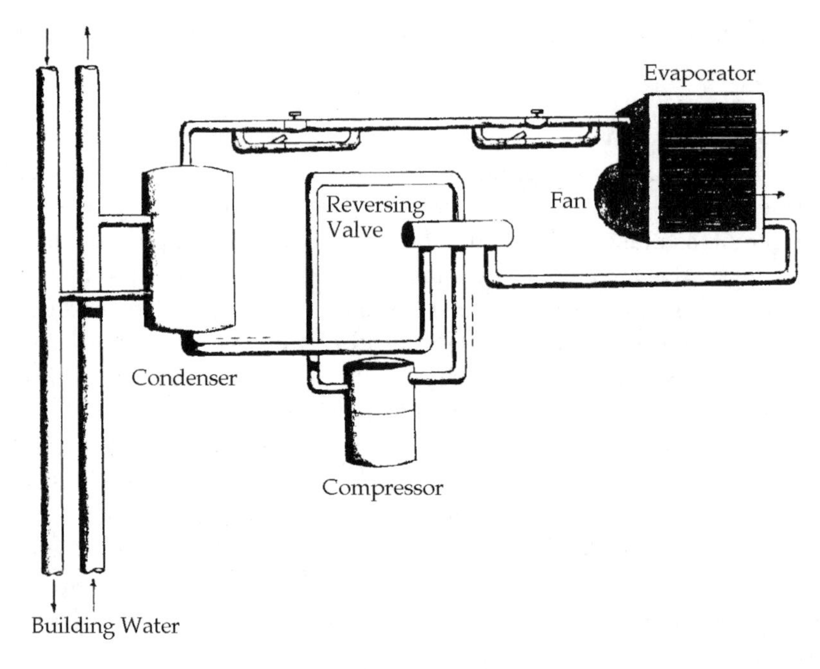

Figure 8-2. Hydronic heat pump.

a constant heat source. Care should be taken to insure that a heat pump connected to such a heat source does not violate ecological interests.

EFFICIENT APPLICATIONS OF REFRIGERATION EQUIPMENT

Liquid Chiller

A liquid chilling unit (mechanical refrigeration compressor) cools water, brine, or any other refrigeration liquid, for air conditioning or refrigeration purposes. The basic components include a compressor, liquid cooler, condenser, the compressor drive, and auxiliary components. A simple liquid chiller is illustrated in Figure 8-3. The type of chiller usually depends on the capacity required. For example, small units below 80 tons are usually reciprocating, while units above 350 tons are usually centrifugal.

A factor which affects the power usage of liquid chillers is the percent load and the temperature of the condensing water. *A reduced condenser water temperature saves energy.* In Figure 8-4, it can be seen that

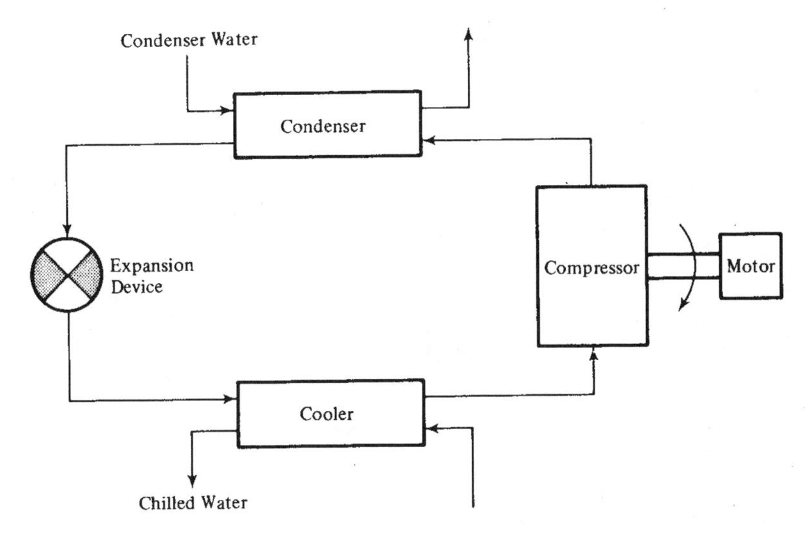

Figure 8-3. Liquid chiller.

by reducing the original condenser water temperature by ten degrees, the power consumption of the chiller is reduced. Likewise, a chiller operating under part load consumes less power. The "ideal" coefficient of performance (COP) is used to relate the measure of cooling effectiveness. Approximately .8 kW is required per ton of refrigeration (0.8 kW is power consumption at full load, based on typical manufacturers' data).

Thus:

$$\text{COP} = \frac{1 \text{ Ton} \times 12{,}000 \text{ Btu/ton}}{.8 \text{ kW} \times 3412 \text{ Btu/kW}}$$

Chillers in Series and in Parallel

Multiple chillers are used to improve reliability, offer standby capacity, reduce inrush currents and decrease power costs at partial loads. Figure 8-5 shows two common arrangements for chiller staging: namely, chillers in parallel and chillers in series.

In the parallel chiller arrangement, liquid to be chilled is divided among the liquid chillers and the streams are combined after chilling. Under part load conditions, one unit must provide colder than designed chilled liquid so that when the streams combine, including the one from the off chiller, the supply temperature is provided. The parallel chillers have a lower first cost than the series chillers counterparts but usually consume more power.

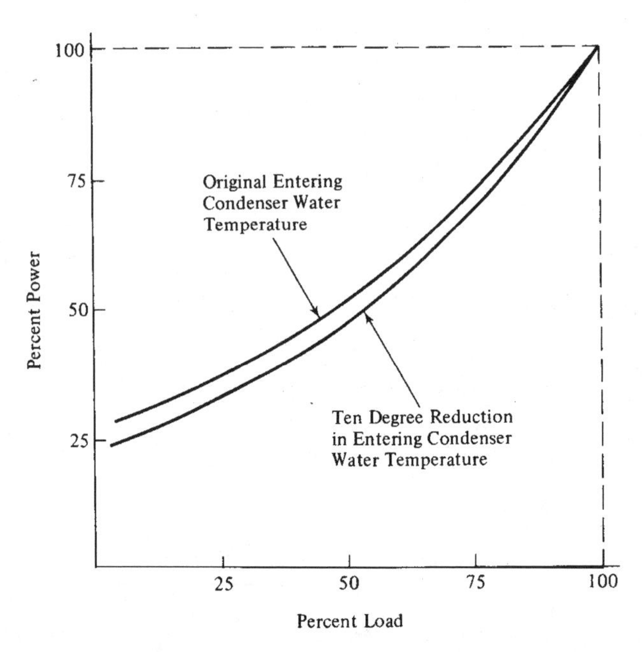

Figure 8-4. Typical power consumption curve for centrifugal liquid chiller.

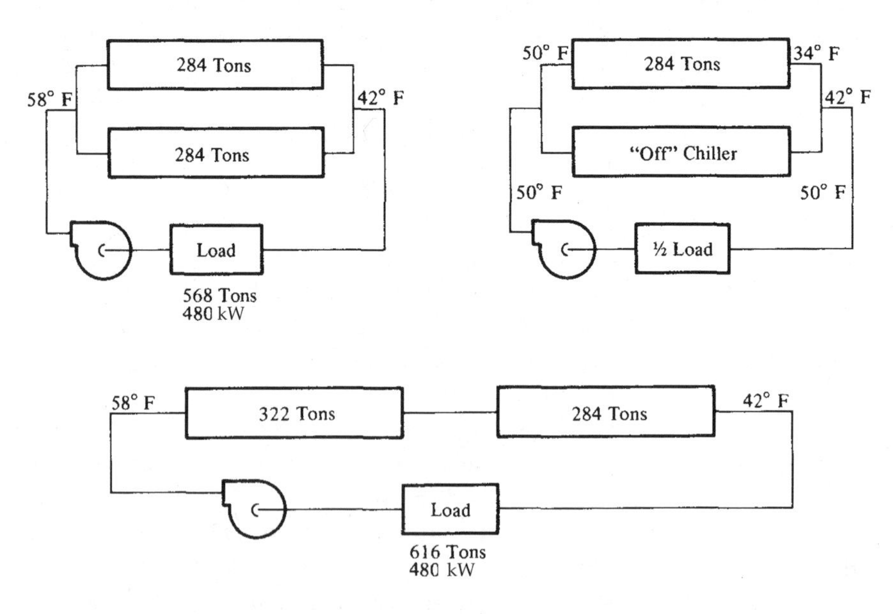

Figure 8-5. Multiple chiller arrangements.

In the series arrangement, a constant volume of flow of chilled water passes through the machines, producing better temperature control and better efficiency under part load operation; thus, the upstream chiller requires less kW input per ton output. The waste of energy during the mixing aspect of the parallel chiller operation is avoided. The series chillers, in general, require higher pumping costs. The energy conservation engineer should evaluate the best arrangement, based on load required and the partial loading conditions.

Absorption System Basics

The absorption cycle begins at the evaporator with heat being absorbed from the air either directly from the air through an air to refrigerant heat exchanger or indirectly through a heat transfer loop. This heat boils the liquid refrigerant (water or R717) which migrates to the absorber where it is absorbed into solution with an absorbent (LiBr or water). This "mixed" fluid (usually called dilute solution) is then pumped to condensing pressure by a small solution pump. The high pressure dilute solution enters the generator section of the unit where heat is added to drive the refrigerant from the solution. The now concentrated absorbent solution returns to the absorber and refrigerant vapor migrates to the condenser where it is liquefied by transferring heat to the outside air (either directly or through a heat transfer loop to a cooling tower) and the refrigerant is then reduced in pressure by an expansion device and returned to the evaporator to begin the cycle again.

A typical schematic for a single-stage absorption unit is illustrated in Figure 8-6. The basic components of the system are the evaporator, absorber, concentrator, and condenser. These components can be grouped in a single or double shell. Figure 8-6 represents a single-stage arrangement.

Evaporator

Refrigerant is sprayed over the top of the tube bundle to provide for a high rate of transfer between water in the tubes and the refrigerant on the outside of the tubes.

Absorber

The refrigerant vapor produced in the evaporator migrates to the bottom half of the shell where it is absorbed by a lithium bromide solution. Lithium bromide is basically a salt solution which exerts a

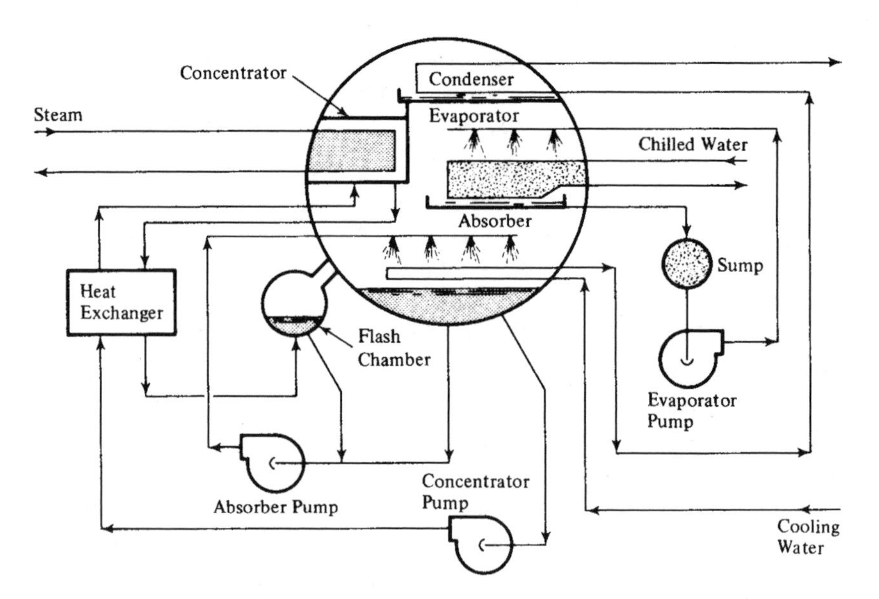

Figure 8-6. One-shell lithium bromide cycle water chiller. (Source: Trane air conditioning manual.)

strong attractive force on the molecules of refrigerant (water) vapor. The lithium bromide is sprayed into the absorber to speed up the condensing process. The mixture of lithium bromide and the refrigerant vapor collects in the bottom of the shell; this mixture is referred to as the dilute solution.

Concentrator

The dilute solution is then pumped through a heat exchanger where it is preheated by hot solution leaving the concentrator. The heat exchanger improves the efficiency of the cycle by reducing the amount of steam or hot water required to heat the dilute solution in the concentrator. The dilute solution enters the upper shell containing the concentrator. Steam coils supply heat to boil away the refrigerant from the solution. The absorbent left in the bottom of the concentrator has a higher percentage of absorbent than it does refrigerant, thus it is referred to as concentrated.

Condenser

The refrigerant vapor boiling from the solution in the concentrator flows upward to the condenser and is condensed. The condensed refrigerant vapor drops to the bottom of the condenser and from there flows to the evaporator through a regulating orifice. This completes the refrigerant cycle.

The single-stage absorption unit consumes approximately 18. 7 pounds of steam per ton of capacity (Steam consumption at full load based on typical manufacturers data.) For a single-stage absorption unit,

$$\text{COP} = \frac{1 \text{ ton} \times 12{,}000 \text{ Btu/ton}}{18.7 \text{ lb} \times 955 \text{ Btu/lb}} = 0.67.$$

The single-stage absorption unit is not as efficient as the mechanical chiller. It is usually justified based on availability of low pressure steam, equipment considerations, or use with solar collector systems.

SIM 8-1

Compute the energy wasted when 15 psig steam is condensed prior to its return to the power plant. Comment on using the 15 psig steam directly for refrigeration.

ANSWER

From Steam Tables 12-20 for 30 psia steam, hfg is 945 Btu per pound of steam; thus, 945 Btu per pound of steam is wasted. In this case where *excess low pressure* steam cannot be used, absorption units should be considered in place of their electrical mechanical refrigeration counterparts.

SIM 8-2

2000 lb/hr of 15 psig steam is being wasted. Calculate the yearly (8000 hr/yr) energy savings if a portion of the centrifugal refrigeration system is replaced with single-stage absorption. Assume 20 kW additional energy is required for the pumping and cooling tower cost associated with the single-stage absorption unit. Energy rate is \$.045 kWh and the absorption unit consumes 18.7 lb of steam per ton of capacity.

The centrifugal chiller system consumes 0.8 kWh per ton of refrigeration.

ANSWER

Tons of mechanical chiller capacity replaced = 2000/18.7 = 106.95 tons.

$$\begin{aligned}\text{Yearly energy savings} &= 2000/18.7 \times 8000 \times 0.8 \times \$.045 - \\ &\quad 20 \text{ kW} \times 8000 \times \$.045 = \$23{,}602\end{aligned}$$

Two-Stage Absorption Unit

The two-stage absorption refrigeration unit as illustrated in Figure 8-7 uses steam at 125 to 150 psig as the driving force. In situations where excess medium pressure steam exists, this unit is extremely desirable. The unit is similar to the single-stage absorption unit. The two-stage absorption unit operates as follows:

Medium pressure steam is introduced into the first-stage concentrator. This provides the heat required to boil out refrigerant from the dilute solution of water and lithium bromide salt. The liberated refrigerant vapor passes into the tubes of the second-stage concentrator, where its temperature is utilized to again boil a lithium bromide solution, which in turn further concentrates the solution and liberates additional

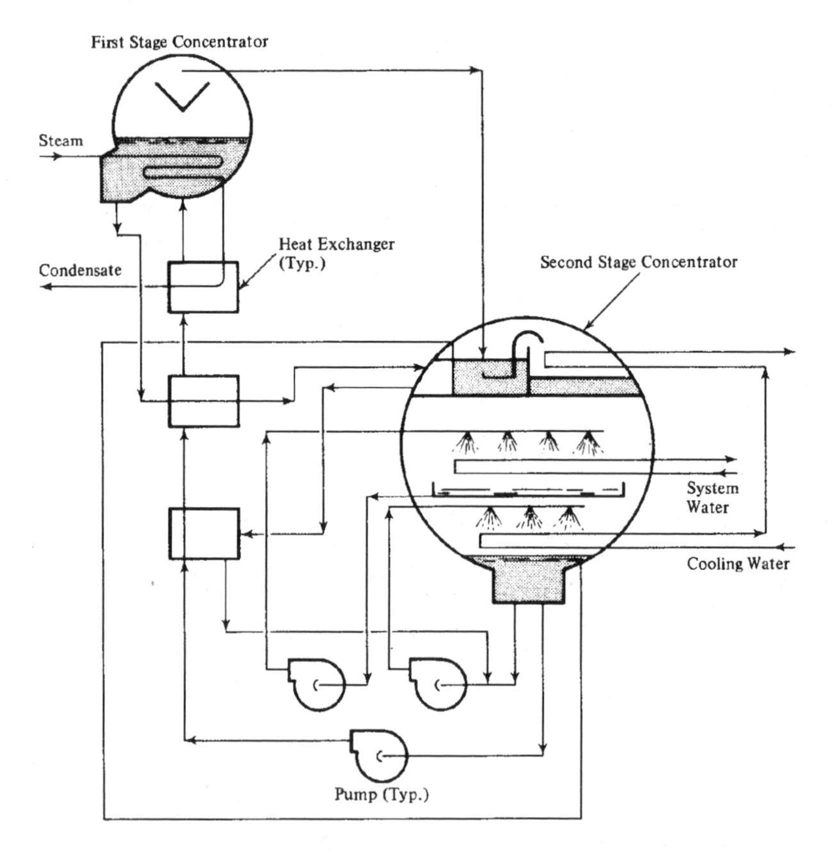

Figure 8-7. Two-stage absorption unit.

refrigerant. In effect, the concentrator frees an increased amount of refrigerant from solution with each unit of input energy.

The condensing refrigerant in the second-stage concentrator is piped directly into the condenser section. The effect of this is to reduce the cooling water load. A reduced cooling water load decreases the size of the cooling tower which is used to cool the water. The remaining portions of the system are basically the same as the single-stage unit.

The two-stage absorption unit consumes approximately 18.7 pounds of steam per ton of capacity; thus, it is more efficient than a single-stage counterpart. The associated COP is

$$\text{COP} = \frac{1 \text{ ton} \times 12{,}000 \text{ Btu/ton}}{18.7 \text{ lb} \times 860 \text{ Btu/lb}} = .67.$$

Either type of absorption unit can be used in conjunction with centrifugal chillers when it is desirable to reduce the peak electrical demand of the plant, or to provide for a solar collector addition at a later date.

GUIDE TO GAS COOLING TECHNOLOGY

The era of deregulation, real time prices, and summer electric price volatility is posing significant risks for the commercial customer. Conventional cooling technologies will leave the customer exposed to the hot weather price swings seen over the last two summers. This chapter summarizes the available and emerging gas technologies that can meet these customers' needs for cost containment and risk reduction, including absorption chillers, engine chillers, desiccant systems, and how such arrangements compare or work with thermal storage and on-site power generating systems.

The Emerging Technologies

Given these market needs, a number of new technologies are emerging, with implication that can significantly change the world of commercial HVAC.

A New Age of Absorption Chillers

They say that if you wait long enough, everything will come around again. Absorption chillers in the new century are one of those things. Manufacturers have prepared an entire new generation of absorption chiller with new features and user conveniences never before available to meet the challenges of electric deregulation.

Absorption is attractive in that it can be driven in so many ways; boiler steam, waste heat, and direct fired from either natural gas or fuel oil. When integrated with engine, steam turbine, or gas turbine generation, electricity and cooling can be supplied simultaneously. When used in conjunction with electric chillers in a "hybrid cooling plant," absorbers offer a fuel fired component to the cooling plant at maintenance costs equivalent to those of an electric chiller.

This last element may surprise many "old hands" in the HVAC design community. Due to the addition of sophisticated electronics, automatic purge, variable speed solution pumps, and a whole host of new features, maintenance contracts on a modern absorption chiller are often quoted at or below that of an electric chiller. In addition, absorbers now feature response times similar to electric chillers, are able to handle variable chiller barrel pumping, can cope with wide and rapid swings in tower water temperature, and often have turndown capabilities that exceed those of electric chillers.

Given these developments, using an absorption chiller for a portion of the capacity can assure the owner of flexibility in energy sourcing without a major change in his overall maintenance cost profile.

Engine Chillers

Another option for new chiller plants is engine driven chillers. These chillers combine the vapor compression systems used in electric chillers with a heavy-duty long-life engine, generally operating on natural gas. These systems feature significantly higher operating efficiencies than absorption chillers and can produce low temperature chilled water if desired. Engine chillers may also be less expensive than absorption chillers, particularly in the popular 100-400RT range. However, engine chillers do have additional maintenance needs, which have to be included in the decision process. However, these chillers have seen rapid market growth in the last 6 years as service structures to handle their particular needs have been established.

Thermal Storage

Another element in the technology mix is thermal storage—usually in the form of ice storage. This can be used to shift cooling loads from day to night power usage, a distinct advantage in areas with wide time-of-use price swings. If used in all-electric plants, thermal storage requires oversizing of the electric equipment to handle the low evaporator temperatures, and substantial floor space for the ice storage system. Using thermal storage to meet peak loads, however, reduces these cost issues. In a previous paper, the author has examined the comparative cost of differing systems and found the most cost effective use of thermal storage to be as a peak load shaver in hybrid gas and electric cooling plants.*

Hybrid Chiller/Generator Products

With the interest in hybrid chiller plants growing, some manufacturers have developed individual "hybrid chillers." This is a single chiller that can operate on two different energy sources alternately as desired by the owner.

The first such unique product available to the market is an engine/electric driven chiller, that features a drive line consisting of an engine, a clutch, and an electric motor, all in drive line, driving the chiller compressor. The engine can drive the compressor or be unclutched freeing the electric motor to drive the compressor. For areas with dramatic swings in time-of-day pricing, this can be highly economical. Depending on the engine for fewer hours of the year reduces maintenance and the electric motor allows the chiller to operate even if the engine is off-line for maintenance.

In time, new equipment of this type should be coming to the market with the added capability of generating electricity by reversing the motor. Using the unit as an engine driven generator allows the chiller to double as a standby generator or as an on-site generator when cooling is not called for.

Desiccant Dehumidification

Another option for thermally activated cooling systems is to relieve the cooling plant of some of the ventilation air cooling loads by using desiccant dehumidification to dehumidify ventilation air.

Desiccant systems have unique capabilities as they can deep dry ventilation air without overcooling and reheating. In high ventilation

*Presented at GLOBALCON 2000, April, 2000 by W.A. Ryan, "Gas Cooling in the Era of Deregulation."

load applications such as schools, theaters, and hospital surgical suites, this can amount to a significant load. In addition, desiccant systems can be operated either by direct fuel firing or from waste heat, such as engine or turbine exhaust heat. This gives the owner an option in integrated cooling, heating, and power systems.

Trigeneration

Trigeneration is a new, more specific term for combining cooling, heating, and power generation systems into one piece of equipment. Development of such comprehensive equipment is under way and should be available to the market in the next few years. In general, the generation source may be an engine, gas turbine or fuel cell with waste heat powering both the facility space heating and cooling system. Integration of water heating into the same system may also be used for applications where water heating is a significant load.

Custom designed trigeneration systems have been applied to large installations in the past. These systems have generally been for such large applications as college campuses, major medical facilities, or major convention centers. The modern concept is to develop integrated packaged trigeneration systems that can be applied to the huge market for individual commercial buildings. This requires smaller capacity, low maintenance, generating options such as microturbines and small fuel cell systems, both in development, and more affordable light capacity thermally activated cooling systems such as small absorption or desiccant systems.

One of the most attractive elements of trigeneration is the "double reduction" of the peak-demand requirements of the facility. By using waste heat to operate the cooling system, the cooling season electric peak is avoided. The power output of the system is then devoted to base loads such as lights, receptacles, and ventilating fans, further reducing the peak electric demand of these base loads. This is more practical than providing a larger power-generation-only system to a facility with all electric chillers to produce the same annual demand limitation. This will be an important issue in the deregulated future, as the facility will need to contract for electric supply service that will handle peak power needs.

District Heating and Cooling Plants

Nearer term is the recent growth in district heating and cooling plants. After decades with little interest being shown in this concept, new facilities have recently been constructed in a number of urban centers. Generally these plants have either been specialized in fuel fired heating and cooling operation or have been electrically driven thermal storage plants, utilizing off-peak power.

These plants provide steam or hot water and chilled water to independently owned buildings in a specific area. This eliminates the summer electric peak at the building, leaving the facility owner with only reasonable steady base loads to cover, once again making the building's load more attractive to electric suppliers.

Since 1990, 172,000,000 sq. ft. of occupied space has been added to district energy systems, with 22,000,000 added in 1999 alone.*

BASICS OF AIR CONDITIONING SYSTEM DESIGN FOR ENERGY CONSERVATION

Chapter 7 illustrated the fundamental aspects of heat gains and losses in buildings. It was seen that several building loads, such as lighting, and equipment losses remain fairly constant with outside temperature. Other loads, caused by the transmission of heat through building walls or by the ventilation of outside air into a building, vary with temperature. Solar radiation can be assumed to be constant with outside temperature.

The following summarizes heat gains and heat losses:

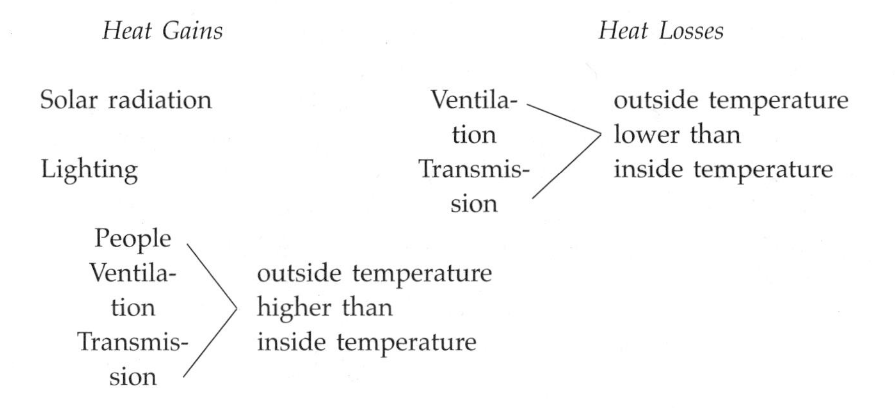

District Energy Now, Feb. 2000, a publication of the International District Energy Association

When heat gains are traded with heat losses, energy savings are realized. The central system is the basic air conditioning system used today. It allows for cooling, dehumidification, and heating equipment to be located in one area. As the central system developed, it became apparent that heat sources varied within the space; thus, the total area had to be divided into zones or sections.

The general categories of cooling systems most commonly used today are:

1. Central units serving individual zones
2. Multi-zone units
3. Double duct systems
4. Reheat systems
5. Variable volume systems
6. Individual room units

There are many variations and combinations, but the above are the basic options available. The major problem facing energy conservation engineering is to avoid generalizing. Each category has its application and economic advantage and should be evaluated based on the economics of the particular systems. In this section, particular applications of air conditioning systems will be discussed.

APPLYING VARIABLE AIR VOLUME SYSTEMS

The Variable Air Volume System has gained popularity in recent years, due to its potential of saving energy. This system is not a cure-all, but will save energy when *properly selected.* Variable Air Volume (VAV) Systems vary the quantity of air at a constant temperature, to match the system load requirements. The energy consumption closely parallels the load in air conditioning systems. This is in contrast to constant volume air systems which control building temperatures by varying the temperature of a constant flow of air. The most popular type of VAV System contains a terminal unit which modulates the air volume between maximum and a predetermined value.

There are four main classes of VAV Systems which encompass 80% of the VAV market. These classes combine the best features of several systems, to obtain maximum efficiency. Figure 8-8 illustrates the four classes.

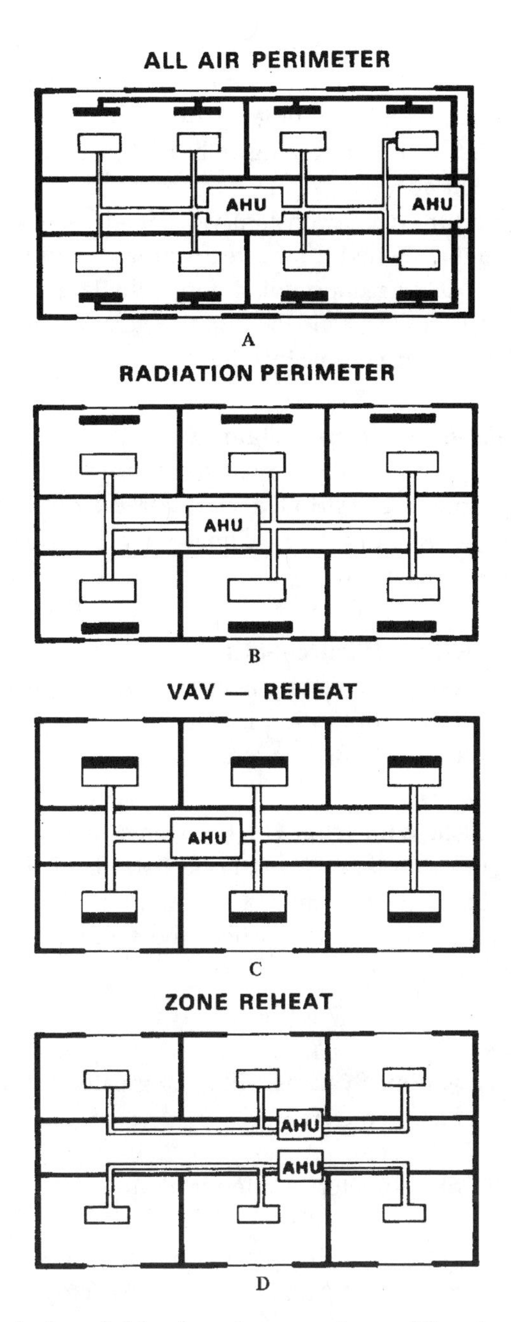

Figure 8-8. Typical variable air volume systems. (Courtesy of the Trane Company.) *Note:* AHU is an air-handling unit.

VAV with Independent Perimeter System (Figure 8-8A)

This system uses a constant volume variable temperature system to offset the transmission losses or gains through the building walls. The *variable volume system* is for cooling only and is used for temperature control of occupied spaces.

The temperature of the constant volume air is controlled from an outside thermostat (shielded from direct sunlight) and the system operates on return air. During unoccupied hours, the larger variable volume system can be shut down, while the building temperature is maintained by the smaller constant volume system.

VAV with Radiation Perimeter (Figure 8-8B)

This system is essentially similar to the previous class except the perimeter temperature is controlled via hydronic or electric heat. This system is mainly used in northern climates where under window heat is needed to offset down draft.

VAV with Zone Reheat (Figure 8-8C)

This system provides heating as well as cooling using a VAV System. This system finds application in smaller buildings, mild climates, and buildings with good construction.

VAV with Reheat or with Dual Duct Systems (Figure 8-8D)

This system modulates the volume to a fixed minimum before activating the reheat coils or mixing warm air at the individual zone. Once the minimum volume is reached, the volume remains constant and heat is added.

Energy Savings

Variable Air Volume Systems will be of greater benefit in buildings having the following characteristics:

1. High ratio of perimeter to interior area

2. Buildings which have high variations in loads, such as schools where the number of people changes regularly

3. Small zone sizes

4. Building faces east and west and has a high percentage of glass (VAV reheat should be investigated for this case)

VAV Systems of the modulating type, which can be used for both perimeter and interior spaces and operating at medium static pressure, are usually the most economical. A true variable volume system has the following characteristic: system volume reduction corresponds to loads, which permits savings in annual fan kWh and mechanical refrigeration.

APPLYING THE ECONOMIZER CYCLE

The basic concept of the economizer cycle is to use outside air as the cooling source when it is cold enough. There are several parameters which should be evaluated in order to determine if an economizer cycle is justified. These include:

1. Weather
2. Building occupancy
3. The zoning of the building
4. The compatibility of the economizer with other systems
5. The cost of the economizer.

What are the Costs of Using the Economizer Cycle?

In life, nothing comes free. Outside air cooling is accomplished at the expense of an additional return air fan, economizer control equipment, and an additional burden on the humidification equipment. Therefore, economizer cycles must be carefully evaluated based on the specific details of the application.

Maximum Savings

The economizer control system is the most significant aspect in achieving maximum savings. Outside air temperature alone does not insure that the economizer cycle is operating at its maximum potential. Ideally, the economizer controls should be based on Btus, not degrees. The enthalpy sensor offers the greatest potential for energy conservation, since it measures the heat content of the outside air. Unfortunately, enthalpy sensors have poor maintenance records and are many times replaced with thermometers.

APPLYING HEAT RECOVERY

Several examples of "heat recovery" have been indicated in this text. Basically, whenever wasted energy is recovered and used, the principle of "heat recovery" is being followed.

Coil Run-Around Cycle

The coil run-around cycle, as illustrated in Figure 8-9A, is another example of heat recovery. The cycle essentially transfers energy from the exhaust stream to the make-up stream. The heat transfer is accomplished by installing coils in each stream and continuously circulating a heat transfer medium, such as ethylene glycol fluid, between the two coils.

In winter, the warm exhaust air passes through the exhaust coils and transfers heat to the ethylene glycol fluid. The fluid is pumped to the make-up air coil where it preheats the incoming air. The system is most efficient in winter operation, but some recovery is possible during the summer.

Heat Wheels

If inlet and outlet exhaust ducts are close to one another, the heat wheel concept can be used. The heat wheel cycle accomplishes the same

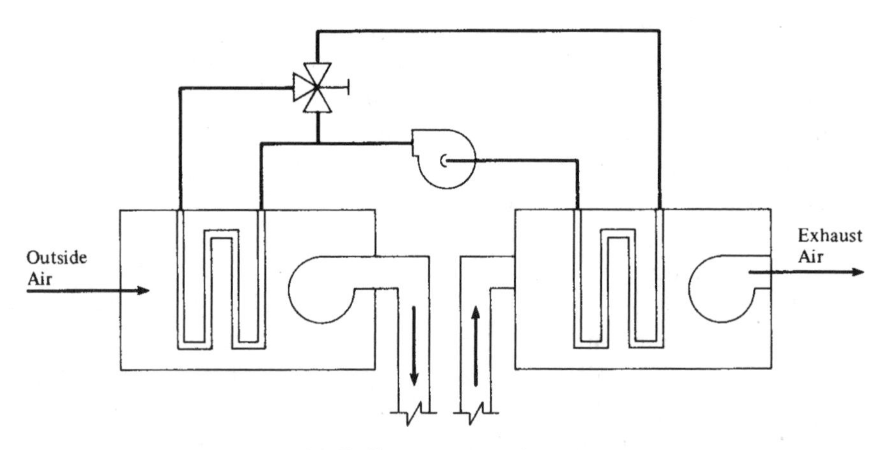

A. Coil-run-around cycle.

Figure 8-9. Heat recovery systems. (Courtesy of the Edison Electric Institute.)

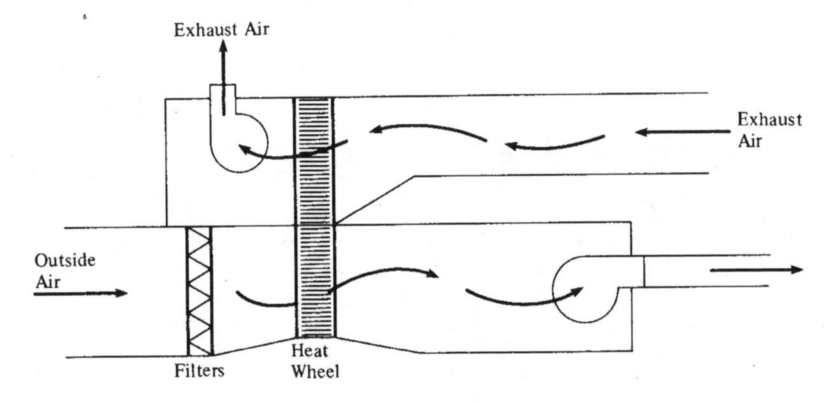

B. Heat wheels.

Figure 8-9. ***(Continued)***

objectives of the coil run-around cycle, as illustrated in Figure 8-9B. The heat wheel can be used to transfer sensible heat or latent heat. It consists of a motor driven wheel frame packed with a heat-absorbing material, such as aluminum, corrugated asbestos, or stainless steel mesh.

To optimize the system in the summer, lithium chloride impregnated asbestos can be used. Since lithium chloride absorbs moisture as well as heat, it can be used to cool and dehumidify make-up air.

Air-to-Air Heat Pipes and Exchangers

Another form of heat recovery is relatively simple, since no moving parts are involved. Either air-to-air heat pipes or exchangers can be used, as illustrated in Figure 8-9C. A heat pipe is installed through adjacent walls of inlet and outlet ducts; it consists of a short length of copper tubing sealed at both ends. Inside is a porous cylindrical wick and a charge of refrigerant. Its operation is based on a temperature difference between the ends of the pipe, which causes the liquid in the wick to migrate to the warmer end to evaporate and absorb heat. When the refrigerant vapor returns through the hollow center of the wick to the cooler end, it gives up heat, condenses, and the cycle is repeated.

The air-to-air heat exchanger consists of an open ended steel box which is compartmentalized into multiple narrow channels. Each passage carries exhaust air alternating with make-up air. Energy is transmitted by means of conduction through the walls.

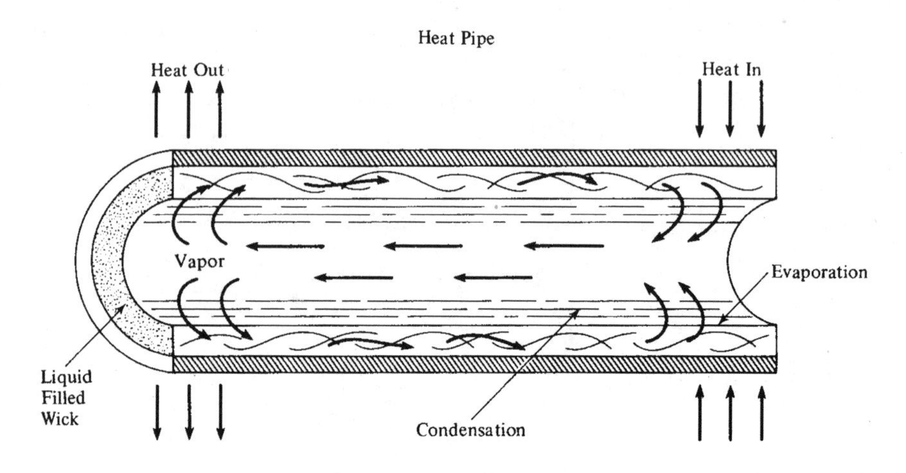

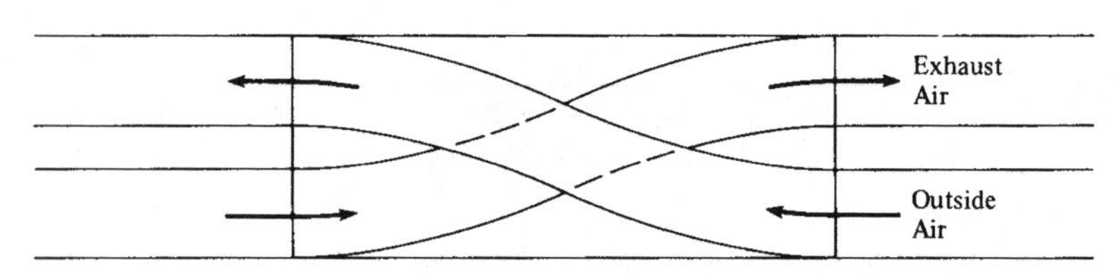

C. Air to air heat pipes and exchangers.

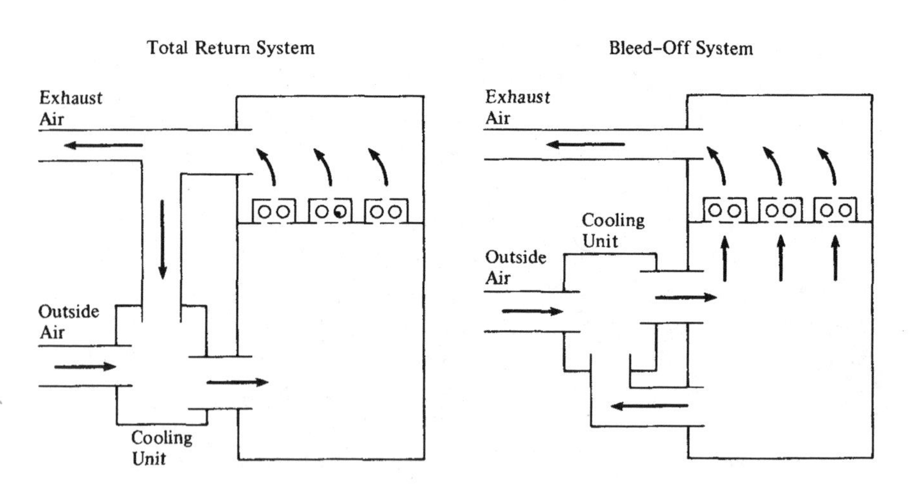

D. Recovery from lighting fixtures.

Figure 8-9. ***(Continued)***

Heat from Lighting Systems

Heat dissipated by lighting fixtures which is recovered will reduce air conditioning loads, will produce up to 13 percent more light output for the same energy input, and can be used as a source of hot air. Two typical recovery schemes are illustrated in Figure 8-9D. In the total return system, all of the air is returned through the luminaires. In the bleed off system, only a portion is drawn through the lighting fixtures. The system is usually used in applications requiring high ventilation rates.

Refrigeration Systems-Double Bundle Condenser

Refrigeration systems using double bundle condensers can be used for heat recovery and offer an economical way to prevent contamination of building water with cooling tower water. Figure 8-9E illustrates the addition of a double bundle condenser to a refrigeration unit. The

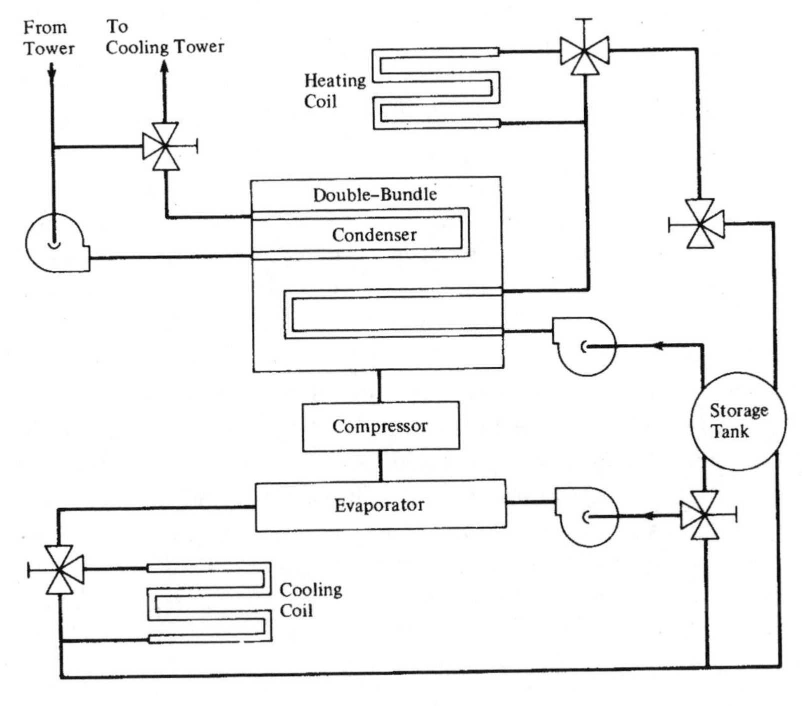

E. Recovery from refrigeration systems.

Figure 8-9. ***(Continued)***

heat rejected by the compressor is now available to the building water circuit. A heating coil has been added to provide supplemental heating. In cases where the amount of heat recovered during occupied hours exceeds the daytime heating requirements of perimeter zones, a heat storage tank can be added.

COOL STORAGE SYSTEM PERFORMANCE

Several utilities are offering substantial discounts to a number of their customers and even waiving demand charges on the use of power by offering time-of-day rates.

Time-of-day rates charge a premium during periods of high demand (on peak) and offer reduced rates during periods of low demand (off peak). The charge may be based on demand, usage, or both. Time-of-day rates have revived the concept of using thermal storage for cooling.

Thermal storage uses large volumes of water or ice in central or modular packages to store refrigeration produced during off peak periods for later use during periods of peak demand. Thermal storage systems can be placed in two general classifications: partial storage and total storage as indicated in Figure 8-10. Partial storage trims the peak refrigeration load only and requires some chiller operation during on peak periods. The storage volumes are manageable and the demand savings are attractive. Total storage systems must produce the entire daily refrigeration load the evening before. No chiller operation is allowed during on peak periods. Total storage systems require enormous thermal storage volumes and are usually attractive only where time-of-day rates offer substantial discounts for off peak power.

Chilled water has been the preferred choice of storage medium. Chilled water storage systems have lower energy requirements when compared to conventional systems. Producing and storing chilled water requires no unusual hardware. However, chilled water storage demands a fair amount of space. Approximately 10 cubic feet of water are required to store one ton-hour of cooling. This storage volume can be reduced by a factor of two to three by using ice storage. Can this space savings justify the additional installed cost and higher energy cost of ice storage? Some recent innovations in packaging give the system designer some options when designing ice storage systems. A review of ice storage systems past and present would be helpful.

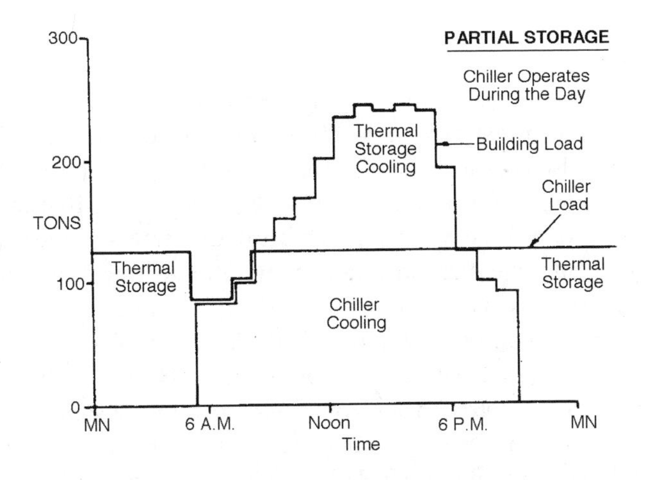

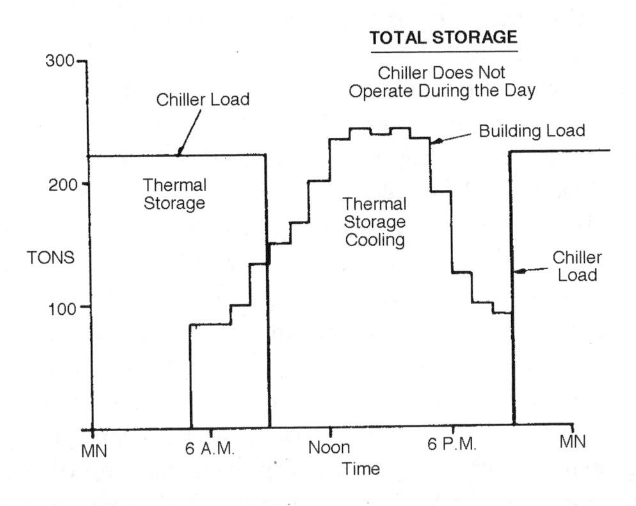

Figure 8-10.

Ice is stored on the heat transfer surface that produces the ice. The size of the ice tank is then a function of the heat transfer efficiency and the arrangement of the heat transfer surface as well as the amount of cooling the designer wishes to store. The ice tank is a large insulated tank containing several feet of steel pipe. The ice is built up on the outside of the pipe while the refrigerant (R22) is circulated inside the pipe. When fully charged, two and a half inches of ice surround the pipe or 13 pounds of ice for each lineal foot of pipe. This equates to five feet of steel pipe for every ton-hour of cooling stored.

There are several methods to circulate the refrigerant within the tubes. Pumping large volumes of Refrigerant 22 is possible, but the pumping system must not allow the premature formation of flash gas in the system. Refrigerant 22 is circulated through the ice tank tubes by direct expansion, liquid recirculation or gravity (flooded systems).

Direct expansion employs the pressure difference between the high side receiver and the suction line accumulator to transport refrigerant through the ice tank. With direct expansion, proper sizing of the suction line will minimize oil return problems. From a design standpoint, direct expansion offers a simple and reliable system. However direct expansion suffers some inefficiencies compared to other systems. Fifteen to 20 percent of the tube surface must be used to provide superheat and is not available for making ice. This results in an installed cost penalty for additional surface area and/or lower efficiencies due to lower suction temperatures.

Liquid recirculation systems use either the compressor or a separate liquid refrigerant feed pump to circulate refrigerant as indicated in Figure 8-11. The recirculation rate is two to three times the evaporation rate. This causes liquid overfeed systems, as they are sometimes called, to return a two-phase mixture from the ice tank to a low pressure receiver. From the low pressure receiver, refrigerant vapor is returned to the compressor while the refrigerant liquid is available for recirculation to the ice tank. The liquid level control meters additional refrigerant from the high pressure receiver to the low pressure receiver, to make up for refrigerant returned to the compressor. The design of the feed pump is critical in liquid recirculation systems. Proper design of the pumping system must be followed to prevent flashing of the refrigerant at the pump suction or in the pump itself. Semihermetic pumps or open pumps with specially designed seals are used to keep refrigerant losses to a minimum.

Pumping drums use hot gas to recirculate refrigerant, thereby eliminating the need for the refrigerant feed pump. The "pumper" is a smaller receiver that is located below the low pressure receiver. When vented to the low pressure receiver, liquid refrigerant drains by gravity into the pumper drum. The pumper is then isolated from the low pressure receiver by means of solenoid valves. Hot gas is used to pressurize the pumper, forcing refrigerant from the pumper through the ice tank. Two pumping drums are used so one is filling while the other is drain-

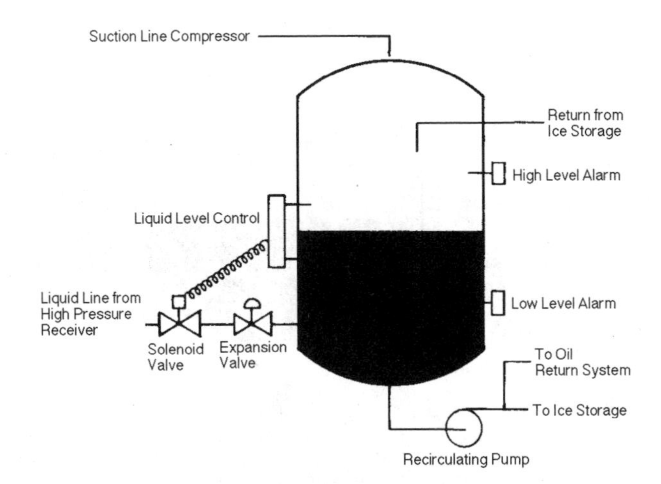

Figure 8-11. Low pressure receiver.

ing. The "double pumper" replaces the expensive and high maintenance liquid feed pump.

Gravity feed systems eliminate the need for pumping refrigerant altogether. With gravity systems, a low pressure receiver, or surge drum, is paired with each ice tank as a coil header. The vertical header carries the refrigerant to the ice tank inlet while a suction line returns refrigerant from the ice tank back to the surge drum. At the surge drum, vapor is returned to the compressor while liquid is recirculated to the ice tank. Due to the large refrigerant charge and the cost of multiple surge drums and controls, gravity flooded systems may be impractical in larger ice storage systems. A horizontal surge drum system is illustrated in Figure 8-12.

Oil return is a special concern with all liquid recirculation systems. At the low pressure receiver, refrigerant vapor is drawn off by the compressor, but the oil remains in the receiver. When a liquid feed pump or pumper drum is used, sufficient pressure drop is available to employ an oil return system. The liquid line to the ice tank is tapped and a small amount of refrigerant is bled off to an expansion valve. The line downstream of the expansion valve is sized to maintain sufficient velocity to carry the entrained oil back to the compressor inlet. With a gravity flooded system, sufficient head is not available and a separate oil recovery system must be used. In addition to an oil return or oil recovery system, an oil separator in the compressor discharge is a must.

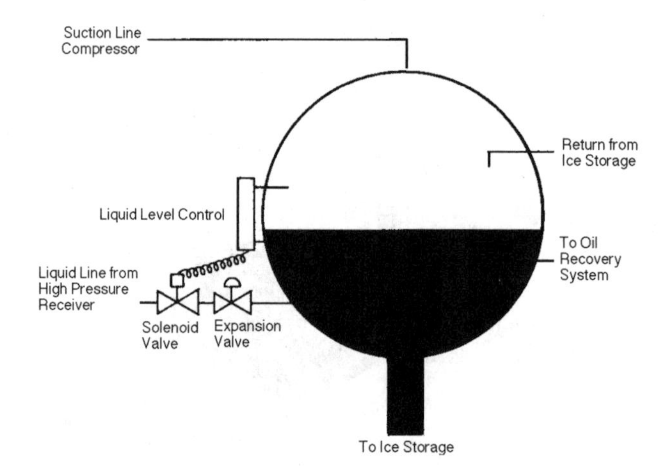

Figure 8-12. Horizontal surge drum.

Finally, there are ice tank accessories. To maintain uniform ice thickness, the water in the ice tank must be thoroughly mixed. An air pump is used to agitate the ice tank water during both the freezing and melting cycles. During the ice melting cycle, chilled water must circulate around the ice covered pipes. An ice thickness probe is required to prevent the ice tank from overfreezing or "bridging." Bridges of ice forming between the ice covered pipes will impede the flow of chilled water through the ice bundle during the melting cycle, resulting in uneven ice melting.

Ice storage systems of this nature require very large refrigerant charges and incur substantial construction costs due to the number of field erected components. Due to these construction costs, many ice systems are difficult to justify except in larger tonnage applications. Recent innovations in the packaging of glycol type ice tanks may make the option of ice storage a greater possibility.

Glycol systems as indicated in Figure 8-13 use an ethylene glycol/water mixture as a low temperature heat transfer medium to transfer heat from the ice storage tanks to a packaged chiller and from the cooling coils to either the ice storage tanks or the chiller. The use of freeze protected chilled water eliminates the design time, field construction, large refrigerant charges and leaks formerly associated with ice systems.

Glycol ice storage tanks create ice by circulating the low temperature fluid through half-inch polyethylene tubing. The polyethylene tubing is coiled inside insulated polystyrene tanks. The ethylene glycol/water

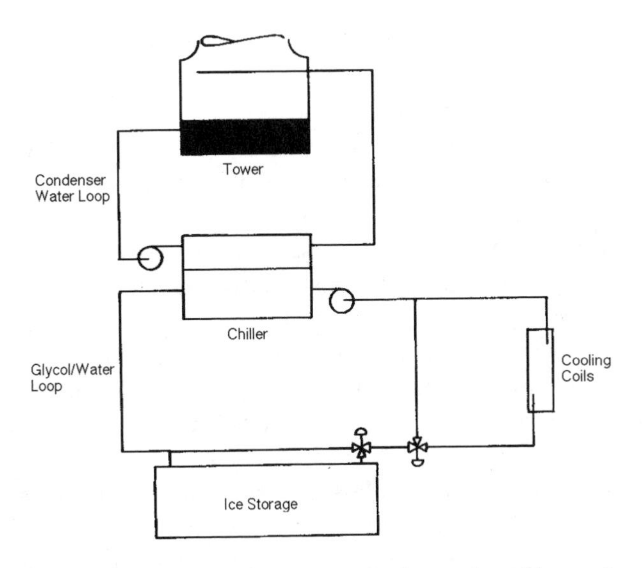

Figure 8-13. Glycol system (ethylene glycol/water).

mixture is also used for cooling by circulating the warm fluid through the tubing, melting the ice. The water that experiences a phase change remains in the tank. Since no water circulates around the tubing, the ice tank may be frozen solid. The problems of ice bridging and an air pump for agitation are eliminated. In this configuration the glycol ice tank is a sealed system similar to a packaged chiller or a car battery.

The heat transfer surface of the glycol ice tank can be increased by four to five times the area used in refrigerant ice tanks due to the low cost of polyethylene. The extended heat transfer area decreases the approach temperature required to make ice. Centrifugal or reciprocating chillers producing 23 to 26 degree glycol are well suited for this application. Centrifugal chillers have an excellent track record in low temperature applications including food processing, cosmetics, pharmaceuticals, clean rooms, other industrial applications and, of course, ice rinks.

THERMAL STORAGE CONTROL SYSTEMS

The control of thermal storage systems and equipment mandates the uses of automated control systems. The complexity of controls should not exceed the complexity of the system. However, more than a time clock and a few extra thermostats are required to insure the maximum benefit of the thermal storage system. Control concepts can range from simple to comprehensive.

The simplest control is chiller priority. This control scheme allows the chiller to accept all of the cooling load until the load exceeds chiller capacity. Excess load is then met by melting ice. Control defaults to limiting chiller capacity by chiller selection or some form of demand limiting. Control is simple, but use of thermal storage is limited to cooling in excess of chiller capacity. This control may not provide the maximum financial return on the thermal storage investment.

There are several variations of ice priority control schemes. The intent of ice priority control is to gain maximum benefit of time-of-day rates by maximizing ice usage each day. During intermediate seasons when the daily cooling load is less than thermal storage capacity, chiller operation can be limited to off peak hours. When daily cooling load exceeds thermal storage capacity, chiller and ice must share the on peak cooling load. The task of assigning cooling load to either ice or chiller is well suited for building automation systems with equation processing capabilities. If ice is used too early or too rapidly, the ice will be depleted before the end of the daytime cooling load, leaving several hours of building load in excess of chiller capacity. On the other hand, if the ice is not used to its maximum potential, the operating cost benefits of thermal storage are not totally realized.

How does the automation system predict the daily cooling load the day before? Most of the required tools are already in place. The automation system has the ability to monitor several building or load producing trends. What the automation system does not possess is the ability to equate these trends to building cooling load.

Fortunately, several HVAC system simulation programs are available for this task. Simulation programs can isolate single load components and plot their dependence to an easily monitored variable. Internal loads are a function of time and calendar. Ventilation and solar loads can be equated to monitored weather variables. A daily cooling load profile can now be constructed by measurement of the monitored variables, profiling those variables for the period of prediction, and summing the results. Automation systems that can produce an expected daily load profile provide the building operator with a very powerful management tool. Only informed and prudent operation of thermal storage systems can guarantee their success.

THE VENTILATION AUDIT*

Indoor Air Quality (IAQ) is an emerging issue of concern to building managers, operators, and designers. Recent research has shown that indoor air is often less clean than outdoor air and federal legislation has been proposed to establish programs to deal with this issue on a national level. This, like the asbestos issue, will have an impact on building design and operations. Americans today spend long hours inside buildings, and building operators, managers and designers must be aware of potential IAQ problems and how they can be avoided.

IAQ problems, sometimes termed "Sick Building Syndrome," have become an acknowledged health and comfort problem. Buildings are characterized as sick when occupants complain of acute symptoms such as headache, eye, nose and throat irritation, dizziness, nausea, sensitivity to odors and difficulty in concentrating. The complaints may become more clinically defined so that an occupant may develop an actual building-related illness that is believed to be related to IAQ problems.

The most effective means to deal with an IAQ problem is to remove or minimize the pollutant source, when feasible. If not, dilution and filtration may be effective.

Dilution (increased ventilation) is to admit more outside air to the building. ASHRAE's 1981 standard recommended 5 CFM/person outside air in an office environment. The newest ASHRAE ventilation standard, 62-2001, now requires 20 CFM/person for offices if the prescriptive approach is used.

Increased ventilation will have an impact on building energy consumption. However, this cost need not be severe. If an airside economizer cycle is employed and the HVAC system is controlled to respond to IAQ loads as well as thermal loads, 20 CFM/person need not be adhered to and the economizer hours will help attain air quality goals with energy savings at the same time. Incidentally, it was the energy cost of treating outside air that led to the latest standard. The superseded 1973 standard recommended 15-25 CFM/person.

Energy savings can be realized by the use of improved filtration in lieu of the prescriptive 20 CFM/person approach. Improved filtration can occur at the air handler, in the supply and return ductwork, or in the

---

*Source: Indoor Air Quality: Problems & Cures, M. Black & W. Robertson, Presented at the 13th World Energy Engineering Congress.

spaces via self-contained units. Improved filtration can include enhancements such as ionization devices to neutralize airborne biological matter and to electrically charge fine particles, causing them to agglomerate and be more easily filtered.

To accomplish an energy audit of the ventilation system the following steps can be followed:

1. Measure volume of air at the outdoor air intakes of the ventilation system. Record ventilation and fan motor nameplate data.
2. Determine code requirements and compare against measurements.
3. If measured ventilation is more than required by code, decrease CFM by changing fan pulley, or by applying a variable speed drive.

Two savings are derived from this change, namely:

- Brake horsepower of fan motor is reduced.
- Reduced heat loss during heating season.

To compute the savings, Equations 8-3 and 8-4 are used. Figure 8-14 can also be used to compute fan power savings as a reslt of air flow reduction.

$$\text{HP (New)} = \text{HP} \times \left(\frac{\text{CFM (New)}}{\text{CFM (Old)}}\right)^3 \qquad (8\text{-}3)$$

$$\text{Q (Saved)} = \frac{1.08 \text{ Btu}}{\text{HR} - \text{CFM} - {}^\circ\text{F}} \times \text{CFM (Saved)} \times \Delta\text{T} \qquad (8\text{-}4)$$

$$\text{kW} = \text{HP} \times 0.746/\eta \qquad (8\text{-}5)$$

where

HP = motor horsepower
CFM = cubic feet per minute
ΔT = average temperature gradient
kW = motor kilowatts (k = 1000)
η = motor efficiency

In addition to reducing air flow during occupied periods, consideration should be given to shutting the system down during unoccupied hours.

If the space is cooled, additional savings will be achieved. The quantity of energy required to cool and dehumidify the ventilated air

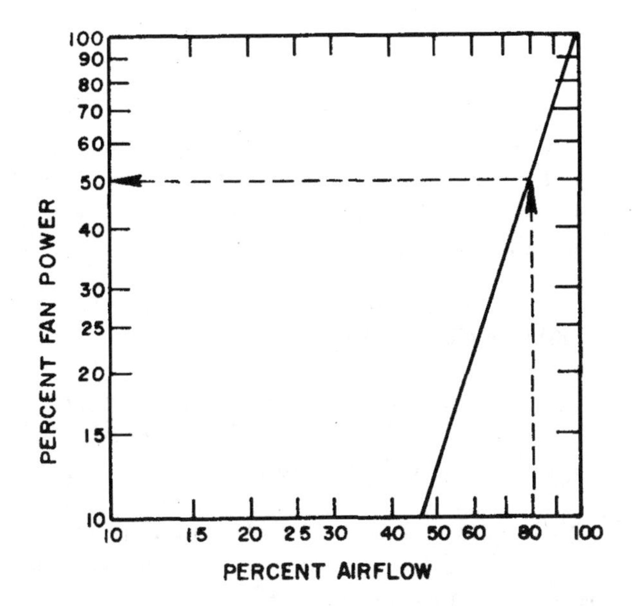

Fig. 8-14. Decrease in horsepower accomplished by reducing fan speed (based on laws of fan performance). (Source: NBS *Handbook 115 Supplement 1.*)

to indoor conditions is determined by the enthalpy difference between outdoor and indoor air.

ENERGY ANALYSIS UTILIZING SIMULATION PROGRAMS

There are a number of software programs which have been successfully used to simulate how a system will perform various load and weather conditions. Before the microcomputer revolution of the 1980s, two of the most widely used public domain programs were DOE-II, developed by Lawrence Berkeley Laboratory for the Department of Energy, and Building Loads Analysis and Steam Thermodynamics, better known as BLAST, developed for the Army and Air Force. In the private sector the TRACE program developed by the Trane Company and Energy Analysis developed by Ross R. Meriwether and Associates dominated the market.

Carrier's Hourly Analysis Program combines the base and hourly load calculation logarithms of the Commercial Load Estimating Program with the sophisticated equipment simulation of the Operating Cost Analysis Program.

The personal computer provides a cost-effective way to analyze energy consumption, make detailed building energy analysis and implement life cycle costing in the decision making process. There are numerous software options available which can help the plant engineer make the best possible choice.

TEST AND BALANCE CONSIDERATIONS

Probably the biggest overlooked low-cost energy audit requirement is a thorough test, balance, and adjust program. In essence the audit should include the following steps:

1. ***Test.*** Quantitative determination of conditions within the system boundary, including flow rates, temperature and humidity measurements, pressures, etc.
2. ***Balance.*** Balance the system for required distribution of flows by manipulation of dampers and valves.
3. ***Adjust.*** Control instrument settings, regulating devices, and control sequences should be adjusted for required flow patterns.

In essence, the above program checks the designer's intent against actual performance and balances and adjusts the system for peak performance.

Several sources outlining Test and Balance Procedures are:

- Construction Specifications Institute (CSI), which offers a specification series that includes a guide specification Document 15050 entitled, "Testing and Balancing of Environmental Systems." Reprints of this paper are available. It explains factors to be considered in using the guide specification for project specifications.

- Associated Air Balance Council (AABC), the certifying body of independent agencies.

- National Environmental Balancing Bureau (NEBB), sponsored jointly by the Mechanical Contractors Association of America and the Sheet Metal and Air Conditioning Contractors National Association as the certifying body of the installing contractors' subsidiaries.

9

Cogeneration: Theory and Practice

Because of its enormous potential, it is important to understand and apply cogeneration theory. In the overall context of energy management theory, cogeneration is just another form of the conservation process. However, because of its potential for practical application to new or existing systems, it has carved a niche that may be second to no other conservation technology.

This chapter is dedicated to development of a sound basis of current theory and practice of cogeneration technology. It is the blend of theory and practice, or praxis of cogeneration, that will form the basis of the most workable conservation technology in the coming years.

DEFINITION OF "COGENERATION" OR COMBINED HEAT AND POWER (CHP)

Cogeneration is the sequential production of thermal and electric energy from a single fuel source. In the cogeneration process, heat is recovered that would normally be lost in the production of one form of energy. That heat is then used to generate the second form of energy. For example, take a situation in which an engine drives a generator that produces electricity: With cogeneration, heat would be recovered from the engine exhaust and/or coolant, and that heat would be used to produce, say, hot water. It is this combination of heat and power which leads to the name CHP.

Making use of waste heat is what differentiates cogeneration facilities from central station electric power generation. The overall fuel utilization efficiency of cogeneration plants is typically 70-80% versus 35-40% for utility power plants.

This means that in cogeneration systems, rather than using energy in the fuel for a single function, as typically occurs, the available energy is cascaded through at least two *useful* cycles.

To put it in simpler terms: Cogeneration is a very efficient method of making use of *all* the available energy expended during any process generating electricity (or shaft horsepower) and then utilizing the waste heat.

A more subjective definition of cogeneration calls upon current practical applications of power generation and process needs. Nowhere more than in the United States is an overall system efficiency of only 30% tolerated as "standard design." In the name of limited *initial* capital expenditure, all of the waste heat from most processes is rejected to the atmosphere.

In short, present design practices dictate that of the useful energy in one gallon of fuel, only 30% of that fuel is put to useful work. The remaining 70% is rejected randomly.

If one gallon of fuel goes into a process, the designer may ask, "How much of that raw energy can I make use of within the constraints of the overall process?"

In this way, cogeneration may be taken as a way to use a maximum amount of available energy from any raw fuel process. Thus, cogeneration may be thought of as *just good design.*

COMPONENTS OF A COGENERATION SYSTEM

The basic components of any cogeneration plant are:

- A prime mover
- A generator
- A waste heat recovery system
- Operating control systems

The prime mover is an engine or turbine which, through combustion, produces mechanical energy. The generator converts the mechanical energy to electrical energy. The waste heat recovery system is one or more heat exchangers that capture exhaust heat or engine coolant heat and convert that heat to a useful form. The operating control systems insure that the individual system components function together.

The prime mover is the heart of the cogeneration system. The three basic types are steam turbines, combustion gas turbines and internal combustion engines. Each has advantages and disadvantages, as explained below.

Steam Turbines

Steam turbine systems consist of a boiler and turbine. The boiler can be fired by a variety of fuels such as oil, natural gas, coal and wood. In many installations, industrial byproducts or municipal wastes are used as fuel. Steam turbine cogeneration plants produce high-pressure steam that is expanded through a turbine to produce mechanical energy which, in turn, drives a device such as an electric generator. Thermal energy is recovered in several different ways, which are discussed in Chapter 5. Steam turbine systems generally have a high fuel utilization efficiency.

Combustion Gas Turbines

Combustion gas turbine systems are made up of one or more gas turbines and a waste heat recovery unit. These systems are fueled by natural gas or light petroleum products. The products of combustion drive a turbine which generates mechanical energy. The mechanical energy can be used directly or converted to electricity with a generator. The hot exhaust gases of the gas turbine can be used directly in process heating applications, or they can be used indirectly, with a heat exchanger, to produce process steam or hot water.

A variation on the combustion gas turbine system is one that uses high-pressure steam to drive a steam turbine in conjunction with the cogeneration process. This is referred to as a combined cycle.

Internal Combustion Engines

Internal combustion engine systems utilize one or more reciprocating engines together with a waste heat recovery device. These are fueled by natural gas or distillate oils. Electric power is produced by a generator which is driven by the engine shaft. Thermal energy can be recovered from either exhaust gases or engine coolant. The engine exhaust gases can be used for process heating or to generate low-pressure steam. Waste heat is recovered from the engine cooling jacket in the form of hot water.

Cogeneration plants that use the internal combustion engine generate the greatest amounts of electricity for the amount of heat produced.

Of the three types of prime movers, however, the fuel utilization efficiency is the lowest, and the maximum steam pressure that can be produced is limited.

AN OVERVIEW OF COGENERATION THEORY

As discussed in the introduction, and as may be seen from Figure 9-1, standard design practices make use of, at best, 30% of available energy from the raw fuel source (gas, oil, coal).

Of the remaining 70% of the available energy, approximately 30% of the heat is rejected to the atmosphere through a condenser (or similar) process. An additional 30% of the energy is lost directly to the atmosphere through the stack, and finally, approximately 7% of the available energy is radiated to the atmosphere because of the high relative temperature of the process system.

With heat recovery, however, potential useful application of available energy more than doubles. Although in a "low quality" form, *all* of the condenser-related heat may be used, and 40% of the stack heat may be recovered. This optimized process is depicted (in theory) as Figure 9-2.

Thus, it may be seen that effective use of all available energy may more than double the "worth" of the raw fuel. System efficiency is in-

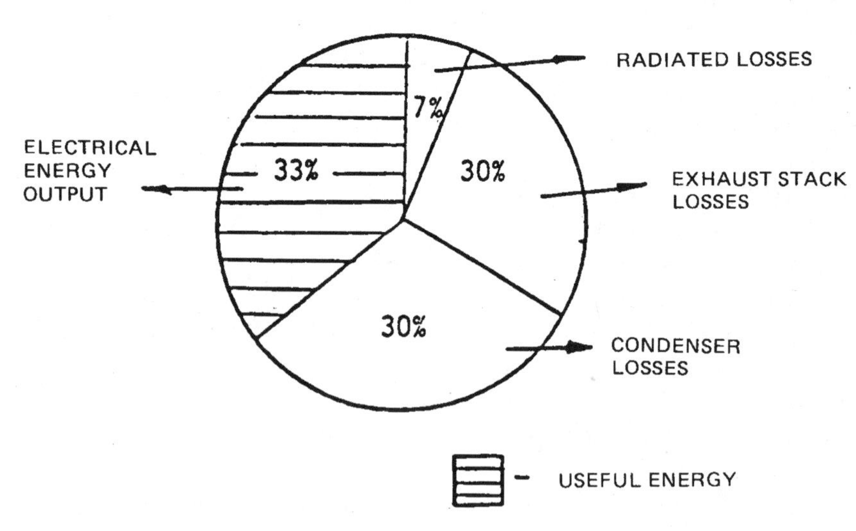

Figure 9-1. Energy balance without heat recovery.

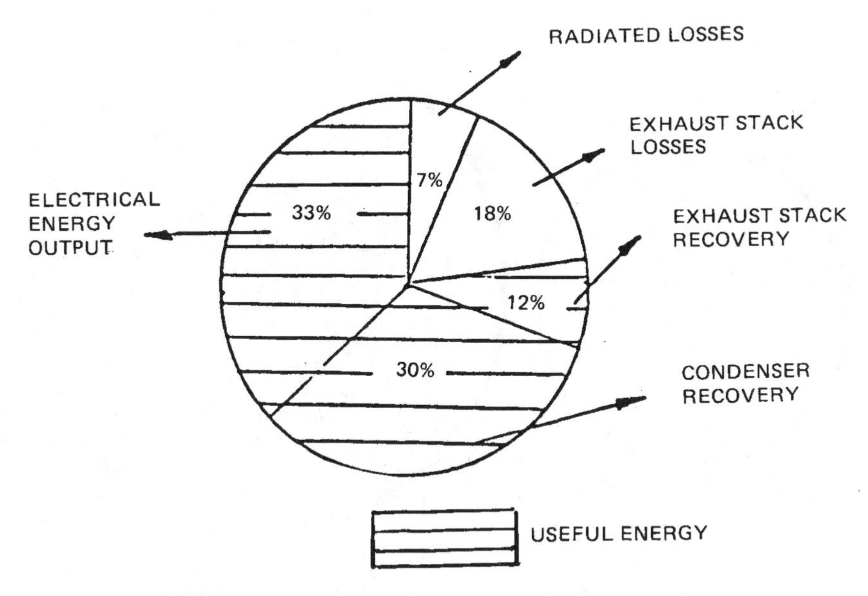

Figure 9-2. Energy balance with heat recovery.

creased from 30% to 75%. This higher efficiency allows the designer to use low grade energy for various cogeneration cycles.

Example Problem 9-1

A cogeneration system vendor recommends the installation of a 20-megawatt cogeneration system for a college campus. Determine the approximate *range* of useful *thermal* energy. Use the energy balance of Figure 9-2.

Analysis

Step 1: "Range" is defined by the best and worst operating *times* of the installed system:

At best, system will operate 365 days per year, 24 hours per day.

At worst, system will operate 5 days per week, 10 hours per day.

Step 2: Perform heat balance. See diagram.

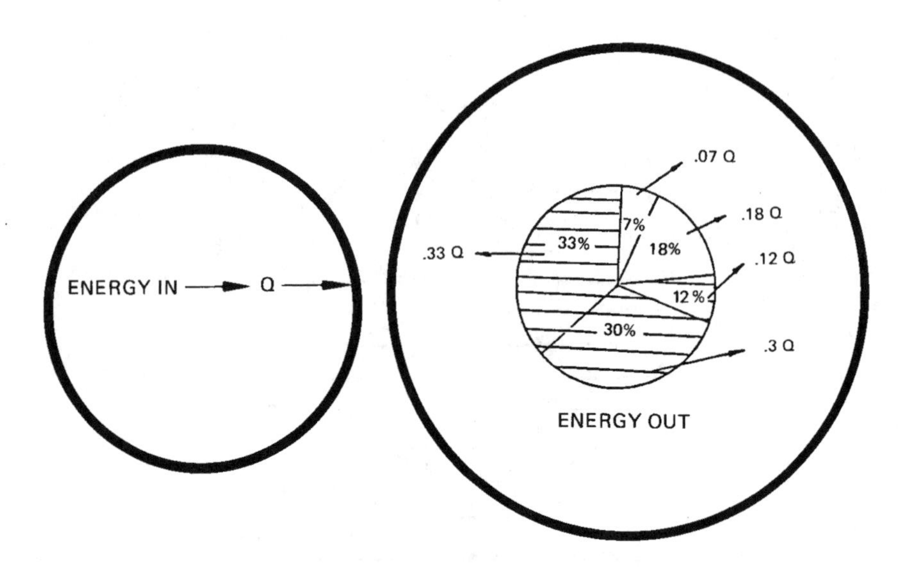

$$Q = E_{out} = E_{electrical} + E_{condenser} + E_{stack} + E_{radiated} \quad (9\text{-}1)$$

Step 3: Calculate available thermal energy.

From Figure 9-2, available energy equals condenser energy and 40% stack energy.

$$E_{available} = E_{condenser} + E_{stack} \quad (9\text{-}2)$$

$$E_{condenser} = .3Q \quad (9\text{-}3)$$

$$E_{stack} = 0.12\ Q$$

$$E_{available} = 0.30\ Q + 0.12Q \quad (9\text{-}4)$$

$$E_{electrical} = .33\ Q \quad (9\text{-}5)$$

$$Q = \frac{E_{electrical}}{.33} \quad (9\text{-}6)$$

By substitution $E_{available} = 0.3 \dfrac{E_{electrical}}{0.33} + 0.12 \dfrac{E_{electrical}}{0.33}$

$$E_{available} = 1.272 \text{ Electrical} \tag{9-7}$$

Calculate available energy

$$\dot{Q} = \dot{Q}_1 \times t = K_1 \, E_{available} \times K_2 \times t$$

K_1 = 3413 Btu/kWh
K_2 = 1000 Kilowatts per megawatt
t = equipment hours of operation

$$Q_1 = 3413 \text{ Btu/kWh} \times (1.272 \times 20 \text{ megawatts}) \times 1000 \text{ kW} \frac{\text{kW}}{\text{MWatt}}$$

$$Q_1 = 86.82 \times 10^6 \text{ Btu/hr}$$

Worst Case: System operates
5 days/wk × 52 wks/yr × 10 hrs/day = 2600 hrs/yr

Best Case: System operates
365 days/yr × 24 hrs/day = 8670 hrs/yr

Thus range is

$$Q = 86.82 \times 10^6 \times 2600 = 225.7 \times 10^9 \text{ Btu/yr}$$
$$Q = 86.82 \times 10^6 \times 8760 = 760.5 \times 10^9 \text{ Btu/yr}$$

APPLICATION OF THE COGENERATION CONSTANT

The *cogeneration constant* may be used as a fast check on any proposed cogeneration installation. Notice, from the sample problem which follows, the ease with which a thermal vs. electrical comparison of end needs may be made.

Example Problem 9-2

A cogeneration system vendor recommends a 20 megawatt installation. Determine the approximate rate of *useful* thermal energy.

Analysis

$$E_{available} = E_{electrical} \times K_c \tag{9-8}$$

E is the cogeneration system electrical rated capacity

K_c is the cogeneration constant

$$E_{available} = 20 \text{ MW} \times 1.272$$
$$= 25.4 \text{ MW of } \textit{useful} \text{ heat}$$

or

$$25.4 \text{ MW} \times 1000 \ \frac{\text{kW}}{\text{MW}} \times = \frac{3413 \text{ Btu}}{\text{kWh}} \times \frac{\text{Therm}}{100{,}000 \text{ Btu}}$$

$$E_{available} = 866.9 \text{ therms/hour}$$

APPLICABLE SYSTEMS

To ease the complication of matching power generation to load, and because of newly established laws, it is most advantageous that the generator operate in parallel with the utility grid which thereby "absorbs" all generated electricity. The requirement for "qualifying facility (QF) status," and the consequent utility rate advantages which are available when "paralleling the grid," is that a significant portion of the thermal energy produced in the generation process must be recovered. Specifically, Formula 9-10 must be satisfied:

$$\frac{\text{Power Output} + 1/2 \text{ Useful Thermal Output}}{\text{Energy Input}} \geq 42.5\% \ \text{(for any calendar year)} \qquad (9\text{-}9)$$

Note that careful application of the *cogeneration constant* will generally assure that the qualifying facility status is met.

BASIC THERMODYNAMIC CYCLES

Bottoming and Topping Cycles

Cogeneration systems can be divided into "bottoming cycles" and "topping cycles."

Bottoming Cycles

In a bottoming cycle system, thermal energy is produced directly from the combustion of fuel. This energy usually takes the form of steam that supplies process heating loads. Waste heat from the process is recovered and used as an energy source to produce electric or mechanical power.

Bottoming cycle cogeneration systems are most commonly found in industrial plants that have equipment with high-temperature heat requirements such as steel reheat furnaces, clay and glass kilns and aluminum remelt furnaces. Some bottoming cycle plants operating in Georgia are Georgia Kraft Company, Brunswick Pulp and Paper Company, and Burlington Industries.

Topping Cycles

Topping cycle cogeneration systems reverse the order of bottoming cycle systems: Electricity or mechanical power is produced first; then heat is recovered to meet the thermal loads of the facility. Topping cycle systems are generally found in facilities which do not have extremely high process temperature requirements.

Figure 9-3 on the following page shows schematic examples of these two cogeneration operating cycles.

A sound understanding of basic cogeneration principles dictates that the energy manager should be familiar with two standard thermodynamic cycles. These cycles are:

1. Brayton cycle
2. Rankine cycle

The Brayton cycle is the basic thermodynamic cycle for the simple gas turbine power plant. The Rankine cycle is the basic cycle for a vapor-liquid system typical of steam power plants. An excellent theoretical discussion of these two cycles appears in Reference 10.

The Brayton Cycle

In the open Brayton cycle plant, energy input comes from the fuel that is injected into the combustion chamber. The gas/air mixture drives the turbine with high-temperature waste gases exiting to the atmosphere. (See Figure 9-4.)

The basic Brayton cycle, as applied to cogeneration, consists of a gas turbine, a waste heat boiler and a process or "district" heating load.

TOPPING CYCLE:

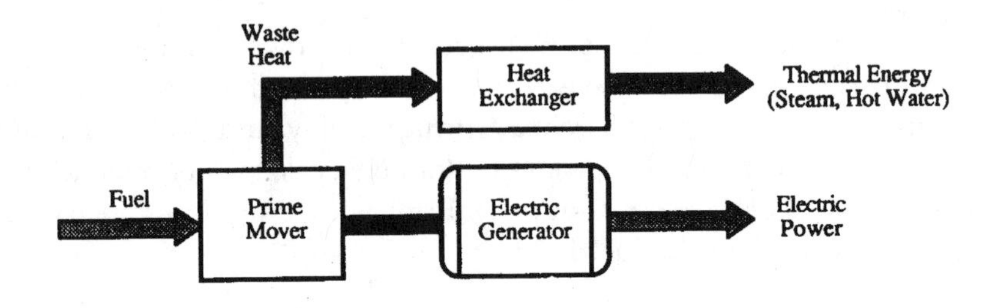

BOTTOMING CYCLE:

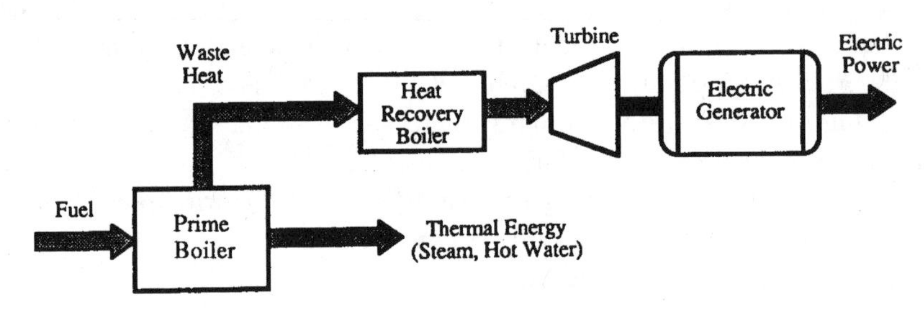

Figure 9-3. Cogeneration operating cycles.

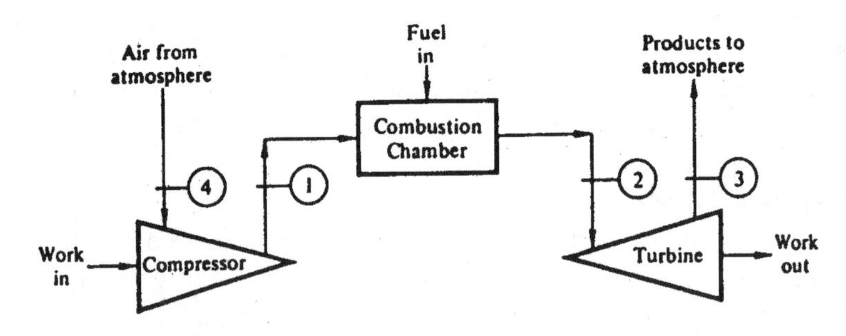

Figure 9-4. Open gas-turbine power plant (Brayton cycle).

This cycle is a *full-load* cycle. At part loads, the efficiency of the gas turbine goes down dramatically. A simple process diagram is illustrated in Figure 9-5.

Because the heat rate of this arrangement is superior to that of all other arrangements at full load, this simple, standard Brayton cycle merits consideration under all circumstances. Note that to complete the loop in an efficient manner, a deaerator and feedwater pump are added.

The Rankine Cycle

The Rankine cycle is illustrated in the simplified process diagram, Figure 9-6. Note that this is the standard boiler/steam turbine arrangement found in many power plants and central facility plants throughout the world.

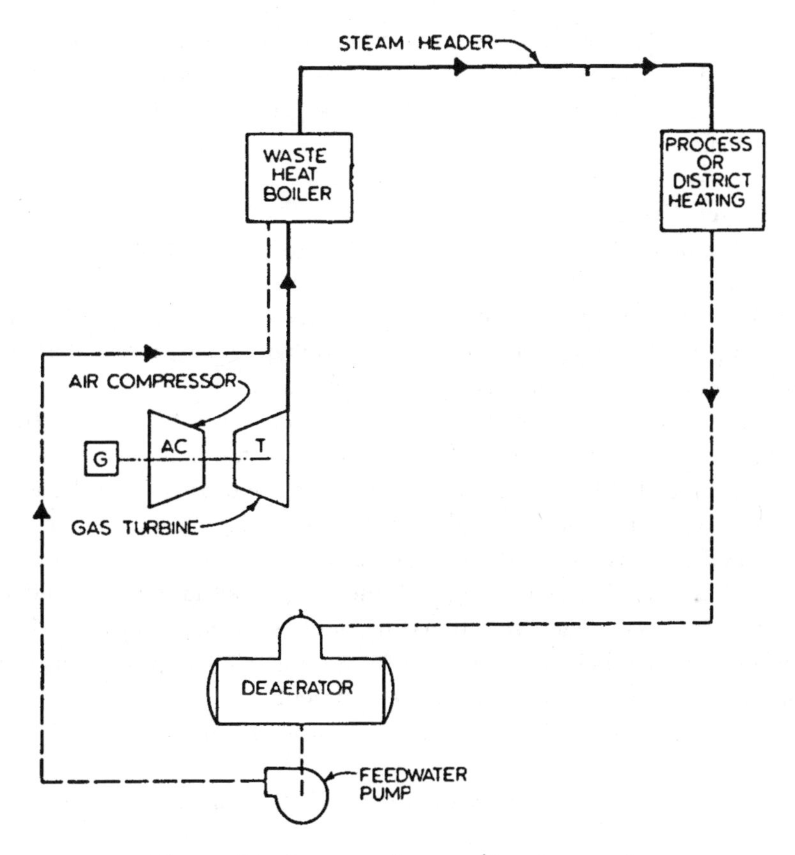

Figure 9-5. Process diagram/Brayton cycle.

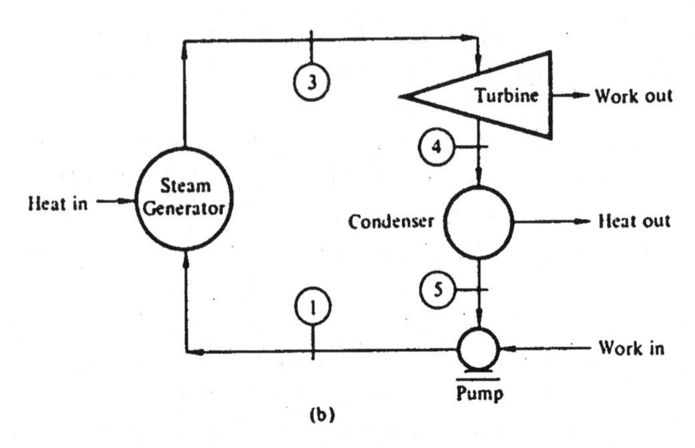

Figure 9-6. Simple Rankine cycle.

The Rankine cycle, or steam turbine, provides a real-world outlet for waste heat recovered from any process or generation situation. Hence, it is the steam turbine which is generally referred to as the topping cycle.

Combined Cycles

Of major interest and importance for the serious central plant designer is the combined cycle. This cycle forms a hybrid which includes the Brayton cycle on the "bottoming" portion and a standard Rankine cycle on the "topping" portion of the combination. A process diagram with standard components is illustrated in Figure 9-7.

The combined cycle, then, greatly approximates the cogeneration Brayton cycle but makes use of a knowledge of the plant requirements and an understanding of Rankine cycle theory. Note also that the ideal mix of power delivered from the Brayton and Rankine portions of the combined cycle is 70% and 30%, respectively.

Even within the seemingly limited set of situations defined as "combined cycle," many variations and options become available. These options are as much dependent on any local plant requirements and conditions as they are on available equipment.

Some examples of combined cycle variations are:

1. Gas turbine exhaust used to produce 15 psi steam for Rankine cycle turbine with no additional fuel burned. This situation is shown in Figure 9-7.

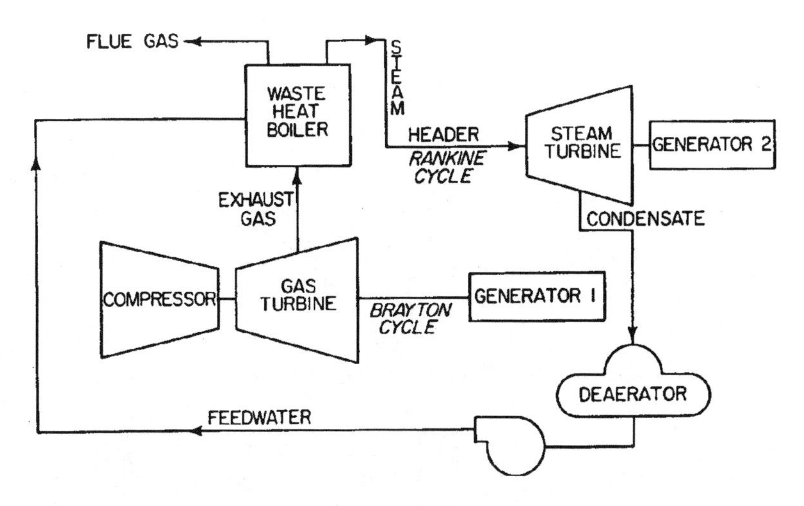

Figure 9-7. Combined cycle operation.

2. Gas turbine exhaust fired in the duct with additional fuel. This provides a much greater amount of produced power with a correspondingly greater amount of fuel consumption. This situation generally occurs with a steam turbine pressure range of 900-1259 psig.

3. Gas turbine exhaust fired directly and used directly as combustion air for a conventional power boiler. Note here that the boiler pressure range may vary between 200 and 2600 psig. (See Figure 9-8.)

One other note to keep in mind is that in any combined cycle case, the primary or secondary turbine may supply direct mechanical energy to a refrigerant compressor. As discussed, the variations are endless. However, a thorough understanding of the end process generally will result in a final, and best, cogeneration system selection.

DETAILED FEASIBILITY EVALUATION*

This section introduces the parameters affecting the evaluation, selection, sizing and operation of a cogeneration plant and shows a means of evaluating those parameters where it counts—on the bottom line.

**Georgia Cogeneration Handbook*, Governor's Office of Energy Resources, August 1988.

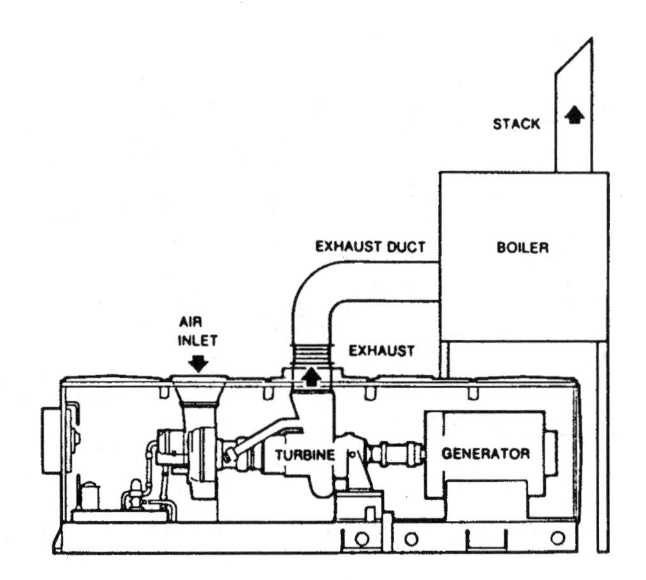

Figure 9-8. Gas turbine used s combustion air.

This section will enable you to answer two basic questions:

1. Is cogeneration technically feasible for us, given our situation?

2. If it is technically feasible, is it also economically feasible, considering estimated costs, energy savings, current utility rates and regulatory conditions in Georgia?

Keep in mind that this chapter is not designed to take the place of a full-scale feasibility study. It is designed to help you and your engineering staff decide—in-house and at minimal expense—whether such a study is warranted. You will reach one of the following conclusions:

- No, cogeneration is not feasible in our case.

- Yes, cogeneration looks promising and a feasibility study is warranted.

- Maybe, but I have some questions.

If your conclusion is "no," you will have saved yourself the expense of a feasibility study. If your conclusion is "yes," you will be assured that the expense of employing outside consultants to perform the study is justified by the potential benefits. If your conclusion is "maybe," you may wish to discuss your situation with qualified engineers before deciding whether to proceed.

Utility Data Analysis

The first step in the feasibility evaluation is the gathering and analysis of utility data. These data are necessary to determine the limits of the thermal and electric energy consumed at your facility. They are also necessary for determining the timing of maximum use of each of these energy sources. From this information, you will also determine the thermal and electric load factors for your facility.

Load Factor Defined

A load factor is defined as the average energy consumption rate for a facility divided by the peak energy consumption rate over a given period of time, and it is an important simplification of energy use data. These values are usually expressed as either a decimal number or a percent. To calculate a load factor you need two pieces of data: the total energy consumption for a given time period and the maximum energy demand observed during that time period.

In the electric power industry, load factors serve as a measure of the utilization of generation equipment. A utility load factor of 0.90, or 90%, indicates very good utilization of power generation facilities, while a load factor of 0.30, or 30%, indicates poor utilization. The lower the load factor, the larger and more expensive your generating equipment must be merely to meet peak power demands for short time periods.

The thermal and electric load factors of your facility are of great importance in sizing a cogeneration plant. In fact, they are even more important to a cogeneration plant than to an electric utility because the cogeneration plant does not have the advantage of an electric power grid to diversify the variations in energy use.

A high thermal or electrical load factor generally indicates that a cogeneration plant would be utilized a major portion of the time and therefore would provide a favorable return on your investment. An ideal situation exists when both load factors are high, indicating that a properly sized cogeneration plant would efficiently utilize most of its output. If both the thermal and electrical load factors are small, you may not be a practical candidate for cogeneration.

In the analysis of your electric and thermal energy consumption data, you will calculate the load factors for your facility. In this evaluation, we will concentrate on annual load factors. It is important to note, however, that monthly and daily load factors may be important in your final analysis.

Electric Energy Consumption Analysis

In order to obtain the necessary energy-use information, you must refer to your monthly electric utility bills for one complete year. From those bills, calculate the following information:

- Determine your annual kWh consumption by tabulating and adding the monthly kWh consumption.

- Tabulate the actual kW demand metered for each month. Be sure to use the actual metered demand and not the billed demand, as the billed demand may be based on time of year and a percent ratchet of the previous year's peak demand.

- Identify the maximum monthly kW demand value as the annual peak kW.

- Determine the annual average kW demand by dividing the annual kWh consumption by 8,760 hours per year.

$$\text{Annual Average kW} = \frac{\text{Annual kWh}}{\text{8,760 hours/year}}$$

- Determine the annual electric load factor by dividing the annual average kW demand by the annual peak kW demand.

$$\text{Annual Electric Load Factor} = \frac{\text{Annual Average kW}}{\text{Annual Peak kW}}$$

Thermal Energy Consumption Analysis

If your boiler plant has an accurate steam or Btu metering system, the following data can be gathered directly from your metered output data. If you do not have this degree of metering, your boiler plant fuel bills for one complete year will be required to obtain the necessary data. Monthly natural gas bills can be analyzed with the same method used for analyzing electrical consumption. If you use fuel oil, propane or coal, make sure that accurate measurements of the reserve supply were taken at the beginning and end of the year.

- Determine your annual fuel consumption. From Table 9-1, select the Btu value per unit of fuel for the type of fuel you use. To obtain the annual fuel Btu input, multiply the annual fuel consumption by this value.

Table 9-1. Typical Fuel Caloric Values (Btu/CF).

Fuel Type	*HHV Higher Heating Value (Approximate)*	*LHV/HHV Lower/Higher Heating Value (Approximate)*
Natural Gas (Dry)	1,000	0.90
Butane	3,200	0.92
Propane	2,500	0.92
Sewage Gas	300-600	0.90
Landfill Gas	300-600	0.90
No. 2 Oil	139,000 Btu/Gal	0.93
No. 6 Oil	154,000 Btu/Gal	0.96
Coal, Bituminous	14,100 Btu/lb	

- Determine your boiler's fuel-to-steam efficiency. An estimated fuel-to-steam efficiency of 78% may be used for a well-maintained boiler plant that operates only enough boilers to keep them well loaded. A poorly maintained boiler plant with some leaks, missing insulation, and oversized boilers that cycle frequently may have a fuel-to-steam efficiency of 60% or lower.

- Determine your annual Btu output by multiplying your annual fuel Btu input by your fuel-to-steam efficiency.

$$\text{Annual Btu Output} = \text{Annual Fuel Btu Input} \times \text{Fuel-to-steam Efficiency}$$

- Determine the annual average Btu/hour demand by dividing the annual Btu output by 8,760 hours per year.

$$\text{Annual Average Btu/Hour Demand} = \frac{\text{Annual Btu Output}}{\text{8,760 hours/year}}$$

If you have metered steam production data, select the maximum value for Btu/hour output that occurred last year as the annual peak Btu/hour demand. If these data are not available, estimate your annual peak Btu/hour demand as a percentage of the maximum possible output of your boiler plant.

NOTE: Boilers that operate at approximately 150 psig and below use the terms "lbs per hour" and "mbh output" interchangeably to represent 1,000 Btu/hour. We will use the unit "mbh" to simplify notation when referring to 1,000 Btu/hour units.

- Determine the annual thermal load factor by dividing the annual average Btu/hour demand by the annual peak Btu/hour demand.

$$\text{Annual Thermal Load Factor} = \frac{\text{Annual Average Btu/hour Demand}}{\text{Annual Peak Btu/hour Demand}}$$

If your annual minimum Btu/hour demand is extremely low or your boiler plant actually shuts down for several months during the summer, your annual thermal load factor is not an accurate indicator of your cogeneration potential. It is probably too large, and a closer examination of the number of hours at the annual minimum Btu/hour demand may be required to assess feasibility. This can be done using monthly and daily load profiles.

If your annual minimum Btu/hour demand is zero for a large number of hours per year (for example, if your boiler plant shuts down during the summer months), you may not be a good candidate for co-generation.

Thermal/Electric Load Ratio

For cogeneration to be feasible, the demands for thermal and electric energy must overlap much of the time. Therefore, once the thermal and electric demands of your facility are known, you must determine the ratio of heat demand to electric demand that may be expected to occur together. This is done by using the thermal/electric (T/E) load ratio. The T/E load ratio is defined as the quantity of heat energy that is coincident with a quantity of electrical energy. In making these calculations, you will attempt to find an optimum match between your facility's T/E load where

$$\text{Thermal/Electric Load Ratio} = \frac{\text{Thermal Demand}}{\text{kW Demand}}$$

If the thermal and electric load factors calculated earlier are high, there is a good possibility that your facility's demand for thermal energy occurs at about the same time as your demand for electric energy. In that case, your facility's annual average thermal/electric load ratio is approximately equal to the annual average Btu/hour demand divided by the annual average kW demand. For convenience, we will use the term "mbh" to represent 1,000 Btu/hr.

$$\text{Annual Average Thermal/ Electric Load Ratio} = \frac{\text{Annual Average mbh Demand}}{\text{Annual Average kW Demand}}$$

If either the thermal or the electric load factor is small, then a worst-case assumption should be made. This assumption is called the minimum demand thermal/electric load ratio and is calculated as follows:

$$\text{Minimum Demand Thermal/ Electric Load Ratio} = \frac{\text{Annual Minimum mbh Demand}}{\text{Annual Minimum kW Demand}}$$

The use of either of the thermal/electric load ratios above must be tempered with the knowledge that these are only approximations of the load correlations and are useful only for a preliminary evaluation. Most facilities will require a more detailed examination of thermal and electric load profiles to obtain the number of hours per year that loads overlap and can be served with a cogeneration plant.

Generally, cogeneration opportunities are good for facilities with T/E load ratios above 5 and are best for ratios above 10, with average annual Btu/hour demand above 10,000,000 Btu/hour, or 10,000 mbh.

If your T/E load ratio is less than two, it is reasonable to assume that you are not a good candidate for cogeneration. This would certainly be the case if your thermal demand is extremely small during the summer months and is only significant during a few winter months. However, if your annual electric load factor is small, you may consider using electric power generation only, in a strategy known as peak shaving. This is discussed in more detail under operating strategies in this chapter.

Equipment Sizing Considerations

For efficient cogeneration, the T/E load ratio in your facility must correspond to the T/E output ratio of the cogeneration systems. T/E output ratios vary for different prime movers and cogeneration plant configurations.

In an analysis of cogeneration options at your facility, you must decide whether to size the cogeneration plant to match your peak electric load, which will produce some waste heat and lower overall plant efficiency, or to size the cogeneration plant to match your heat load and supplement your electric needs with more expensive purchased power. If your load factor is small for either thermal or electric demand, you should size toward the minimum value of that demand.

Operating Strategies

Electric Dispatch

One method of operating a cogeneration plant is to supply your facility's total electrical requirements as a first priority and generate steam as a second priority. This mode of operation is referred to as "electric dispatch."

Under the electric dispatch mode, the cogeneration plant is sized for annual peak kW demand, with some additional capacity for growth. A cogeneration plant sized in this manner can operate totally independently of the electrical utility company.

Standby electric service from your local utility is required for scheduled and unscheduled equipment shutdown. The cost of this standby power can be very expensive, especially if demand is high or if the utility company must provide additional generation or transmission equipment to serve your facility. Total on-site backup using standby emergency power generating equipment is rarely economical.

Electric dispatch requires that you operate your cogeneration plant to meet your electric load demand requirement. Operation of the major electric loads of your facility should be coordinated with the cogeneration plant to prevent sudden load spikes or irregular, repetitive load spikes that could exceed the capacity of your generating plant.

Also, if your annual electric load factor is low, your cogeneration equipment will be oversized for your facility load most of the time. This will reduce efficiency, increase maintenance costs and lower the return on investment.

Peak Shaving

Peak shaving is the practice of selectively dropping electric loads or generating on-site electricity during periods of peak electric demand. This procedure is not cogeneration; normally, no heat recovery equipment is installed and no heat is recovered from the generating process. Peak shaving is most commonly used to reduce annual electric utility costs. Peak shaving with an electric generator has a lower initial cost than a true cogeneration system since no heat exchangers or associated recovery equipment are installed.

Thermal Dispatch

Thermal dispatch operation is the complement to electric dispatch. In this mode of operation, the cogeneration plant is sized to meet your facility's annual peak Btu/hour (or mbh) demand. You will then purchase a part of your electric power from the local utility and cogenerate the rest. For this mode of operation to be successful, thermal load requirements should parallel each other in the same way required for electric dispatch.

For maximum savings on electric energy, electric generation should be near its maximum capacity during the summer peak electric demand period. This means that the ideal candidate for thermal dispatch operation would have maximum thermal loads occurring during summertime electric peak periods. If this is not the case, you will save less on electric energy.

Hybrid Strategy

A "hybrid strategy" utilizes the best features of the electrical and thermal dispatch strategies. As the name suggests, this operating strategy is a hybrid of the electric dispatch and thermal dispatch operating strategies periodically adjusted to minimize operating costs and maximize return on investment.

With the hybrid strategy, the cogeneration plant is sized smaller than for electrical or thermal dispatch. The hybrid strategy calls for the plant to be operated to achieve maximum savings during electric peak demand periods. At other times, it would be operated in a thermal load following mode. To satisfy the total thermal load requirements, it may be necessary to maintain existing boilers or install additional ones. As in the thermal dispatch strategy, your facility still must purchase a portion of its electric energy from the utility.

Prime Mover Selection

Choosing the most appropriate prime mover for a cogeneration project involves evaluation of many different criteria, including:

- Hours of operation
- Maintenance requirements
- Fuel requirements
- Capacity limits

Following are some general guidelines for evaluating each of the different prime movers with respect to these criteria.

Hours of Operation

Most cogeneration plants are designed for continuous operation, with the exception of some small reciprocating engine packages. Large plants are not economical if they cycle on and off on an hourly or even daily basis.

Because cogeneration equipment is relatively expensive as compared with boilers of similar thermal capacity, the equipment must be operated as much as possible to achieve an acceptable return on investment.

If an analysis of your facility shows a low electric load factor, indicating relatively few hours of high electric demand, cogeneration is probably not a cost-effective option. Instead, you may wish to examine the potential of peak shaving with reciprocating engine generators.

Maintenance Requirements

All cogeneration equipment requires some type of periodic maintenance. The frequency and amount of maintenance varies considerably between prime movers.

RECIPROCATING ENGINE GENERATORS have the highest maintenance requirements. Like automotive engines, diesel engines require routine maintenance such as oil and filter changes as often as once a week. However, natural gas fired reciprocating engines require more overhauls of the head and block, and they have a shorter life expectancy.

GAS TURBINE ENGINES are mechanically simpler and require less frequent maintenance. They can operate for longer periods between major maintenance intervals, and the major maintenance is usually simpler, such as bearing inspection and replacement.

STEAM TURBINES require even less mechanical maintenance than gas turbines since they have no combustion equipment attached. Occasional bearing inspection and replacement is generally the extent of steam turbine maintenance. The cost is a function of the size and number of turbines.

Fuel Requirements

RECIPROCATING ENGINES are not flexible with respect to their fuel requirements. Generally, you buy either a diesel engine or a natural gas fired engine. Special design adaptations can produce engines that burn other fuels, such as low Btu gas or heavier oils. Occasionally, natural gas engines can be adapted to switch between natural gas and propane as an alternate fuel. Diesel engines are more efficient in their fuel-to-energy conversion. However, air pollution regulations may impose greater constraints on these engines than gas turbines, usually requiring catalytic converters and carburetion limits.

GAS TURBINES can be switched from natural gas to diesel fuel to take advantage of price fluctuations and backup fuel requirements. In addition, they can be adjusted for other fuel oils, low Btu gas, and, occasionally, organic byproducts of industrial processes such as "black liquor" from pulp and paper miffs.

STEAM TURBINES are limited only by the fuel for their steam source. In addition to the above fuels, coal, wood, waste, peanut shells, any suitable biomass, and incinerated municipal waste can be used to generate steam for steam turbines.

Capacity Limits

The three prime movers discussed here have separate and distinct capacity ranges. Reciprocating engine generators range in size from about 40 kW to over 3,000 kW. Generally, small electric demand plants with still smaller heat requirements can be satisfied with reciprocating engines. The quality of the heat recovered from these engines can be a limitation. Only about 35% of the recoverable heat is available as 125 psig steam; the rest of it is available only as 180°F hot water.

Almost all of the gas turbine heat is recoverable as 125 psig steam. The lower limit of gas turbine equipment size is about 480 kW, and the upper limit is over 30,000 kW. Combined-cycle gas and steam turbine plants can produce over 100,000 kW.

Steam turbines are the most limited with respect to power generation. Their practical lower limit is about 1,000 kW. Their conversion efficiency for power generation is below 15% if practical upper limits of 200 psig of superheated inlet steam and 100 psig of saturated outlet steam are maintained. This efficiency can be improved by increasing the inlet pressure and temperature, which dramatically increases the cost of the steam production plant. Another alternative for improving electric generating efficiency is to use a condensing turbine, but this lowers the temperature of the outlet steam to the point that it cannot be used for much more than low-temperature hot water production.

Figure 9-9 on the following page shows the thermal and electric energy output typically available from several different cogeneration system configurations.

Energy Savings Analysis

To arrive at an estimate of the energy savings potential of cogeneration at your facility, you will need to make two simplifying assumptions. For this analysis we will assume that the proposed cogeneration system is sized such that 100 percent of its thermal and electric output is utilized. We will also assume that the plant will operate at or near full output capacity at all times. These assumptions represent a "best case" operating scenario for your cogeneration system. In reality, your plant probably will not operate in this manner; however, these assumptions are necessary here for the purpose of this evaluation.

In addition to the above assumptions, you will need to obtain the following information on your proposed cogeneration plant:

- The thermal and electric output capacities of the cogeneration equipment

- The fuel consumption rate of the cogeneration equipment

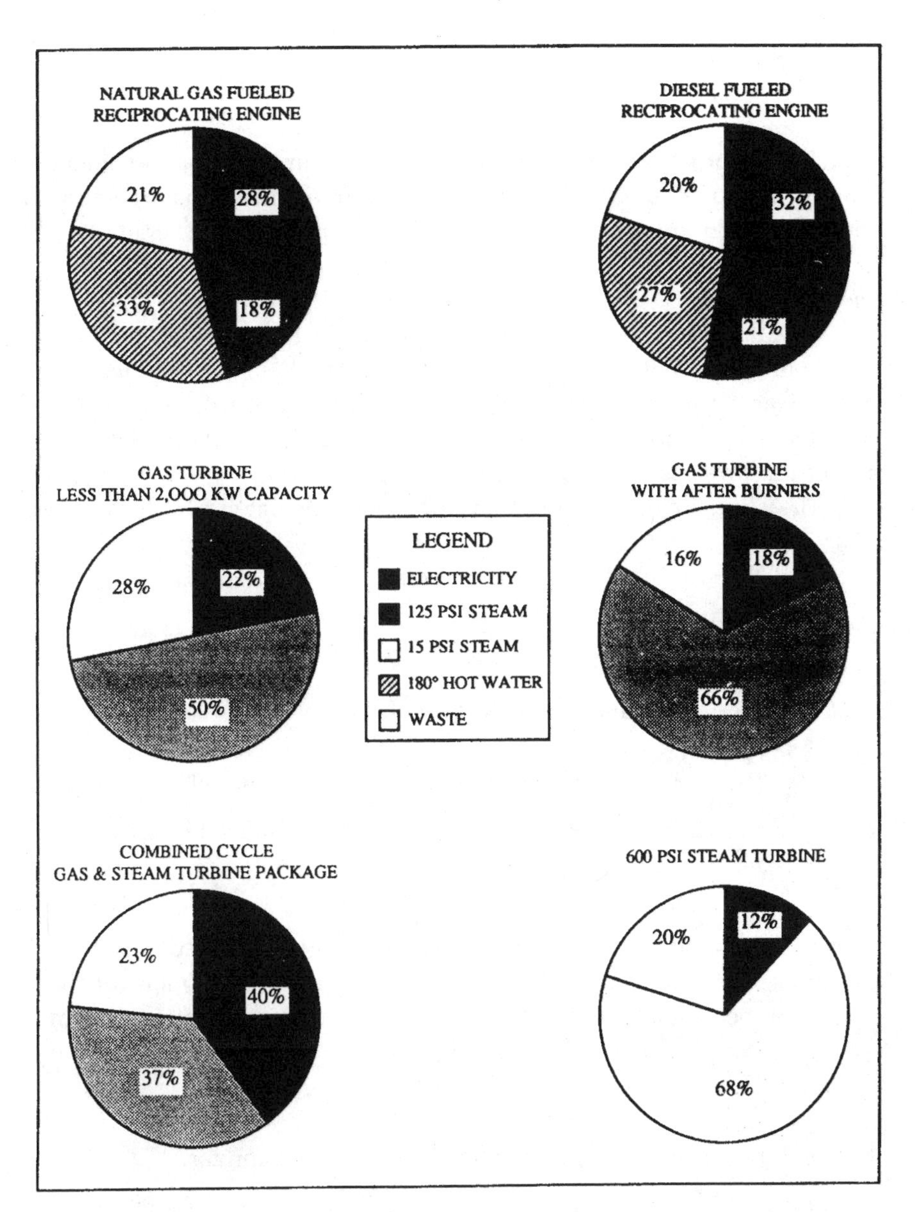

Figure 9-9. Energy production in cogeneration systems.

- The number of hours per year that you plan to operate the cogeneration system

This information can be obtained from equipment data available from the manufacturer. You will need to determine the annual operating hours based on knowledge of your facility operating characteristics.

Energy Production

The thermal and electric energy produced by your cogeneration plant will be used at your facility to offset thermal energy produced by your boiler plant and electricity purchased from your local utility. The next step is to determine the annual thermal and electric energy production of the proposed cogeneration equipment.

The annual electric energy production is calculated as follows:

$$\text{Annual Electric Energy Production} = \text{Electric Output Capacity} \times \text{Annual Operating Hours}$$

The annual thermal energy production is calculated as follows:

$$\text{Annual Thermal Energy Production} = \text{Thermal Output Capacity} \times \text{Annual Operating Hours}$$

Fuel Consumption

To determine the fuel use requirements of your proposed cogeneration plant, you will need to know the rate of fuel consumption for the equipment being evaluated here. This information can be obtained from the equipment manufacturer. Your cogeneration plant fuel consumption is calculated as follows:

$$\text{Cogeneration Fuel Consumption} = \text{Equipment Fuel Consumption Rate} \times \text{Annual Operating Hours}$$

Since you have assumed that all of the output will be used by your facility, you will now calculate the value of the cogenerated energy based on your current utility costs.

Current Utility Costs

To determine your annual average cost per kWh, first tabulate electric energy costs from the bills that you used for gathering the electric demand and consumption data. (Generally, a more detailed analysis of your electric rate structure is needed to develop an accurate unit cost of the purchased power you will offset with cogeneration. For this evaluation, we will assume the unit cost of electricity is the same as the annual average unit cost of electricity.) Next, divide your annual electric cost by your annual electric consumption.

$$\text{Annual Average Cost/kWh} = \frac{\text{Annual Electric Cost}}{\text{Annual Electric Consumption}}$$

You can use the following method to determine your thermal energy cost:

- Tabulate monthly fuel costs from the bills used to obtain the fuel consumption data earlier.

- Determine your average fuel unit cost by dividing the annual fuel cost by your annual fuel consumption.

$$\text{Fuel Unit Cost} = \frac{\text{Annual Fuel Cost}}{\text{Annual Fuel Consumption}}$$

- Determine your current cost per thermal MMBtu of steam or hot water. To do this, first multiply the fuel unit cost by the Btu/unit value from Table 9-1, then divide by your boiler's fuel-to-steam efficiency. Divide the result by 1,000,000 to convert to MMBtu units.

$$\frac{\text{Current Cost}}{\text{Thermal MMBtu}} = \frac{\text{Fuel Unit Cost} \times \text{Btu/Fuel Unit}}{\text{(Fuel-to-steam Efficiency) (1,000,000)}}$$

Annual Savings

The annual energy savings of your cogeneration plant are calculated as the savings from electric and thermal energy production less the cost of operating the plant. The annual electric energy savings can be determined from the following equation:

$$\text{Annual Electric Energy Savings} = \text{Annual Electric Energy Production} \times \text{Annual Average Cost/kWh}$$

The annual thermal energy savings is given by the following:

$$\text{Annual Thermal Energy Savings} = \text{Annual Thermal Energy Production} \times \text{Current Cost/Thermal MMBtu}$$

Operating and Maintenance Costs

Operating and maintenance costs are, to a large degree, dependent on plant operating hours and, therefore, proportional to fuel consumption. For the purpose of this evaluation, maintenance costs can be approximated as 15% of fuel costs for reciprocating engines and 7% of fuel costs for gas or steam turbines. Use a lower figure for steam turbine maintenance costs if some boiler maintenance is already included in your operating expenses.

The cost of the fuel to operate your cogeneration plant can be determined using the following equation:

$$\text{Cogeneration Fuel Cost} = \text{Cogeneration Fuel Consumption} \times \text{Fuel Unit Cost}$$

You now have all the information necessary to determine your annual savings from the operation of a cogeneration plant. This is calculated as follows:

$$\text{Annual Savings} = \text{Annual Electric Energy Savings} \times \text{Annual Thermal Energy Savings} - \text{Cogeneration Fuel Cost} - \text{Operating and Maintenance Cost}$$

At this point, it is important to note that the annual savings derived from this analysis is only a preliminary estimate of the savings potential at your facility and is based on simplifying assumptions you have made.

Economic Analysis

Initial Cost

To develop the total initial cost of the cogeneration plant you are evaluating, first refer to Table 9-2 below for approximate equipment cost ranges per kW of installed cogeneration electric capacity.

Table 9-2.
Approximate Cost Per kW of Cogeneration Plant Capacity.

Reciprocating Engine Packages (High Speed)	
Large 900 to 3,000 kW packages	$600 to $400/kW
Medium 400 to 800 kW packages	$500 to $700/kW
Small 45 to 300 kW packages	$700 to $1,200/kW
Gas Fired Turbine Engine Packages	
Large 4,000 to 10,000 kW	$800 to $1,000/kW
Medium 500 to 4,000 kW	$1,200 to $1,800/kW
*Steam Powered Turbine Engines**	
Less than 125 psig inlet turbines	$100 to $130/kW
Less than 250 psig inlet turbines	$90 to $120/kW

*Note initial costs for steam powered turbines are highly dependent on the entering and leaving pressures of the turbine. The best efficiencies are for 1,000 to 3,000 psi superheated steam. The cost of small-scale steam boilers in this range is prohibitive, and the cost of condensing steam turbines is not listed since considerable extra equipment is required. The costs as given do not include the cost of heat recovery equipment.

The cost figures in Table 9-2 do not include the cost of new buildings to house the equipment, nor do they include allowances for the electrical wiring necessary to connect your facility to your utilities. If additional space must be constructed or major work is required to connect the utilities, these expenses must be added to the equipment cost estimates to determine your total initial cost. Also, these costs were assembled at an earlier time and may need to be revalidated for your current analysis.

Investment Analysis

The final decision to build a cogeneration plant is usually based on investment analysis. Broadly defined, this is an evaluation of costs versus savings. The costs for cogeneration include the initial capital cost for the equipment or the cost of operating and maintaining that equipment, costs; fuel costs, finance charges, tax liabilities, and other system costs, that may be specific to your application. Savings include offset electrical power costs; offset fuel costs; revenues from excess power sales, if any; tax benefits; and any other applicable savings that are offset by the cogeneration plant, such as planned replacement of equipment.

A number of economic analysis techniques are available for comparing investment alternatives. The simple payback period is the most commonly used and the least complex, and is usually adequate for a go, no-go decision at this stage of project development. It is the period of time required to recover the initial investment cost through savings associated with the project. Simple payback period is calculated as follows:

$$\text{Simple Payback Period} = \frac{\text{Initial Cost, \$}}{\text{Annual Savings, \$/Yr}}$$

The simple payback period will give you an indication of the attractiveness of cogeneration at your facility and will help you decide whether a full-scale feasibility study is warranted.

Figure 9-10 is a summary of the information used in this evaluation of cogeneration at your facility.

It is important to note that many of the variables used in this evaluation are assumed values based on your knowledge of the operating characteristics of your facility and the simplifying assumptions made for this preliminary analysis. In a full-scale cogeneration feasibility study, you will want to evaluate several equipment types and sizes in conjunction with different operating strategies for each system configuration. This type of evaluation should be performed for you by an experienced consulting engineer.

UTILITY DATA ANALYSIS

Electric Energy Consumption		________	kWh/year
Annual Peak Electric Demand		________	kW
Electric Energy Cost	$	________	/year
Thermal Energy Consumption		________	Btu/year
Annual Peak Thermal Demand		________	Btu/hour
Annual Fuel Cost	$	________	/year
Annual Electric Load Factor		________	%
Annual Thermal Load Factor		________	%
Thermal/Electric Load Ratio			

ENERGY ANALYSIS

Cogeneration Plant Capacity			
Electric Output		________	kW
Thermal Output		________	Btu/hour
Annual Operating Hours		________	hours/year
Annual Electric Energy Production		________	kWh/year
Annual Thermal Energy Production		________	Btu/year
Cogeneration Plant Fuel Consumption		________	Btu/year
Current Utility Costs			
Average Annual Electric Cost	$	________	/kWh
Average Thermal Energy Cost	$	________	/MMBtu
Annual Savings			
Electric Energy Savings	$	________	/year
Thermal Energy Savings	$	________	/year
Cogeneration Fuel Cost	$	________	/year
Operating and Maintenance Costs	$	________	/year

ECONOMIC ANALYSIS

Annual Savings	$	________	/year
Initial Cogeneration Equipment Cost	$	________	
Simple Payback Period		________	years

Figure 9-10. Cogeneration feasibility evaluation summary.

10

Establishing a Maintenance Program For Plant Efficiency And Energy Savings

GOOD MAINTENANCE SAVES $

Energy losses due to leaks, uninsulated lines, dirt buildup, inoperable furnace controls, and other poor maintenance practices is directly translated into additional energy costs. Good maintenance saves in the plant's yearly operating costs.

In this chapter, you will see the results of a survey on maintenance effectiveness, look at ways to turn around the maintenance program, and learn to apply a preventive maintenance program to energy conservation.

WHAT IS THE EFFECTIVENESS OF MOST MAINTENANCE PROGRAMS?

Poor maintenance can be readily translated into inefficient operation of systems. Inefficient operation of systems spells energy waste. The problem is magnified by a survey of 1,000 plant engineers and senior maintenance supervisors that indicates the sad state of maintenance management. Findings of the survey indicate:

- Maintenance is not receiving management's attention

- Planning procedures are performed by:
  - Superintendents—35%
  - Maintenance foremen—35%
  - Planners, schedulers—30%

- Few maintenance departments issue regular reports to management

- Few use work order (maintenance request) system

- Few maintenance departments analyze the cause of equipment backlog

- Most companies surveyed had started preventive maintenance programs, but most were dissatisfied with their progress

- Most companies had poor control over material and spare parts account

The aspect of specific interest to energy conservation is that preventive maintenance is not occurring satisfactorily. Only 55 percent of the maintenance departments have a tickler file or some other filing system for automatic work order generation for filter changes and other periodic preventive maintenance practices. Another factor which affects the energy conservation program is that most managers have little or no control over maintenance work backlog. This means that maintenance work tends to happen rather than being planned.

HOW TO TURN AROUND THE MAINTENANCE PROGRAM

Management Attention

The results of this survey indicate that the sad state of maintenance has been perpetuated because management has paid little attention to it. If management wants to conserve energy, it must give maintenance top priority. Management has historically devoted much of its efforts to improving manufacturing productivity, and in the reporting, measuring and analyzing of manufacturing costs. Similar efforts to improve maintenance productivity are needed. Plant engineers need to evaluate their present practices to determine the status of their maintenance program. Periodic reports, quantitatively stating key maintenance performance indicators, need to be issued to management.

Work Order Request

Effective maintenance insures that the activity is being controlled. A work order request system needs to be established and used. Work orders serve as an historical record of repairs and alterations made to key equipment. Since the work order is used to authorize maintenance work, it is the prime requisite for any maintenance control system.

Maintenance Planning

Planning of maintenance activity needs to come under the jurisdiction of a maintenance manager. The maintenance manager must have the same status within the organization as manager of production. In planning maintenance work, the following needs to be identified:

Scope of the job

Location of the job

Priority

Cause (why the work must be done)

Methods to be used

Material requirements

Manpower requirements (number and crafts).

In addition, the following should be implemented:

- Labor time standards should be established to cover recurring or highly repetitive tasks.
- Reliable estimates should be established which cover nonrecurring (less than once a month) maintenance work.
- Procedures for issuing maintenance reports should be established.
- Reports should include productivity of labor force.

Applying a Preventative Maintenance Program to Energy Conservation

In the beginning of this chapter, we discussed first the sad state of maintenance and indicated some ways to turn this situation around. *Only* then can maintenance for energy conservation be considered.

The first step is to identify energy wastes that can be corrected by maintenance operations. Table 10-1 indicates an Energy Savings Survey which should be made. Let's look at some specific areas of maintenance as it applies to energy conservation.

STOP LEAKS AND SAVE

Steam Leaks

Leaks not only waste energy; they cause excessive noise as well. A case in point was a noise citation a plant received for steam discharges. The plant solved its noise problem by using steam traps and, in addition, it saved energy which paid for the installation.

Figure 10-1 illustrates the annual heat loss from steam leaks.

SIM 10-1

A maintenance survey of the plant indicates that steam leaks exist from 100 lines operating at 100 psig. The average size of the leak is estimated to be 1/8 in. diameter. The value of steam in the plant is $\$4/10^6$ Btu. What is the annual savings from repairing the leaks?

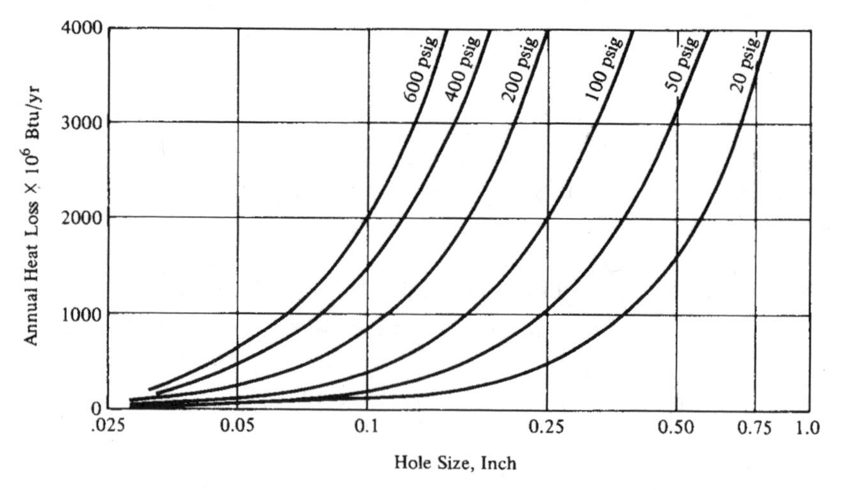

Figure 10-1. Annual heat loss from steam leaks.

Table 10-1. Energy saving survey checklist.

Department ______________________
Date ______________________
Supervised By ______________________

Location #1	Location #2
Fuel, Gas or Oil Leaks	
Steam Leaks	
Compressed Air Leaks	
Condensate Leaks	
Water Leaks	
Damaged or Lacking Insulation	
Leaks of, or Excess HVAC	
Burners Out of Adjustment	
Steam Trap Operation Note: Each trap is to be tagged with date of inspection	
Cleanliness of Heating Surfaces Such as Coolers, Exchangers, etc.	
Hot Spots on Furnaces to Determine Deterioration of Lining	
Bad Bearings, Gear Drives, Pumps, Motors, etc.	
Dirty Motors	
Worn Belts	
Proper Viscosity of the Lubricative Oils for Large Electric Drives, Hydraulic Pumps (Proper Viscosity Minimizes Pump Drive Slippage)	
Cleanliness of Lamps	

ANSWER
From Figure 10-1, the steam loss per line is 540 × \$6/10^6 Btu. The annual savings is then

$$540 \times 100 \times 6 = \$324{,}000 \text{ per year.}$$

Steam leaks can be detected visually or by using acoustic or temperature probes. Leaks should also be checked at valves, heaters, and other equipment. Leaking or stuck traps, stuck by-pass valves discharging to the sewer and condensate systems should also be checked. A program should be established to regularly inspect for leaks, to shut off steam to equipment taken out of service, to repair steam leaks promptly, and to route piping such that leaks that develop are visible.

Leaks In Combustible Gas Lines

As with steam, leaks of combustible gases such as natural gas, methane, butane, propane, hydrogen, etc., are a direct waste of valuable energy, and causes the hazard of fire or explosions. Tables 10-2 and 10-3 summarize losses for various combustible gas piping systems.

SIM 10-2

A natural gas leak is detected in above ground piping. The natural gas pressure is 100 psig and the leak is from a 1/8 in. diameter hole. What is the annual loss, considering the value of natural gas at \$6.00 per million Btu?

ANSWER
From Table 10-3, leakage is 13,500 × 10^3 SCF. Since the heating value of natural gas is 1000 Btu/CF, then the heating value of the leak is:

$$13{,}500 \times 10^6 \text{ Btu}$$

The corresponding yearly energy waste is \$81,000.

Leaks in Compressed Air Lines

Leaks in compressed air lines waste energy and cost yearly operating dollars. An air compressor supplies the plant air; thus, any air leak can be translated into wasting air compressor horsepower. Table 10-4 illustrates losses for various hole diameter openings.

Table 10-2. Natural gas loss per year from leaks in *underground* pipelines at various pressures. (KSCF).

Corrosion Hole Diameter, in	Line Pressure (psig) 0.25	5	25	60	100	300	500
1/64	1	4	10	20	30	80	140
1/32	2	6	20	35	60	160	250
1/16	10	40	100	200	320	900	1,500
1/8	50	200	600	1,200	1,800	5,000	8,200
1/4	250	1,200	3,300	6,500	10,300	28,300	46,500
1/2	1,400	6,600	18,800	37,300	58,000	156,500	263,000

Adapted from *NBS Handbook 115.*

Table 10-3. Natural gas loss per year from leaks in *above ground* pipelines at various pressures.

Corrosion Hole Diameter, in	Line Pressure (psig) 0.25	5	25	60	100	300	500
1/64	5	26	69	136	212	581	953
1/32	21	102	277	544	846	2,330	3,810
1/16	85	409	1,110	2,180	3,390	9,300	15,300
1/8	341	1,640	4,430	8,700	13,500	37,200	61,000
1/4	1,360	6,540	17,700	34,800	54,200	149,000	244,000
1/2	5,450	26,200	70,900	139,000	217,000	595,000	977,000

Adapted from *NBS Handbook 115.*

SIM 10-3
A maintenance inspection of the plant compressed air system revealed the following leaks in the 70 psig distribution piping system:

Number of Leaks	Estimated Diameter In
5	1/4
10	1/8
10	1/16

What is the yearly energy loss, assuming an energy cost of $.060 per kWh?

ANSWER
From Table 10-4,

Number of Leaks	Estimated Diameter	Cost of Power Wasted $/yr
5	1/4	17,550
10	1/8	8,882
10	1/16	2,199
Total Energy Loss		$ 28,631

Air leaks occur at fittings, valves, air hoses, etc. A common way of detecting leaks is by swabbing soapy water around the joints. Blowing bubbles will indicate air leaks.

From SIMs 10-1, 10-2, and 10-3, the total energy waste is $433,631.00. *Only* when maintenance personnel *understand* that stopping leaks saves a significant amount on utility costs will the program become effective.

PROPERLY OPERATING STEAM TRAPS SAVE ENERGY

Many existing plants remove condensate from steam lines by using tubing connected to the steam piping and discharged directly to the atmosphere. Figure 10-1 shows the loss of steam through a hole. Thus, all steam leaks, whether purposeful or not, should be avoided. A steam trap permits the passage of condensate, air, and noncondensible gas from the steam piping and equipment while preventing (trapping)

Table 10-4. Annual heat loss from compressed air leaks.

Hole Diameter, in	Free Air Wasted (a), cu ft per year, by a Leak of Air at:	Fuel Wasted (b), MBtu/yr	Cost of Power Wasted (c), $/yr at Unit Power Cost of		
	100 psig		$0.020/kWh	$0.040/kWh	$0.060/kWh
3/8	79,900,000	2190	4370.00	8740	13110
1/4	35,500,000	972	1940.00	3880	5820
1/8	8,880,000	243	486.00	972	1458
1/16	2,220,000	60.6	121.00	242	363
1/32	553,000	15.1	30.30	60.6	90.90
	70 psig		$0.020/kWh	$0.040/kWh	$0.060/kWh
3/8	59,100,000	1320	2650.00	5300	7950
1/4	26,200,000	587	1170.00	2340	3510
1/8	6,560,000	147	294.00	588	882
1/16	1,640,000	36.6	73.30	146.60	219.90
1/32	410,000	9.2	18.40	36.80	55.20

(a) Based on nozzle coefficient of 0.65
(b) Based on 10,000 Btu fuel/kWh
(c) Based on 22 brake horsepower per 100 cu ft free air per min for 100 psig air and 18 brake horsepower per 100 cu ft free air per min for 70 psig air

the loss of steam. Air and noncondensible gases are undesirable in steam systems since they act as insulating blankets and reduce the equipment efficiency. Therefore, the use of steam traps is a must. Steam traps should be properly sized and maintained. An oversized or poorly functioning steam trap causes steam to be wasted.

Traps should be inspected once a week to determine the following:

1. Is the trap removing all of the condensate?

2. Is tight shut-off occurring after operation?

3. Is the frequency discharge in an acceptable range? (Too frequent discharge indicates possible under-capacity, while too infrequent discharge indicates possible over-capacity and inefficiency.)

Check All Steam Traps

Why? To save steam . . . and to prepare your tracing and heating systems for winter operation.

By Whom? Plant Operators, Mechanics, and Fitters from the Field Service Units.

Frequency of Testing: Once/shift by plant operators not unreasonable, daily, weekly, etc.

Safety: Check all steam and condensate lines for insulation. When testing steam traps protect yourself from thermal burns.

Testing: Bucket Traps – Strong No. 141 or Armstrong 811

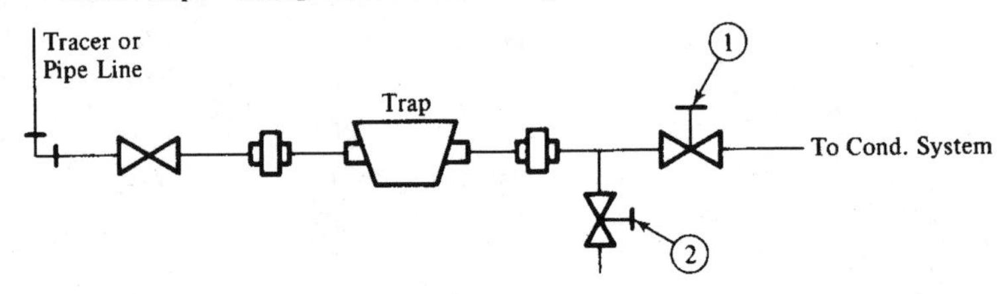

Testing: Thermodynamic or Disc Trap. "Sarco" 52, Yarway 29 Series or strong DD-70

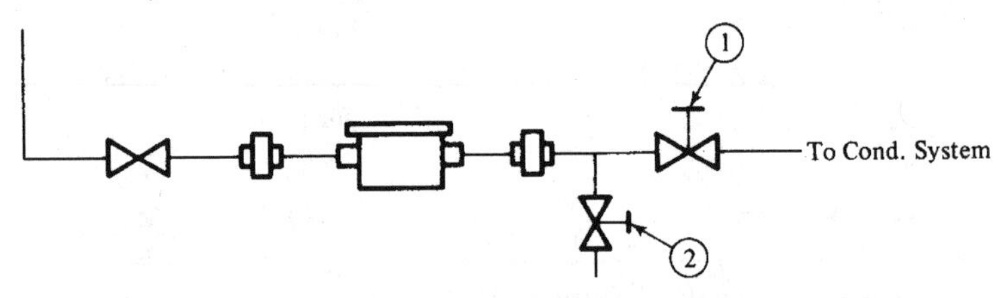

Close valve No. 1 – Open valve No. 2 – Trap should cycle and close. Should trap fail to operate, close valve No. 2 – open valve No. 1. Red Tag for further check. Should trap blow continuously, it could be worn out, undersized, on a cold system, or process start-up.

Blowing Traps Waste Steam!!

Figure 10-2. Check all steam traps.
***Source:* Federal Power Commission, position paper No. 17.**

SIM 10-4

A plant maintenance inspection indicated 10-100 psig steam traps were stuck open. The orifice in the trap is 1/4 inch. Compute the yearly energy savings by repairing the trap, assuming the value of steam at \$6.00 per 10^6 Btu.

ANSWER

From Figure 10-1, the steam loss is 2100×10^6 Btu/yr per trap.

Thus: energy savings = $2100 \times 10 \times 6.00$ = \$126,000 per year.

Reduce Energy Wastes By Insulating Bare Steam Lines

Inspecting an existing plant will reveal that lack of insulation and damaged insulation are common on steam piping. In some instances, the fine was originally insulated, but due to piping repairs or rerouting, the insulation was discarded. From Figure 10-3, the energy loss due to bare steam lines is apparent.

SIM 10-5

A maintenance inspection indicates uninsulated steam lines as follows:

Length	Size	Pressure
300	8″	100 psig
500	6″	50 psig

What is the yearly energy savings resulting from insulating the bare steam line? Assume a steam cost of \$6.00 per 16^6 Btu.

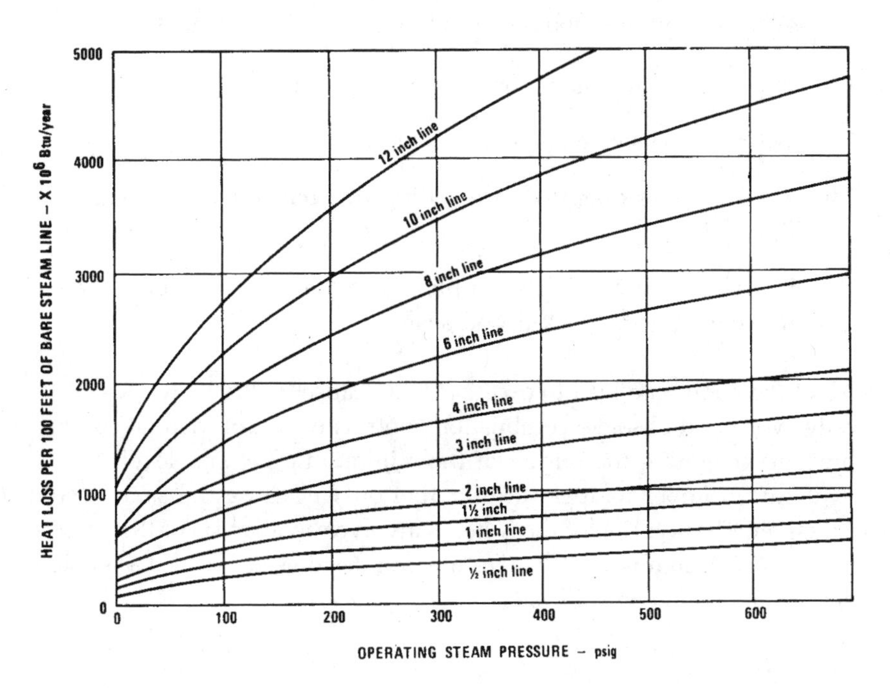

Figure 10-3. Heat loss for bare steam lines. (Adapted from *NBS Handbook 115*.)

ANSWER
From Figure 10-3

Length	Size	Loss	Energy Savings By Insulating Pipes
300	8″	5700×10^6 Btu	34,200
500	6″	5500×10^6 Btu	33,000
		Total yearly savings	$ 67,200

Don't Overlook Steam Tracing

Steam tracing must be maintained as well as turned off when not required, such as in summer periods. Either automatic controls or improved operator attention is required in order to minimize steam waste.

EXCESS AIR CONSIDERATIONS

What is common about the following?

1. Improper temperature for fuel oil atomization.
2. Worn and obsolete combustion controls.
3. The burner is operated at a setting different than the design setting.
4. Shortage of natural gas supplies forces a burner to be operated manually on the back-up oil system.

All of the above can lead to excess air and cause fuel to be wasted. Basically, worn and obsolete combustion controls or use of manual controls prevents operating the burner at the minimum excess air levels.

An improper temperature for fuel oil will cause the same result: excess air. The solution of these items involves an analysis of the present operations, changing the temperature and design settings and a possible new combustion control system.

SIM 10-6

An existing natural gas burner is operating at a stack temperature of 800°F with a 5.5% oxygen content. A maintenance check indicates that the controls are worn and that excess air is difficult to control. What annual (8000 hr/yr) savings can be achieved by installing a $35,000 oxygen analyzer and control system to reduce the oxygen content to 2 percent? The flue gas flow is 180,000 scfh (standard cubic feet per hour). Assume a 10 year life and a rate of return before taxes of 15%. The price of natural gas is $6.00 per million Btu.

ANSWER

Using Figure 5-8, the operating conditions are drawn. Figure 5-8 illustrates a 4% fuel savings.

From Table 1-4, the heating value of natural gas is 1,000 Btu/cf.

$$q = 8000 \text{ hr/yr} \times 180{,}000 \text{ cf/hr} \times 1000 \text{ Btu/cf} \times 0.04$$

$$57{,}600 \times 10^6 \text{ Btu per year}$$

Thus, the annual savings are:

$$57{,}600 \times \$6.00 = \$345{,}600$$

$$\text{Annual owning cost} = \$35{,}000 \times CR \quad = \$35{,}000 \times 0.2 = \$7000.00$$

$$i = 15\%$$

$$n = 10$$

Annual savings before taxes = $338,600

DIRT AND LAMP LUMEN DEPRECIATION CAN REDUCE LIGHTING LEVELS BY 50%

When lighting systems are designed, a light loss factor is used to compensate for lamp lumen depreciation based on lamp burning hours and dirt buildup on lamps and reflecting surfaces. This means that lighting levels will initially be higher than needed and decrease to levels below requirements based on the maintenance program.

Figure 10-4 illustrates the effect of dirt on a reflector. Light must first pass through the layer of dirt to reach the reflector and then pass through again to escape from the luminaire to the working plane. The object of the lighting maintenance program is to use the light available to get the maximum output for the system. This means that lamps should be replaced *prior* to burning out. A group relamp program at 70-80% of lamp life and a lighting cleaning program should be initiated.

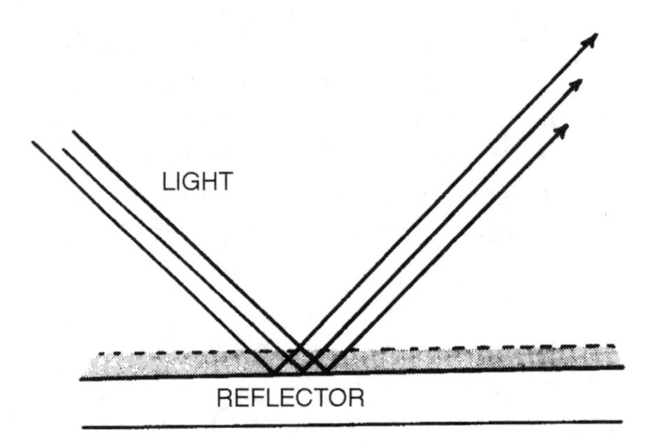

Figure 10-4. The effect of dirt on a reflector.

The results will be that luminaires can be removed from service without lowering the overall lighting level.

SUMMARY

How Much Is Saved By A Good Maintenance Program?

From the discussion in this chapter, it is evident that *good maintenance* spells *energy savings*. Savings in excess of one million dollars per year is common if the maintenance program outlined in this chapter is followed. A good maintenance program provides for a quiet, safer, energy efficient plant where production downtime is minimized. Thus, the cost for the program will more than pay for itself.

11

Managing an Effective Energy Conservation Program

ORGANIZING FOR ENERGY CONSERVATION

By now it should be clear that energy management affects almost every major activity of a plant. It is involved in:

Electrical Engineering
Control Systems Engineering
Utility Engineering
Piping Design
Mechanical Engineering
Chemical Engineering
Heating, Ventilation, and Air Conditioning Engineering
Building Design
Environmental Engineering
Operations
Maintenance
Accounting and Financial Management

Each plant has assigned individuals who are responsible for one or more of the above functions. The problem facing management is how to organize the energy conservation activity so that all functions are moving in a common direction. The situation becomes more complex when several plants or outside consultants are involved. Direction and coordination for the program need to be provided. In most organizations, the function of the energy conservation coordinator/committee emerges.

In this chapter, you will see how to establish energy conservation goals, learn how to set priorities and how to improve communications and coordination, see how to use the critical path method to schedule energy conservation activities, and discover how to encourage the creative process.

TOP MANAGEMENT COMMITMENT

Energy conservation requires top management commitment. Table 11-1 indicates a checklist for top management. Formulating a committee and assigning a coordinator does not solve the energy conservation problem, but it does make individuals responsible for energy conservation. An individual does not become committed to a goal unless he believes in the cause.

Table 11-1. Checklist for top management.

A. Inform line supervisors of:
   1. The economic reasons for the need to conserve energy
   2. Their responsibility for implementing energy saving actions in the areas of their accountability

B. Establish a committee having the responsibility for formulating and conducting an energy conservation program and consisting of:
   1. Representatives from each department in the plant
   2. A coordinator appointed by and reporting to management
   Note: In smaller organizations, the manager and his staff may conduct energy conservation activities as part of their management duties.

C. Provide the committee with guidelines as to what is expected of them:
   1. Plan and participate in energy-saving surveys
   2. Develop uniform record keeping, reporting, and energy accounting
   3. Research and develop ideas on ways to save energy
   4. Communicate these ideas and suggestions
   5. Suggest tough, but achievable, goals for energy saving

6. Develop ideas and plans for enlisting employee support and participation
7. Plan and conduct a continuing program of activities to stimulate interest in energy conservation efforts

D. Set goals in energy saving:
   1. A preliminary goal at the start of the program
   2. Later, a revised goal based on savings potential estimated from results of surveys

E. Employ external assistance in surveying the plant and making recommendations, if necessary

F. Communicate periodically to employees regarding management's emphasis on energy conservation action and report on progress

---

Adapted from *NBS Handbook 115.*

In the face of frozen fuel allotments, shortages of raw materials and supplies, and the increasing fuel costs, commitment develops quickly out of the necessity of "survival."

WHAT TO CONSIDER WHEN ESTABLISHING ENERGY CONSERVATION OBJECTIVES

Energy conservation requirements must be translated into clear goals. Since funds are limited, the financial goals as well as the energy goals need to be defined. Typical energy conservation goals are illustrated in Table 11-2.

How To Set Priorities of Energy Conservation Projects

When setting priorities, it is necessary to make sure that competing energy conservation projects are evaluated on the same basis. This means that all life cycle cost analysis should be based on the same fuel costs and the same assumptions of escalation. In addition, the financial factors, such as depreciation method and economic life, need to be evaluated the same way. Table 11-3 summarizes the information to be furnished *prior* to the start of the life cycle cost analysis.

Competing projects can be ranked in order of the best rate of returns on investment, best payout period, or best ratio of Btu/year savings to capital cost.

Table 11-2. Typical energy conservation goals.

1. Overall energy reduction goals
   (a) Reduce yearly electrical bills by ___%.
   (b) Reduce steam usage by ___%.
   (c) Reduce natural gas usage by ___%.
   (d) Reduce fuel oil usage by ___%.
   (e) Reduce compressed air usage by ___%.
2. Return-on-investment goals for individual projects.
   (a) Minimum rate of return on investment before taxes is ___________.
   (b) Minimum payout period is ___________.
   (c) Minimum ratio of $\dfrac{\text{Btu/year savings}}{\text{capital cost}}$ is ___________
   (d) Minimum rate of return on investment after taxes is ___________.

Other factors enter into the decision making process. For example, if natural gas supplies are curtailed, projects reducing natural gas supplies will be given priority.

The Vital Elements—Communication and Coordination

As obvious as it may seem, a company which fosters communication and coordination of energy conservation goals and information has a good chance of significantly reducing energy consumption.

A central energy conservation committee or plant conservation coordinator can only do so much. It is up to the individuals who are familiar with the detailed operations to make the greatest contributions. They need to be informed and to understand the energy conservation goals and the potential for saving yearly operating expenses. They need to know that energy conservation directly affects the profitability of the company.

Another aspect involved in coordination of energy conservation information is the flow of information generated by the project team assigned to a specific plant expansion. The project team is comprised

Table 11-3. Information required to set energy projects on the same base.

Fuel	Cost At Present	Estimated Cost Escalation Per Year	Energy Equivalent
1. *Energy equivalents and costs for plant utilities.*			
Natural gas	$_____/1000 ft^3	$_____/1000 ft^3	_____Btu/ft^3
Fuel oil	$_____/gal	$_____/gal	_____Btu/gal
Coal	$_____/ton	$_____/ton	_____Btu/lb
Electric power	$_____/kWh	$_____/kWh	_____Btu/kWh
Steam			
_____ psig	$_____/1000 lb	$_____/1000 lb	_____Btu/1000 lb
_____ psig	$_____/1000 lb	$_____/1000 lb	_____Btu/1000 lb
_____ psig	$_____/1000 lb	$_____/1000 lb	_____Btu/1000 lb
Compressed air	$_____/1000 ft^3	$_____/1000 ft^3	_____Btu/1000 ft^3
Water	$_____/1000 lb	$_____/1000 lb	_____Btu/1000 lb
Boiler make-up water	$_____/1000 lb	$_____/1000 lb	_____Btu/1000 lb

2. *Life Cycle Costing Equivalents*

A) After tax computations required ____________

Depreciation method ____________

Income tax bracket ____________

Minimum rate of return ____________

Economic life ____________

Tax credit ____________

Method of life cycle costing (annual cost method, payout period, etc.) ____________

of a project engineer and supporting design disciplines. An energy conservation coordinator assigned to the project insures that energy conservation goals are being implemented for the project. Figure 11-1 illustrates the coordination of energy conservation data.

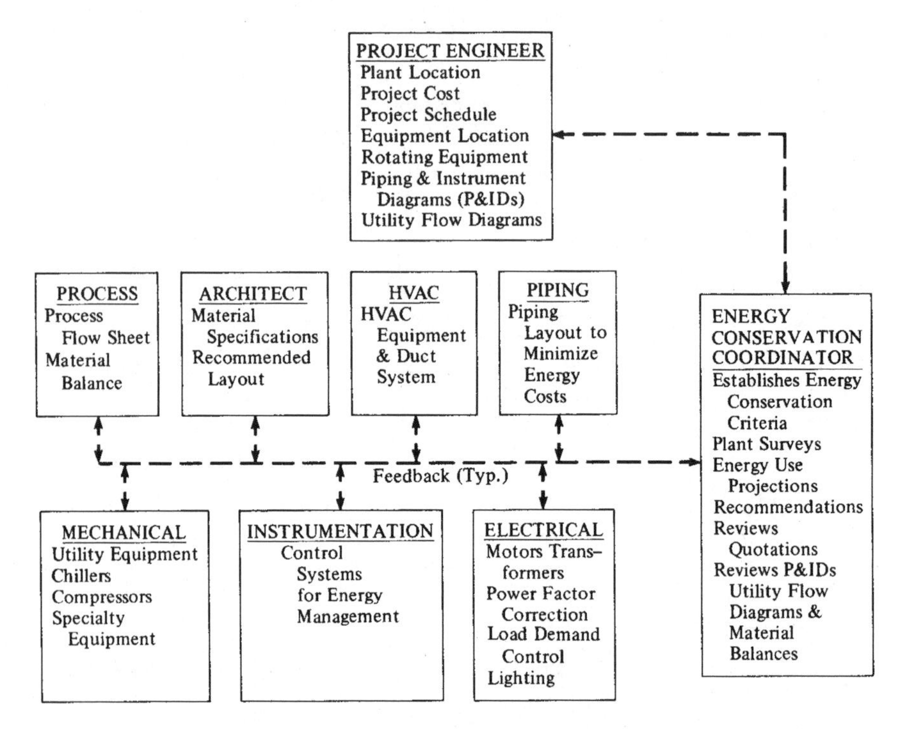

Figure 11-1. Coordination of energy conservation information.

USING THE CRITICAL PATH SCHEDULE OF ENERGY CONSERVATION ACTIVITIES

The critical path schedule for energy conservation takes into account that certain activities should take place prior to others. For example, before economic alternatives (life cycle cost analysis) is made, the objectives, financial data, utility rates, and plant survey need to have been completed. This schedule contains activities which affect plant expansions. For example, prior to purchasing equipment, energy consumption per item needs to be evaluated. Traditionally, bids received by competitive equipment manufacturers were compared on a first cost basis only. Today bids should be evaluated on a life cycle costing basis.

Key items to evaluate are:

- Base price
- Delivery

- Technical specifications
- Cost for pollution compliance
- Yearly energy and maintenance cost

This means that during the quotation phase, equipment such as fans, air conditioners, pumps, compressors, etc., need to be evaluated with respect to energy consumption. This simple step will point up inefficient equipment: the penalty of operation prior to purchase.

ELECTRICAL SCHEDULING OF PLANT ACTIVITIES

As indicated in Chapter 4, the load demand controller is used to reduce peak electrical demand.

Scheduling of plant operations requires more than an electrical black box. As an example, the startup of a plant expansion can involve several start-up engineers working on several systems. A peak load demand can occur by testing several large motors at the same time. A little coordination can save substantially on the energy demand charge.

After the plant is in operation, the electrical energy curves should be analyzed. Careful analysis can indicate ways in which peak demands can be lowered by rescheduling plant operations.

AN EFFECTIVE MAINTENANCE PROGRAM

Plant maintenance cannot be overemphasized in any energy conservation program. As indicated in Chapter 10, management needs to take an active role.

CONTINUOUS CONSERVATION MONITORING

As discussed in Chapter 1, an energy conservation program must be continuously monitored in order to be controlled. This means that each operator needs to manage his own fuel usage. Adequate instrumentation is essential to the monitoring process.

Operator accountability can exist only if all utilities are metered. In planning what point in the process will be metered, consideration should be given to how the process will be operated and by whom. As an example, the fuel oil tank would come under the jurisdiction of the boiler plant operator. Any fuel take-offs to the process operator should be separately metered. This will enable each operator to control his usage, independent of other operations.

ARE OUTSIDE CONSULTANTS AND CONTRACTORS ENCOURAGED TO SAVE ENERGY BY DESIGN?

The type of contract may well influence the type of energy savings which will be realized. As an example, a competitive "lump sum" contract encourages the consultant to achieve the lowest first cost and ignore additional expenditures which will minimize future operating costs. This type of contract will not achieve the energy conservation goals unless the owner specifies the energy conservation requirements during the bid phase. Thus, the owner must be careful not to penalize creative energy conservation design during the conceptual phase.

ENCOURAGING THE CREATIVE PROCESS

Energy conservation engineering relies on creatively applying existing technology in new ways. Many examples were illustrated in this text, but this in no way can be considered the entire gambit. "Why" questioning must be promoted. A team of engineers representing several disciplines should meet periodically to "brainstorm" energy conservation concepts. "Do it the same way as the last project, use rules of thumb, or experience tells us..." should be replaced with "*Why*" questioning.

Examples are:

1. Why can't vacuum pumps be used in this process instead of steam vacuum jets?

2. Can conventional warehouse heaters be replaced with infrared heaters?

3. Can air coolers be used instead of water coolers?

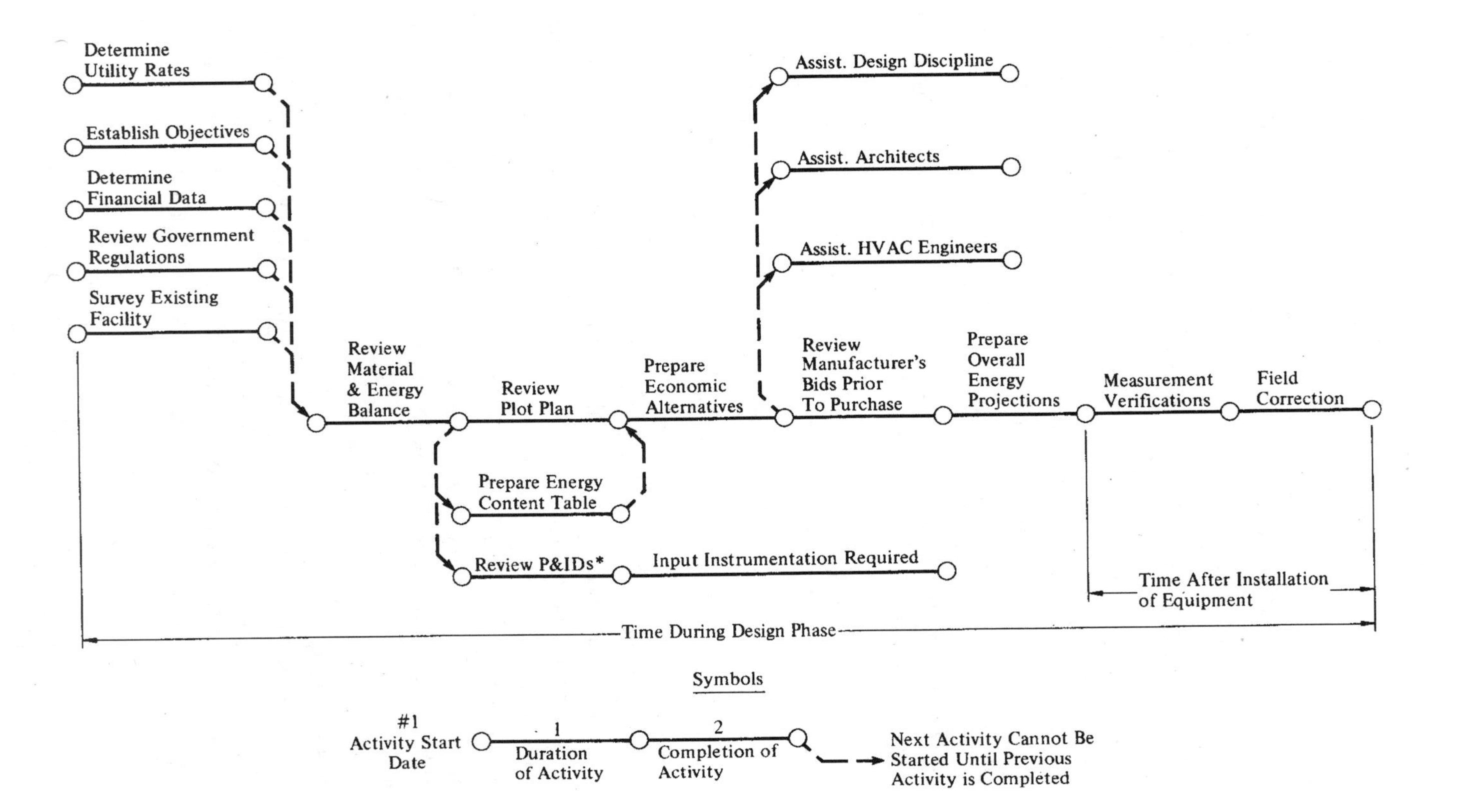

Figure 11-2. Energy conservation engineering schedule.

4. Can no-vent equipment, such as vertical motors with special packing, be used to permit process steam to be returned to the boiler without steam traps?

5. Can loads be re-adjusted so that equipment runs at optimal levels of efficiency?

6. Can short runs on thermal process equipment be avoided?

7. Can spot cooling for personal cooling be used instead of area cooling?

8. Can waste heat be recovered by means of systems employing organic working fluids?

9. Can smooth coated pipe be used to reduce friction loss, thereby reducing horsepower?

10. Can pumps be eliminated by locating equipment to take advantage of gravity?

ENERGY EMERGENCY AND CONTINGENCY PLANNING

Energy managers need to deal with many unknowns. Experience of the last ten years indicates coal strikes and fuel allocations do occur. How to plan for emergencies such as these is an important aspect of the energy manager's role.

In November of 1979, President Carter signed Public Law 96-102, Emergency Energy Conservation Act, giving the President the authority to establish energy demand targets for each state in case of a national emergency.

Risk Assessment

In order to establish the contingency plan, the energy manager needs to establish the risk associated with various scenarios.

There is a cost to business for doing nothing in the event an emergency does occur and no plan was put into effect. This cost is computed according to Equation 11-1.

$$C = P \times C_1 \tag{11-1}$$

where

C = cost as result of emergency where no contingency plans are in effect

P = the probability or likelihood the emergency will occur

C_1 = the loss in dollars as a result of an emergency where no emergency plans are in effect.

Thus risk assessment can help the energy manager determine the effect, money, and resources required to meet various emergency scenarios.

An example of a simple contingency plan would be in the justification of a backup fuel system. The risk assessment would indicate the probable cost of losing the primary system and the expected duration of the loss. The key is to develop the plan prior to the emergency occurring.

CONCLUSION

The energy conservation program, when properly managed, will pay for itself. Engineering "rules of thumb" are no longer sufficient because of increases in the cost of energy and shortages of raw materials. The decision making process is based on economics and local, state, and federal regulations. The guidelines and practices presented in this text will prove to save energy without decreasing the function of the system. The energy manager's role must be understood by the public and the government.

The answer to energy management is *not* to decrease the quality of life by living in darkness, but rather to apply the technologies available to increase the efficiency of the systems. The challenge facing the energy engineer/manager is great and the stakes are high.

12

Electric Motors

Electric motors are the rotating machines that drive today's industrial manufacturing. A typical manufacturing facility may operate from hundreds to thousands of motors. The U.S. Dept. of Energy has reported that electric motors use over half of the electricity consumed in the United States in a given year.

Many plant managers do not maintain a motor inventory and only examine a motor once it has failed. This reflects positively on the reliability of electric motors. Unfortunately, this approach to motor management can greatly affect the efficiency of motor operations in a facility.

MOTOR TYPES

There are three basic motor designs found in most industrial facilities. The first type is the AC synchronous motor. This design offers a fixed speed, rotating shaft that is useful for mechanical work. It also provides power factor correction capability.

AC Synchronous Motors

The AC synchronous motor is fed from two sources. The outer stationary shell of the motor (called the stator) contains coils of wire that are fed from three phase electric power. When energized, they create a rotating magnetic field within the motor that can induce the inner shaft of the motor (known as the rotor) to begin spinning. Attached to the rotor, are magnetic coils that are fed from a separate DC source which creates a fixed magnetic field within the machine that interacts with the magnetic field created by the stator.

During start-up, the stator circuit on a synchronous motor, which is also called the armature circuit, is first energized while the rotor circuit is not engaged. In fact, it is actually short-circuited. The circuit breaker feeding the armature circuit is often labeled as the "start" breaker. This allows the motor to begin spinning as an induction motor while it slowly gains speed. Once it reaches a steady-state speed, the field circuit is engaged which causes the motor to lock into a fixed speed that is a function of the number of magnetic poles in the motor and the electrical frequency being applied to the stator. The breaker on this circuit is often labeled the "run" breaker. The fixed operational speed is known as synchronous speed and is calculated as

$$Nsynch = (120 * f_e)/\#\ of\ magnetic\ poles$$

f_e is the electrical frequency

Both the electrical frequency and the number of poles can be found on the motor nameplate.

Synchronous motors found in industrial facilities are often fairly large and typically only account for one percent of the motors found on site. They are similar in construction to induction motors, but they are far more expensive and require more maintenance. The reason for this is that to energize the field circuit, one must supply energy from an external DC source to a rotating shaft. To accomplish this, a series of slip rings are installed on the shaft and then they form a connection by resting on carbon brushes which connect back to the external DC source. The extra copper and wiring adds weight and cost while the use of brushes in frictional connections creates maintenance issues.

The primary benefit in using a synchronous motor is that the magnetism induced in the rotor circuit allows the operator to adjust whether the motor consumes reactive power or supplies reactive power. Induction motors always operate at a lagging power factor. Elevating a synchronouos motor's excitation allows it to supply reactive power to other induction motors thus elevating the overall power factor of the facility in the eyes of the local power utility.

DC Motors

A second type of motor found at some facilities is the DC motor. A DC motor is powered by direct current being applied to magnetic poles located on its inner shell which is called the stator. These poles establish the primary magnetic field in the machine and are called the field poles. They could be permanent magnets which would identify the machine as a permanent magnet DC machine.

The rotating shaft on a DC motor also has its own magnetic poles.

This circuit is energized through a frictional connection made of small copper connection points on the rotor contacting carbon brushes that are wired to an external DC source. The small copper connection points form a segmented ring around the shaft which is called the commutator.

One of the benefits of a DC motor is that the speed can be tightly controlled by varying either the voltage applied to the rotor or the voltage applied to the field circuit on the rotor. Prior to the development of the variable frequency drive, DC motors were the only motors available which allowed for speed variation without the use of pulleys, belts or gears.

The drawbacks to DC machines are the extra costs due to the large amount of copper used in their construction along with the increased maintenance costs due to the use of brushes and a commutator. Less than 5% of the motors in industrial facilities today are DC.

AC Induction Motors

Induction motors make up the majority of motors used today. Their general construction consists of a three phase set of armature poles being formed on the stator. They are energized by an external, three phase source. The rotor of the motor appears like a small cage similar in shape to an exercise wheel found in a gerbil cage. In fact, a common induction motor design is called the squirrel cage induction motor. When three phase power is applied to the stator, a rotating magnetic field is created within the motor which causes the rotor to begin rotating.

The primary benefits of induction motors are that they are inexpensive and offer a wide range of selection for various tasks. The savings in price over DC and AC synchronous machines come from the fact that there is no electrical connection to the rotor. This results in less copper and less maintenance issues. It is common for a three phase induction motor operated within limits to last thirty years. One drawback of induction motors is that they only operate on a lagging power factor.

MOTOR EFFICIENCY AND POWER FACTOR

With the wide range of motors available to the typical plant engineer today, one needs to closely examine the needs of the application to ensure that motors are operating at their peak efficiency and provide the necessary torque-speed properties with the proper insulation to meet the load.

The details of the motor nameplate are extremely important in determining whether a motor is correct for its application. It should be noted that motor nameplate values represent full load conditions. The motor power rating is a measure of the rated output horsepower delivered to the motor shaft. The motor efficiency is the ratio of mechanical output power to the electrical input power at full load.

When sizing a motor, the first question that must be asked is if a motor can provide enough output horsepower to meet the requirements of the mechanical load. If a motor is overloaded, it will draw too much current which generates heat as a function of the current squared times the armature resistance. Excessive heat will breakdown the motor insulation and cause premature failure of the motor. Typically, a motor will include a service factor which represents the acceptable overload percentage before damage begins to occur.

The alternate situation, which is very common, is where the motor is oversized for its application. This occurs because engineering practice is conservative in nature. While an oversized motor can clearly meet the needs of a light load, the operational efficiency suffers. Figure 12-1 illustrates the relationship between motor efficiency and percentage load for a motor.

As Figure 12-1 demonstrates, motor operational efficiency drops off significantly below 40% load. To accurately estimate motor load, a three phase load logger can be attached to the electrical feed for the motor. A current probe can be used as a rough estimate by comparing input current to the nameplate rated current. If the input current is below 40% of the nameplate rated current, then the motor should be replaced with a smaller unit that will be able to operate at a higher efficiency.

Figure 12-1 also demonstrates how power factor is affected by motor load. When a motor operates, it consumes both real power (kW) and reactive power (kVAR). The combination of these two values is known as the apparent power demand (kVA). The reactive power is consumed by the coils of wire that use magnetism to convert electrical energy to mechanical torque at the shaft.

For a motor to begin to turn, it consumes reactive power to establish a magnetic field. In this process, there are losses from hysteresis and

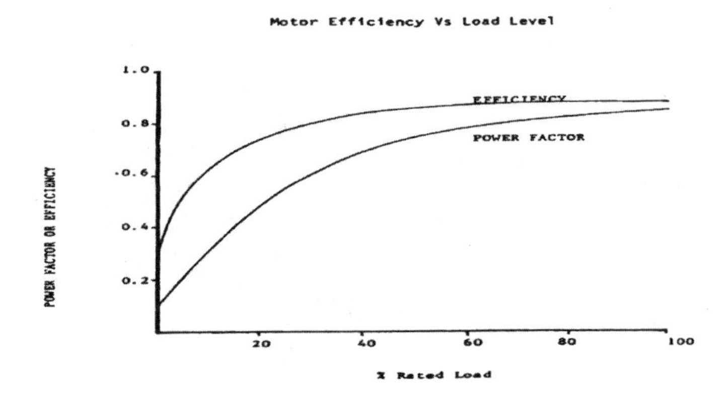

Figure 12-1. Motor efficiency vs. load level. Source: Electrical and Energy Management, IEES, Ga. Tech., Atlanta, GA.

the eddy currents that circulate in the magnetic poles. These losses exist regardless of the motor load. As motor load increases, the real power consumption increases but the reactive power increases only slightly. Power factor is defined as the ratio of real power consumption to the overall apparent power consumption (kVA). When a motor operates at full load, it consumes a large amount of real power (kW) with respect to its overall apparent power load (kVA) which equates to a power factor value close to 100%. At light load, the real power load is significantly reduced but the reactive power consumption only reduces slightly. Thus, the power factor drops significantly. This is important to the plant engineer since power utilities typically charge extra for poor power factor.

MOTOR VOLTAGE

The rated voltage should be noted since electrical voltage can vary significantly within an industrial facility. As applied voltage varies above or below rated voltage for the motor, motor performance will be altered. Figure 12-2 demonstrates the effects of voltage variation on motor performance.

Besides voltage magnitude variation, motor performance is also affected by variation in voltage between the three electrical phase volt-

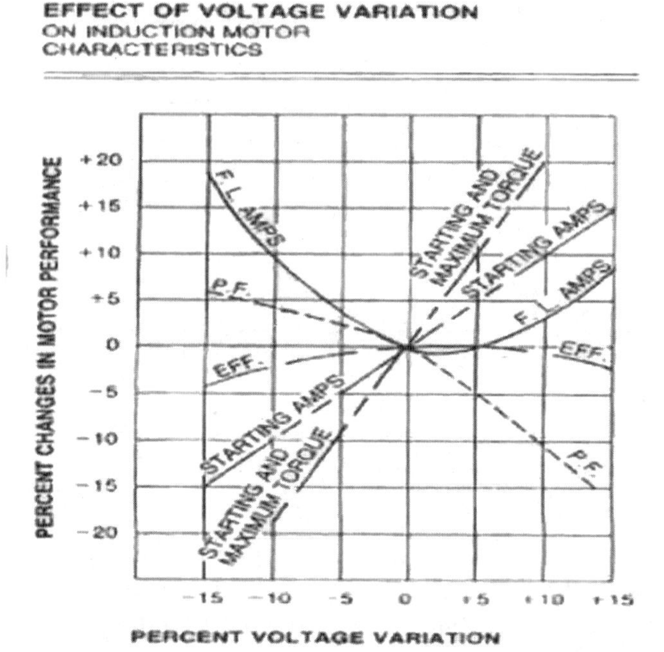

Figure 12-2. Motor performance vs. voltage variation. Source: http://www.reliance.com/mtr/pwrspymn.htm.

ages. Ideally, a three phase motor should be fed from a balanced source with all three phase voltages equal to the same value. However, single phase loads within a facility can vary the load drawn from each phase which can create a voltage imbalance between phases. Voltage imbalance will increase motor losses. Figure 12-3 illustrates the relationship between voltage imbalance and motor losses.

MOTOR REWINDS

When a motor fails, a plant manager must decide to either replace the motor with a new unit or take the damaged motor to a motor rewind shop. If a new motor is not in stock in the plant or in the inventory at the local motor supplier, the only option may be a rewind.

A motor rewind involves heating the internal parts of the motor to high levels to break down the resin binding copper conductors to the

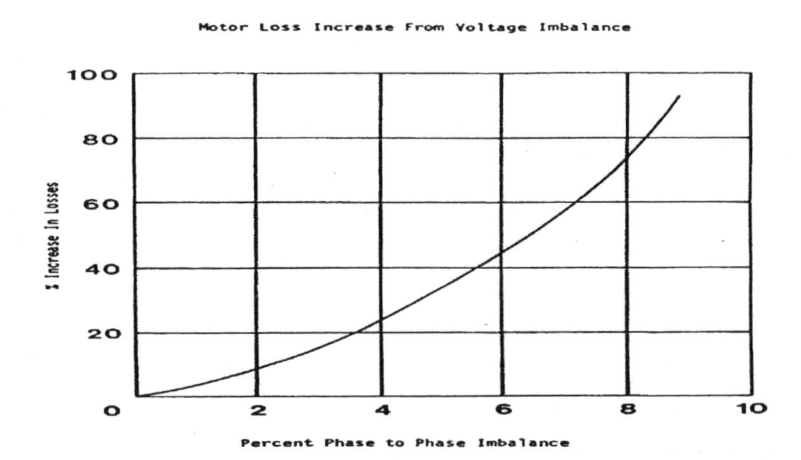

Figure 12-3. Motor losses vs. voltage imbalance. Source: Electrical and Energy Management, IEES, Ga. Tech., Atlanta, GA.

motor stator or rotor. Once the parts are separated and cleaned, new windings are installed and the motor is tested before returning it back to the customer. When proper procedure is followed, a motor that has been rewound may be returned to service in pristine condition. However, the heat exposure to the core can affect its magnetic characteristics over time. This may result in efficiency losses of 1% with each rewind. For larger motors, it is usually more economical to rewind a motor rather than replacing it. For motors less than 50 hp, the economics seem to favor replacement.

13

Reliable and Economic Natural Gas Distributed Generation Technologies

Distributed generation (DG) has emerged as a viable alternative to conventional central station power production. This chapter discusses natural gas-fueled distributed generation technologies, focusing on advanced industrial turbines and microturbines, reciprocating engines, and fuel cells. Each of these systems has the capability to improve power reliability and reduce environmental impacts at lower overall costs. Market barriers, such as the lack of standardized utility interconnection protocols, environmental permitting complexities, and public unfamiliarity with distributed generation technologies, have interfered with successful market applications. However, studies indicate that there is substantial growth potential for DG in the steel, petroleum, chemical, forest products, and other industries, as well as commercial buildings, government facilities, hospital complexes, industrial parks, multi-family buildings, and school campuses. This chapter will show that with continued support distributed generation will play a key role in energy production in the next millennium.

INTRODUCTION

Distributed generation (DG) involves small, modular electricity generation at or near the point of use. Utilities or customers can own DG systems, but there is a growing trend towards third party ownership. DG differs fundamentally from traditional central station generation and

---

Presented at Strategic Energy Forum by William Parks, Patricia Hoffman, Debbie Haught, Richard Scheer and Brian Marchionini

delivery in that it can be located near end users, in an industrial area, inside a building, or in a community (as illustrated in Figure 13-1).

Distributed generation includes renewable and natural gas-fueled equipment. Photovoltaics, concentrating solar power, solar buildings, wind, hydro, and geothermal plants are all examples of renewable DG. Enabling technologies such as energy storage systems, interconnection equipment, and sensors and controls help DG systems to integrate into utility or customer operations effectively. The focus of this chapter is on gas-fired technologies, including advanced turbines, microturbines, fuel cells, and reciprocating engines. Because of recent technological advancements and the relatively low cost of natural gas, industrial and commercial end users have expressed particular interest in these technologies. In addition, the Department of Energy (DOE) provides R&D funding for technologies in order to encourage DG as a reliable power option in a competitive market.

There are many different terms being used to describe the concept of distributed generation, including distributed resources, distributed power, and distributed utilities. While the terms differ, the concept they describe is fundamentally the same. In this era of increasingly competitive markets for electricity and natural gas, consumers have greater

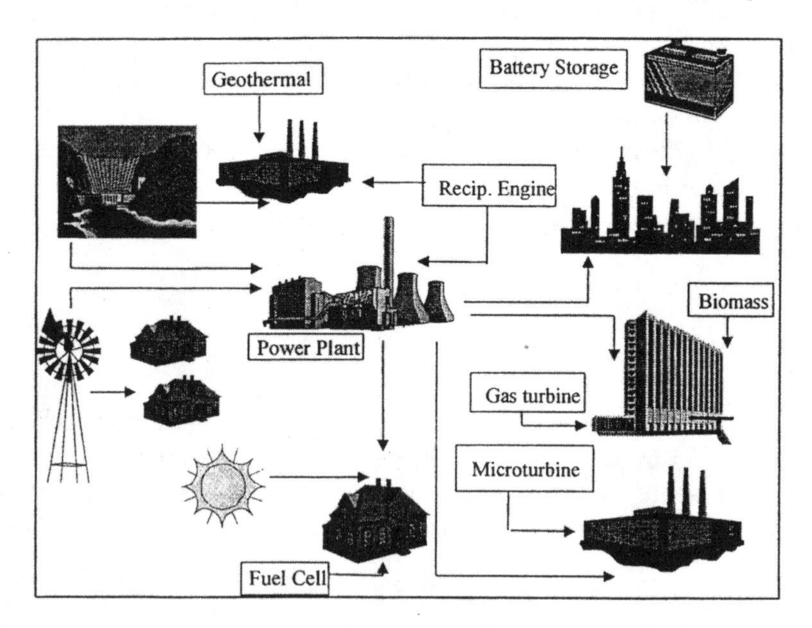

Figure 13-1. Distributed generation.

choices than ever before for satisfying their needs for energy services. Certain industrial and commercial customers find having greater access to and control over their sources of energy to be advantageous. With uncertainty about the future utility system and concern about the reliability of energy services, distributed generation offers customers an additional energy option.

Combined heat and power (CHP) is an important DG application. CHP systems capture the waste heat from power generation that is normally vented to the atmosphere and use it for productive purposes, such as making hot water and steam, industrial process heat, driving chillers and other mechanical devices, or regenerating desiccants for humidity control. CHP systems can use turbines, engines, or fuel cells and couple them with heat recovery equipment. While the most efficient electricity generation systems are able to convert 40-50 percent of the fuel into useful products, CHP systems can convert as much as 80 percent of the fuel into useful work.[1] This increase in efficiency lowers overall costs to users and reduces emissions of air pollutants and greenhouse gases.

ELEMENTS OF DG

Partnerships

There are several participants involved with the development and deployment of natural gas-fired DG systems (see Figure 13-2). Industrial and commercial end users demand power that is clean, economical, and reliable. Power outages and interruptions can be costly.[2] These needs are recognized by federal and state government agencies, who are leading a national effort to develop cleaner, more reliable, efficient, and affordable energy systems. Equipment manufacturers are involved in the research and development efforts. Given favorable market conditions, equipment manufacturers can and do engage in R&D activities without the government's encouragement. Energy service providers, such as electric and gas utilities, work with the equipment manufacturers to market DG technologies and bring the power to the user. In addition, energy service providers are often responsible for interconnecting DG systems to the grid. Project developers include energy service companies, engineering companies, and design firms. Their role is siting, permitting, and financing projects.

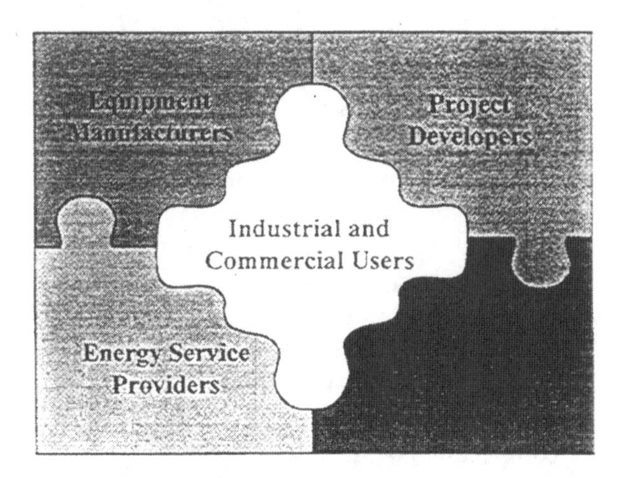

Figure 13-2. Key participants in DG.

The Department of Energy is actively involved in research and development and supporting activities to accelerate DG system implementation into the market. DOE invests more than $250 million annually in research and development programs for advanced distributed generation technologies, including renewable and natural gas-fueled systems, as well as supporting technologies in transmission and distribution, sensors and controls, combustion, advanced materials, and energy storage systems. DOE has also proposed comprehensive electricity restructuring legislation that calls for opening markets across the country to retail competition.

In addition to these research and development efforts, the Office of Energy Efficiency and Renewable Energy has launched the Combined Heat and Power Challenge which has the goal of doubling U.S. CHP capacity by 2010 and then double it again by 2020. By conducting outreach activities across the nation, the Department hopes to encourage CHP use by educating legislators, policy officials, and end users of the potential benefits of utilizing distributed generation technologies in CHP applications. Additional information about these activities is available on the industrial power homepage (www.eren.doe.gov/der) and CHP Challenge website (www.eren.doe.gov/der/chp).

Benefits

According to the North American Electric Reliability Council (NERC), the current economy is expected to drive long-term demand growth for electricity faster than projected; planned generating capacity will not be able to keep pace with this growth[3]. DG offers an additional option to meet this load growth.

The restructuring of electricity and natural gas markets across the country is primarily the result of action by states. Twenty-four states have enacted comprehensive restructuring legislation or regulatory orders that will enable customers to select alternative service providers. As a result of the implementation of retail choice legislation, there will be a wider range of products available for end users. These will include options for third parties to own and operate on-site generation. Distributed generation has many potential benefits in a more competitive electricity market.

From an economics standpoint, DG offers benefits to both end users and utilities (see Figure 13-3). Distributed generation becomes viable when it can produce power at costs equal to or less than what is available from the grid. As a result of electric restructuring, there is a trend towards greater use of time of day pricing. For some customers this has meant an increase in electricity costs. Customers can install DG systems as a means of lowering their electricity bills.

Additional savings are realized when consumers utilize CHP applications because these systems use heat that is normally vented to the atmosphere or discharged into lakes or rivers. Additionally, DG can

	Benefit	¢/kWh
Customer	Increased Power Reliability	0.6
	Reduced Energy Cost for Thermal Energy Loads	2.1
	Decreased Exposure to Electricity Price Volatility	1.0
Utility	Avoided Increases in System Capacity	1.0
	VAR Support	0.7
	Reduced T&D Losses	0.9
	T&D Upgrade Deferral	0.6

Source: Arthur D. Little

Figure 13-3. DG cost savings.

offer a new source of revenue if the customer can sell excess power or ancillary services to power markets. DG at the customer's site can also provide many benefits to the electric utility. Utilities can defer investments in distribution system upgrades by using DG to relieve congestion at key locations in the network. For example, the addition of a new feeder line at a substation can be avoided if DG systems are used to meet the load growth. By reducing the need for transmission and distribution services, DG reduces the stress on these systems during peak periods and reduces losses. In addition, strategically located DG systems can be used by grid operators for frequency control, voltage, and reactive power support.

Another attractive benefit that DG can offer the customer is increased power reliability. By operating a system that is on-site or near the end-user, DG can avoid or reduce the impacts of grid power outages that cause operational downtime and health and safety concerns. By contributing to a less complex transmission and distribution network, DG can reduce or eliminate grid voltage variation and harmonics that negatively affect a customer's sensitive loads, resulting in improved power quality.

The increased power efficiency of on-site CHP systems is an added benefit to DG. Currently, the electricity consumed in the United States is produced from large power plants owned by independent power

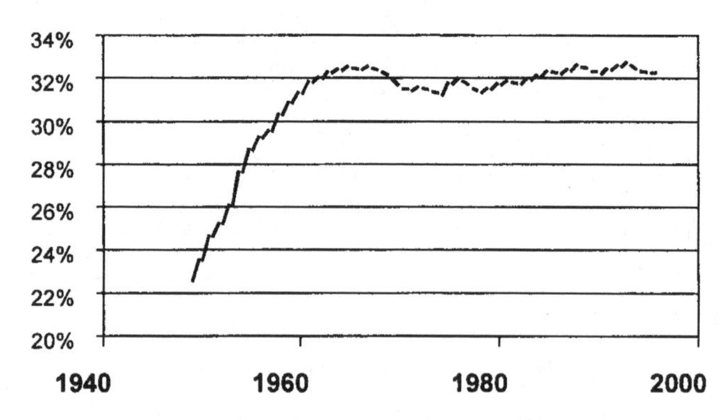

Fossil Electric Generation Efficiency (at plant, W/O T&D)
Source: Annual Energy Review (International Energy Agency, 1996)

Figure 13-4. Efficiency has reached a stagnant level.

producers, utilities and large industrial facilities. Over the past thirty years, there has been a slight increase in overall efficiency of the U.S. electric system. However, it has reached a plateau of around 33% (see Figure 13-4). Fossil and nuclear plants vent a large amount of heat into the atmosphere and lakes and rivers each year. Although all DG technologies have different efficiencies, when utilized in a combined heat and power application, they are more efficient than conventional power production methods.

Environmental quality is also a major benefit of DG. Some DG systems substantially reduce air pollutants. Tightened emission standards for SO_x and NO_x and increased concern over greenhouse gases have made low-emission DG systems more appealing for power production. Another benefit of some DG systems is size: fuel cells and microturbines can be the size of an air conditioning unit. The small size and modularity support a broad range of customer and grid-sited applications by providing electricity and thermal power in a precise way where central power plants would be impractical.

Barriers

In order to achieve an expanded role for DG, several market and regulatory barriers will have to be overcome. These barriers include utility practices and electricity rate designs that discourage on-site generation and lengthy and costly environmental permitting and siting processes.

Non-standardized interconnect requirements have been a barrier to widespread deployment of DG technologies. Interconnect requirements vary by state and/or utility and are often not based on state-of-the-art technology or data. Compliance often requires custom engineering and lengthy negotiations that add cost and time to system installation. Non-standardized requirements also make it difficult for equipment manufacturers to design and produce modular packages, hampering their ability to realize economies of scale.

Typically, customer-sited generation requires a backup source of power to meet load requirements during generation outages or routine maintenance periods. Utility charges for this standby/backup power are not always cost-justified. In a restructured market, the generation backup charge will be negotiated between the user and the generation supplier and the distribution charge will be negotiated with the utility. A competitive means for supplying these services is essential to ensure economic application of DG. However, state regulators, struggling with the more contentious issues of restructuring legislation, are often unaware of the importance of standby fees and backup charges on the economic viability of on-site generation.

Treatment of DG projects by air quality and air permitting agencies varies widely among jurisdictions, ranging from straightforward and low cost to expensive and difficult. Making the challenge even greater is the process itself. Local environmental permitting of power facilities has been structured to handle a few installations. For DG, a rapid, predictable and inexpensive permitting process is a prerequisite for the project to be economically viable. One way to achieve this is through precertification of certain types of DG facilities. Precertification involves the evaluation of DG projects in advance. If the application equipment meets predetermined requirements, the system is precertified and is able to move through the permitting process more quickly. DG would also benefit if output-based standards were adopted. By using the full electric and thermal output to set standards, applications for DG projects with high efficiency and pollution-preventing characteristics could also move through the permitting process more quickly.

One of the most significant impediments to deployment of on-site generation is the inconsistent and location-specific rules, regulations, and procedures affecting general siting and permitting. In order to obtain a permit, a variety of additional criteria must be addressed, including water impacts, noise, land use, visual impacts, fire, safety, fuels, and hazardous materials. Present permitting requirements have been developed either for backup generators or large baseload projects and are often inappropriate for DG applications. Lack of familiarity with DG technologies and applications, and a lack of a precertification options result in site-specific negotiations that can be needlessly time-consuming and costly.

TECHNOLOGIES

Distributed generation encompasses many distinct power generation technologies that vary by size, application and efficiency. Gas turbines and reciprocating engines have been on the market for decades, while smaller-scale technologies, such as microturbines and fuel cells, are in the early stages of commercialization. If these technologies continue to advance and prove commercially viable and regulations and markets evolve to encourage their acceptance, these new options for supplying power will take hold.[4]

Advanced Turbines

Initiated by the DOE, the Advanced Turbine System Program focused on completing the development and demonstration of ultra-high efficient natural gas turbine systems. The goal was to develop turbines that are not only more efficient but also cleaner and less expensive to operate than first generation turbines. The target was to develop gas turbines with >60% efficiency and <10 ppm NO_x emissions. The goals wer exceeded.

Simple cycle gas turbines that are less than 20 MW are being improved upon in the Advanced Turbine Program. A gas turbine is a heat engine that uses a high-temperature, high-pressure gas as the working fluid. Part of the heat supplied by the gas is converted directly into the mechanical work of rotation. In most cases, hot gases reduced to operating a gas turbine are obtained by burning a fuel in air, which is why gas turbines are often referred to as "combustion" turbines. Because turbines are compact, lightweight, and simple to operate, gas turbines have been widely used in electricity generation. Gas turbines can be used in a cogeneration application in industrial and utility settings to produce electricity and steam. (Many industrial processes require steam in addition to electricity.) In such cases, "simple cycle" gas turbines convert a portion of input energy to electricity and use the remaining energy to produce steam in a steam generator.

Solar Turbines, a major manufacturer of turbine systems, has teamed up with the DOE to develop an advanced turbine, and is also involved in improving the efficiency of their existing turbines with ceramic retrofits. Malden Mills, a textile company in Massachusetts, installed two turbine engines with ceramic retrofits, as shown on the next page, and at full production will save over $1 million annually (illustrated in Figure 13-5).

Microturbines

The U.S. Department of Energy established the Advanced Microturbine Program as part of its Distributed Energy Program. The federal government invested $60 million over a six-year period to further

Figure 5. Malden turbine.

develop units that are cleaner, more fuel efficient and more reliable.

Microturbines are machines ranging in size from 30 kW to 500 kW, which includes a compressor, combustor, turbine, alternator, recuperator, and generator. They have the potential to produce power for sites that have space limitations, such as cramped locations in factories or basements of buildings (see Figure 13-6). Waste heat recovery can be used with these systems to achieve efficiencies greater than 80%. Microturbines offer a number of potential advantages compared to other small-scale power generation technologies, including a small number of moving parts, compact size, and light weight, as well as opportunities for greater efficiency, lower emissions, lower electricity costs, and the use of waste fuels. Fuel flexibility allows microturbines to use high- and low-pressure natural gas, propane, and diesel. Microturbine technology is still relatively young and testing is not at a point where precise emissions and efficiency can be derived. Also, there are no long-term data available to determine the reliability and durability of the application.

Microturbine systems are just entering the market; manufacturers are targeting applications in the industrial and buildings sectors, including CHP, backup power, continuous power generation, and peak shaving, to reduce costs during peak demand periods. So far, four U.S. manufacturers have made commitments to enter the microturbine market. Honeywell (Allied Signal) is offering a 75 kW product, Capstone has a 30 kW product, Elliott has 45 and 80 kW products, and Northern Research and Engineering Company will have several products in the 30 to 250 kW size range.

Fuel Cells

Fuel cell power systems are an emerging class of technologies that convert chemical energy directly into electricity, producing almost no pollution. Heat is a by-product of the reaction and can be recovered in much the same way as in combustion-based systems. Fuel cells offer the potential advantage of lower maintenance requirements because there are no moving parts in the power generating stacks. There are four types of fuel cells: Proton Exchange Membrane (PEM), Phosphoric Acid Fuel Cell (PAFC), Molten Carbonate Fuel Cell (MCFC), and Solid Oxide Fuel Cells (SOFC). The main differentiation among fuel cell types is in the electrolytic material. Each electrolyte has advantages and disadvantages based on cost, operating temperature, efficiency, power-to-weight ratio, and other operational considerations. PEM fuel cells are being tested in homes to support residential loads (see Figure 13-7). These fuel cells are also most commonly used in testing for transporta-

Figure 13-6. Capstone Model 330. *Courtesy Capstone Turbine Corporation.*

tion because they provide a continuous electrical energy supply from fuel at high levels of efficiency and power density. The PEM's power-to-volume ratio makes it appealing for hybrid electric vehicle research and development. The PAFC and MCFC fuel cells are primarily used for larger-scale power production.

The economics of distributed generation continue to become more advantageous as technology and industry partnerships evolve.

Figure 13-7. PEM fuel cell.

Reciprocating (Recip) Engines

The DOE is in the process of developing an advanced Reciprocating Engine Program based on natural gas. The program is still in the planning stage and several meetings were held last year to discuss the scope and direction of the program.

Reciprocating engines (also known as internal combustion engines) were first commercialized over 100 years ago and have long been used for electricity generation (see Figure 13-8). There are three major operational components to reciprocating engines: fuel, air, and an ignition source. The two types of ignition sources are spark ignition (SI), which is common in conventional and gasoline or gaseous-fueled engines, and compression ignition (CI), which is used in conventional diesel-fueled engines.

Recips used for backup power are the fastest selling distributed power technology in the world today. Engine manufacturers are enhancing production capabilities in anticipation of new orders as distributed generation technologies are chosen by more manufacturers both here and abroad. Institutions, large industrial establishments, and commercial buildings have used reciprocating engines with fractional horsepower all the way to 60 MW. Existing engines achieve efficiencies in the range of 30% to greater than 40%. Further improvements are possible in efficiencies and lowered emissions, and there are op-

Figure 13-8. Recip engine. ***Courtesy Caterpillar.***

portunities to use bio-based fuels in place of petroleum and liquids and gases derived from natural gas.

Valley Medical Center cogeneration plant, located outside of Seattle, operates four General Electric reciprocating engines to cover the hospital's electrical demand of 3.6 MW. The steam generated from utilizing the thermal output is used for space heating. The hospital had initially decided to install the reciprocating engine plant to reduce the uncertainty of the power supply while decreasing a hospital's operating costs. The system saves the hospital more than $50,000 per month on energy bills and will make the payback period five years. Besides the economic benefit of the system, the plant has also significantly improved power quality and reliability as it now has the ability to operate during emergency situations.[5]

MARKET POTENTIAL

Forecasts indicate that approximately 40% of all new power generation worldwide over the next decade will be provided by distributed generation. This translates into a $4-5 billion global market for distributed generation.[6] Forecasts also indicate that the United States could generate up to 5 GW of DG power a year in the next fifteen years (see Figure 13-9).

Reciprocating engines are the fastest selling distributed generation technology on the market today. Advanced turbines are moving into commercial production with the expectation that sales will be brisk. Microturbines are emerging on the market, while high-tempera-

	U.S.	Rest of World
Installed Capacity (GW)	770	2310
Average Growth Rate (%)	1.6	5.0
Potential DG Market Share—15-year average (GW/y)	3-5	23-46

Source: Solar Turbines

Figure 13-9. DG market potential.

ture fuel cells and fuel cell hybrids are already commercially available. Nevertheless, these early, first generation products are expected to improve rapidly as increasing production leads to low-cost, more robust devices. There is a great deal of interest in fuel cells but there are a limited number of products on the market and their costs remain higher than the other options.

The Department of Energy supports various programs that aim to significantly improve the resource efficiency and productivity of energy- and waste-intensive industries in the United States. USDOE is helping to develop technology solutions for critical energy and environmental challenges. The aluminum, chemical, forest product, glass, metal casting, mining, petroleum, and steel industries will have large future power demands and, given an aggressive research and development scenario, are expected to use microturbines, fuel cells and reciprocating engines for power solutions (see Figure 13-10). By the year 2020, these technologies are forecast to supply over 5 GW of power with energy cost savings of $600 million. Additionally, CO_2, SO_2, and NO_x emissions will be reduced significantly. DG is predicted to have a large impact particularly in the steel, petroleum, chemical, and forest products industries.

In addition, natural gas DG technologies used in commercial, office, institutional, government, and multifamily buildings will play a large role in the market. DG used in a building application provides for electricity, heating, cooling, and humidity control while addressing air quality issues. DG potential is greater in new buildings where the building can be designed around the DG system. However, the market

Technology	Market Penetration (MW)	Energy Cost Savings ($Million)	CO_2 Displaced (kTons)	SO_2 Displaced (kTons)	NO_x Displaced (kTons)
Microturbines	2,200	270	12,800	44	38
Reciprocating Engines	1,300	130	7,300	25	17
Fuel Cells	1,600	200	12,600	36	29
Total	**5,100**	**600**	**32,700**	**105**	**84**

Source: Arthur D. Little

Figure 13-10. Net impacts of microturbines, recips, and fuel cells by 2020.

for retrofit applications is still strong. There are no published market studies available for DG technologies in buildings as of yet. However, experts believe that the market for buildings is at least as big as the industrial market.

CONCLUSION

Distributed generation could play a significantly larger role in the utility market if research and development efforts are successful in reducing costs, improving performance, and addressing the regulatory and institutional barriers. Institutional, commercial, and industrial facilities could be served by DG and will proliferate as awareness grows and environmental permitting and interconnection becomes more streamlined. With the onset of utility restructuring comes a competitive and innovative market that will demand lower electricity prices with augmented energy efficiency and environmental performance. Favorable regulations, combined with economic benefits and technological advances, will promote the use of advanced turbines, microturbines, fuel cells, and reciprocating engines. If these changes come about, DG will change the way we receive and use power in the new millennium.

References

1. Energy and Environmental Analysis, Inc., USEPA Combined Heat and Power Partnership, Sept. 2007.
2. R. Hof, *Business Weekly,* The 'Dirty Power' Clogging Industry's Pipeline, April 8, 1991, p82.
3. Arthur D. Little, White Paper, *Distributed Generation: Understanding the Economics,* 1999, p4.
4. Arthur D. Little, White Paper Distributed Generation: Understanding the Economics. 1998
5. Robert Swanekamp, *Power Magazine,* May/June 1998 Vol. 142 No. 3, page 43.
6. *Electric World,* 7/97

14

Financing Energy Efficiency Projects

INTRODUCTION

Every day spent by an organization without having installed the appropriate energy efficiency measures means lost savings and lost opportunities. Performance contracting is a sophisticated solution to this problem. As with any sophisticated system, there are elements of complexity to be managed. Therefore, communication, knowledge and experience are essential for successful project completion.

DEFINITIONS AND CLARIFICATIONS

For the purposes of this chapter, it is assumed that the energy services company (ESCO) is providing the energy conservation measures (ECMs) such as audit, design, installation, monitoring and maintenance for the customer and that a separate third party, such as a bank or investment company, is providing the capital for the project. Often, energy services companies market themselves to customers as providing financing. In many cases, there is an independent financing source involved in the background. Alternatively, an ESCO can be a utility subsidiary which uses the utility's shareholder money to finance projects. For simplicity's sake, we will treat the ultimate source of capital for projects as a separate lender with its own guidelines.

FINANCING ALTERNATIVES

There is a variety of options for financing energy conservation projects. Some of the most common are:

General Obligation Bond

Specifically applicable to municipalities, these bonds are based on the general credit of a state or local government. The process is long and complicated, but interest rates are low.

Municipal Lease

Specifically applicable to municipalities, state entities and local entities, the lessor earns tax-exempt credit and the borrowing entity pays low interest rates.

Commercial Loan

This is a loan to the customer from a conventional bank based on the customer's assets and credit quality. This form of financing is rarely offered by typical finance companies or financial institutions for energy conservation projects requiring less than $5 million of capital.

Taxable Lease

There are a number of leasing vehicles with a variety of names such as: operating lease, capital lease, guideline lease, tax-oriented lease, and non-tax-oriented lease. The tax ramifications of leasing are often not well understood. Figure 14-1 will help clarify some of those distinctions. Figure 14-2 will show the dynamics of the parties involved.

It is important to note that in all of the categories listed above, the customer is directly obligated to make payments relating to the installed energy-saving measures REGARDLESS OF PERFORMANCE of either the equipment or the ESCO. The customer may well have recourse under a separate contract to the service provider or equipment manufacturer, but will still owe under the financing instrument. The interest rate and any additional loan terms are based almost exclusively on the creditworthiness of the customer.

RELATIVE BENEFITS OF PROJECT FINANCING

Many companies and government entities are undergoing severe budget cutbacks. When these entities are approached by an ESCO offering virtually no up-front investment and a guaranteed amount of savings, new possibilities are created for energy projects to be developed and completed. If structured properly, performance contract financing

Figure 14-1. Leasing options.

Guideline Lease/True Lease/Tax-Oriented Lease

Compliance with all of the IRS guidelines including those listed below is required:

(a) The total lease term (including extensions and renewals at a predetermined, fixed rate) must not exceed 80% of the estimated useful life of the equipment at the start of the lease, i.e., at the end of the lease the equipment must have an estimated remaining useful life equal to at least 20% of its originally estimated useful life. Also, this remaining useful life must not be less than one year, thereby limiting the maximum term of the lease.
(b) The equipment's estimated residual value at the expiration of the lease term must equal at least 20% of its value at the start of the lease. This requirement limits the maximum lease term and the type of equipment to be leased.
(c) No bargain purchase option.
(d) The lessee cannot make any investment in the equipment.
(e) The equipment must not be "limited use" property. Equipment is "limited use" property if no one other than the lessee or a related party has a use for it at the end of the lease.
(f) Tax-oriented leasing is 100% financing. Guideline leases (tax-oriented leases) may be either a capital lease or an operating lease for reporting purposes under Financial Accounting Standards Board Rule #13.

Capital Lease

A capital lease is one that fulfills any ONE of the following criteria:

(a) The lease transfers ownership of the property to the lessee by the end of the lease term.
(b) The lease contains a bargain purchase option (less than fair market value).
(c) The lease term is equal to 75% or more of the estimated economic life of the leased property.
(d) The present value of the minimum lease payments equals or exceeds 90% of the fair value of the leased property.

The capital lease shows up on the lessee's balance sheet as an asset and a liability.

Operating Lease

An operating lease does not meet ANY of the above criteria for a capital lease.

An operating lease is not booked on the lessee's balance sheet but is recorded as a periodic expense on the income statement.

Sources

Accounting For Leases, Financial Accounting Standards Board
Leasing and Tax Reform: A Guide Through The Maze, General Electric Credit Corporation
Handbook of Leasing, General Electric Credit Corporation

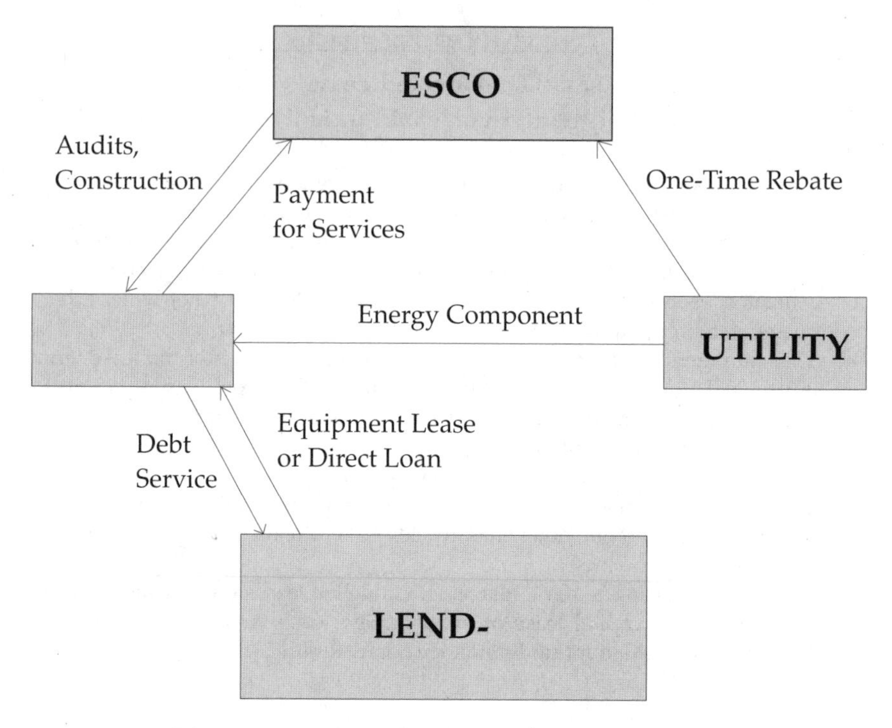

Figure 14-2. Equipment lease/direct financing.

has the benefits to the customer of an operating lease, i.e., off-balance-sheet treatment, with the added benefit of being non-recourse to the customer. By maintaining this financing out of its balance sheet, the customer can retain the use of available credit for expansion, research and development, additional inventory or business emergencies.

BASIC STRUCTURE

Performance contract financing from a lender's perspective is a challenging combination of business credit analysis and project evaluation. The project financing is structured as follows:

(1) The ESCO contracts with the customer to provide energy conservation measures (lighting, variable speed drives, etc.) as well as ongoing services which may include warranties, handling and disposal of wastes, operation and maintenance of installed equip-

ment, repair and replacement of measures, and measurement, monitoring and verification of savings. The Energy Services Agreement (ESA) is the foundational document upon which a lender will rely to confirm that the customer and ESCO have a clear understanding of all aspects of the ECMs being installed. The ESA incorporates a payment from the customer which is designed to cover the costs to finance the project as well as paying the ESCO for the services provided. (A sample ESA is included in Figure 14-4. Note that it is a guideline only. Be sure to have your legal counsel review any document before you authorize it.)

(2) The lender lends directly to the ESCO. The ESCO uses the funds to recoup its project development expenses and purchase and install the equipment. Repayment to the lender is made by the ESCO out of the funds paid to it by the customer. One variation is to create a single-purpose entity exclusively to hold the assets of the project financing. Because the loan is technically made to this entity, it has the added benefit of keeping the transaction off the balance sheet of the ESCO as well as the customer (see Figure 14-3). In addition, the lender can lend up to 95% of the project amount.

PROPOSAL REVIEW

The lender will analyze a proposed project with these questions in mind:

- Will the revenues generated by the energy savings, utility rebates and customer payments be sufficient to cover debt service?
- Can the ESCO perform as required under its contract with the customer and/or the utility for the term contract?
- Are the risk allocations among all parties fair from both business and legal perspectives?
- Will the customer and/or the utility be able to meet its payment obligations for the term of the financing?

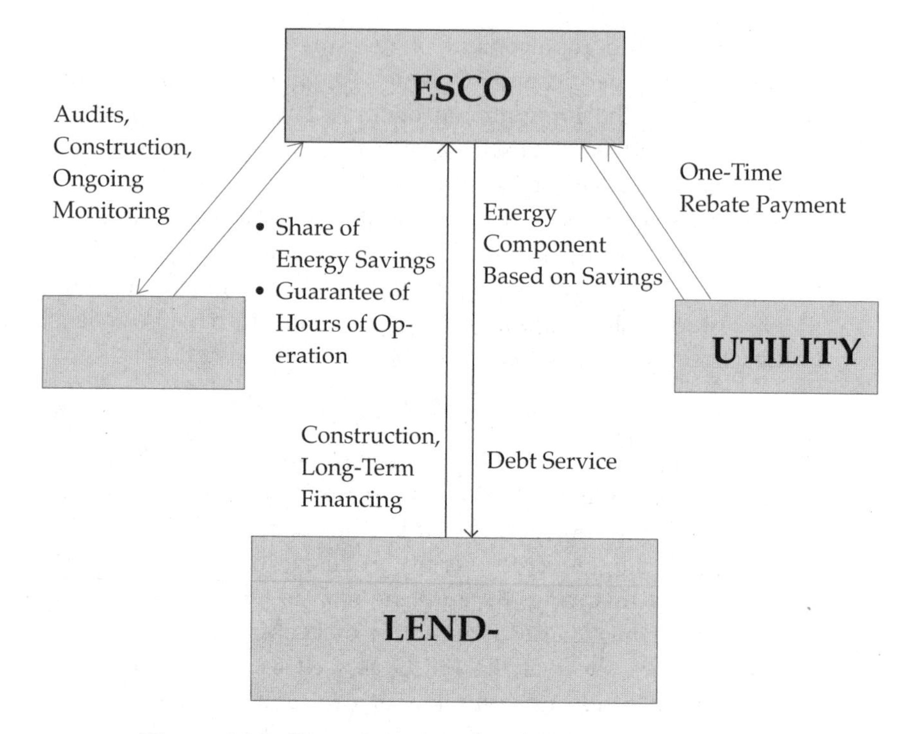

Figure 14-3. Shared savings/performance contracting.

Figure 14-4. Sample energy services agreement.

Note: Consult your legal counsel before authorizing any legal agreement.

I. TERM SHEET.

The purpose of the Agreement is the evaluation, engineering, design, procurement, installation, financing and monitoring by **ESCO** of Energy Conservation Measures ("ECMs") at **Customer's** facility(ies) identified in Appendix A attached hereto ("Premises").

ESCO and **Customer** (the "Parties") agree to the following terms pursuant to which this Agreement shall be performed:

1. EXECUTION DATE:

2. TERM OF AGREEMENT:

   (years after ECM Commencement Date)

3. LOCATION:

4. OWNERSHIP OF PREMISES:

PROJECT FIGURES

	Preliminary Estimates	Final Installed Figures
5. **ESCO** PERCENTAGE OF ENERGY SAVINGS:	________	________
6. PROJECT TOTAL CAPITAL COST ($ 000):	________	________
7. ECM COMMENCEMENT DATE:	________	________
8. VALUE OF FIRST YEAR ENERGY SAVINGS ($):	________	________
9. **CUSTOMER** FIRST YEAR PAYMENT TO **ESCO** (line 5 × line 8) ($):	________	________
10. **CUSTOMER** FIRST YEAR MONTHLY PAYMENT (line 9 ÷ line 12) ($):	________	________
11. **CUSTOMER** FIXED MONTHLY PAYMENT (OPTIONAL):	________	________

Please Initial: ESCO ________

CUSTOMER ________

The lender will want to see three years of audited financial statements: balance sheet, income statement and cash flow statement with explanatory notes from the customer. These will be used to make trend and financial ratio analyses. The ESCO's financial statements are also important and will receive a similar review.

With utility payments and/or customer payments based on actual savings, the lender needs to make a thorough analysis of the projected savings of the energy conservation measures installed. The project figures (simple payback, types of equipment, maintenance savings, if any, construction period) will be carefully reviewed to determine how the savings will cover the loan payments and to determine the effects of factors such as energy price fluctuations and inflation.

Any information which the parties can provide early on in the proposal negotiation to make the lender feel more comfortable with these issues will save a tremendous amount of time, money and effort to all parties.

CONCLUSION

The key point here is that if an event of default occurs in the ESA between the ESCO and the customer which gives the customer the right to reduce or cease its payment obligation, it is the lender which is most at risk of suffering a loss. The lender must either be comfortable that the ESCO will cure the default or be confident that it can hire another ESCO to cure the project default and force the customer to resume payments. The lender is also at risk of the energy savings not being estimated or measured properly. Despite all of this, good projects are being funded and customers are extremely satisfied with the resulting benefits. Final Installed Figures appearing in the right-hand column above shall be completed in accordance with Section V and the Parties shall then re-execute the Agreement below and enter their initials above.

As indicated in term #11 above, upon completion of any determinations required by Section V, **Customer** shall have the option to fix its monthly payment to **ESCO**, as required under Section VI, by multiplying One Hundred and ______ Percent (1____%) of **ESCO**'s Percentage of Energy Savings by the Final Installed Figure for the Value of First Year Energy Savings and paying such product to **ESCO** each year in twelve (12) monthly payments ("**Customer** Fixed Monthly Payment"). **Customer** shall indicate its exercise of such option by initialing the appropriate line below. Except as set forth in this Agreement, such monthly payments shall be due and payable each month of this Agreement from and after the month in which the Commencement Date occurs and shall be made by **Customer** without regard to the amount of Energy Savings in any such month or year. Any excess of Energy Savings over the Final Estimated Value of First Year Energy Savings shall be retained by **Customer. Customer**'s Fixed Monthly Payment shall not be revised except as may be specifically required in accordance with the terms of this Agreement.

ESCO: **Customer:**

____

____

(business organization) (business organization)

By:____________________________________
By:____________________________________
(entity) (entity)

By:____________________________________
By:____________________________________
(name) (name)

Its:____________________________________
Its:____________________________________
(title) (title)

ACCEPTANCE OF FINAL INSTALLED FIGURES

ESCO: **Customer:**

____

____

(business organization) (business organization)

By:____________________________________
By:____________________________________
(entity) (entity)

By:————————————————————————————
By:————————————————————————————
(name) (name)

Its:————————————————————————————
Its:————————————————————————————
(title) (title)

CUSTOMER ACCEPTANCE OF FIXED MONTHLY PAYMENT

Accepted Not Accepted

—————— —————— (please initial)

II. DEFINITIONS.
When used in this Agreement, the following terms shall have the meaning specified:

2.1 **Agreement:** This Agreement between **Customer** and **ESCO**.

2.2 **ESCO**'s **Percentage of Energy Savings:** The percentage of Energy Savings **ESCO** shall receive as compensation for its services under this Agreement, subject to **Customer**'s option to fix monthly payments set forth in Section I, paid to **ESCO** in accordance with Section VI and Section VII.

2.3 **Current Market Value of Energy Savings:** The total market value rate expressed in r/kWh of electrical energy use and/or $/kW of electrical demand imposed by the Utility company in the current monthly period then occurring, or in any future monthly period then being considered, including applicable taxes, surcharges and franchise fees, if applicable. The **Current Market Value of Energy Savings** shall be determined in accordance with Appendix C.

2.4 **ECM Commencement Date:** The ECM Commencement Date shall be the date on which the installation of the ECMs is substantially complete. Prior thereto, **ESCO** shall give **Customer** a Notice of Substantial Completion and shall therein identify the ECM Commencement Date, which shall occur no sooner than fifteen (15) days after such notice.

2.5 **Energy Audit Report ("EAR"):** The analysis performed by **ESCO** of the electric energy use by **Customer** at the Premises, and the potential for electric energy savings. Such analysis includes, without limitation, the ECMs recommended by **ESCO** and agreed to by the Parties for installation at the Premises and the Measurement Plan for measuring the savings estimated to result from such ECMs, all as attached hereto and made a part hereof as Appendix B.

2.6 **Energy Conservation Measure ("ECM"):** The various items of equipment, devices, materials and/or software as installed by **ESCO** at the Premises, or as repaired or replaced by **Customer** hereunder, for the purpose of improving the efficiency of electric consumption, or otherwise to reduce the electric utility costs of the Premises.

2.7 **Energy Savings:** Electric energy reduction (expressed in kilowatt-hours of electric energy and/or kilowatts of electric demand and measured in accordance with the Measurement Plan) achieved through the more efficient utilization of electricity resulting from the installation of the ECMs agreed to by the Parties under this Agreement.

2.8 **Measurement Plan:** The plan for measuring Energy Savings under this Agreement, which shall be in accordance with the requirements of the Utility Agreement and shall be a part of the Energy Audit Report attached as Appendix B hereto.

2.9 **Monthly Period:** A span of time covering approximately 30 days per month, corresponding to **Customer**'s billing period from its electric utility.

2.10 **Party: Customer** or **ESCO**. **Parties** means **Customer** and **ESCO**.

2.11 **Premises:** The buildings, facilities and equipment used by **Customer**, as identified in Section I and as more fully described in the attached Appendix A, where **ESCO** shall implement the Project under this Agreement.

2.12 **Project:** The complete range of services provided by **ESCO** pursuant to this Agreement, including evaluation, engineering, procurement, installation, financing and monitoring of ECMs at the Premises.

2.13 **Uncontrollable Circumstances:** Any event or condition having a material adverse effect on the rights, duties or obligations of **ESCO**, or materially adversely affecting the Project, if such event or condition is beyond the reasonable control, and not the result of willful or negligent action or omission or a lack of reasonable diligence, of **ESCO**; provided, however, that the contesting by **ESCO** in good faith of any event or condition constituting a Change in Law shall not constitute or be construed as a willful or negligent action, or a lack of reasonable diligence. Such events or conditions may include, but shall not be limited to, circumstances of the following kind:

a. an act of God, epidemic, landslide, lightning, hurricane, earthquake, fire, explosion, storm, flood or similar occurrence, an equipment failure or outage, an interruption in supply, an act or omission by persons or entities other than a Party, an act of war, effects of nuclear radiation, blockade, insurrection, riot, civil disturbance or similar occurrences, or damage, interruption or interference to the Project caused by hazardous waste stored on or existing at the Project site;

b. strikes, lockouts, work slowdowns or stoppages, or similar labor difficulties, affecting or impacting the performance of **ESCO** or its contractors and suppliers;

c. a change in law or regulation or an act by a governmental agency or judicial authority.

2.14 **Utility Agreement:** The agreement entered into by **ESCO** with ________, a _____ public utility company ("Utility"), pursuant to which **ESCO** is required to install certain ECMs at facilities such as **Customer**'s Premises and in accordance with the terms of which **ESCO** has entered into this Agreement.

III. ECM COMMENCEMENT DATE AND TERM OF AGREEMENT. The term of this Agreement shall commence as of the date on which this Agreement is executed and shall continue, unless sooner terminated in accordance with the terms hereof, for the period of years after the ECM Commencement Date set forth in Section I.

Upon receipt of the Notice of Substantial Completion identifying the ECM Commencement Date, **Customer** shall provide **ESCO**, within fifteen (15) days, any comments and requests for work or corrections. **ESCO** shall make all commercially reasonable efforts to respond to such comments and requests within thirty (30) days of the ECM Commencement Date, which shall occur on the identified date.

Upon the expiration or termination of this Agreement the provisions of this Agreement that may reasonably be interpreted or construed as surviving the expiration or termination of this Agreement shall survive the expiration or termination for such period as may be necessary to effect the intent of this Agreement.

At the end of the term of this Agreement, **Customer** shall purchase the ECMs and **ESCO** shall transfer title to **Customer**, free and clear of all liens and encumbrances, all as set forth in Section 6 of the General Terms and Conditions.

IV. SCOPE OF ESCO'S SERVICES.

Subject to and in accordance with the terms and conditions of this Agreement, **ESCO** shall provide the evaluation, engineering, design, procurement, installation, financing and monitoring of the ECMs set forth in the Energy Audit Report. The ECMs in the EAR may be modified pursuant to Section V. **ESCO** shall use reasonable commercial efforts to achieve the ECM Commencement Date estimated in Section I, line 7, column entitled, "Engineering Estimates."

ESCO agrees to extend its funds and install such ECMs in return for **Customer**'s agreement to perform hereunder and in particular, to pay **ESCO**'s Percentage of Energy Savings pursuant to Section VI. The Parties anticipate the measurement of Energy Savings, but notwithstanding this expectation or any provision of this Agreement which may suggest to the contrary, in light of the factors affecting savings which are beyond **ESCO**'s reasonable control, **ESCO** assumes no obligation that any particular level of savings shall materialize due to its services hereunder.

ESCO warrants that the ECMs which have a lifetime greater than one (1) year shall be, and shall remain, free of defects for one (1) year after the ECM Commencement Date. In addition, **ESCO** agrees to assign all manufacturers warranties for such ECMs to **Customer** for the period of manufacturer warranty, subject to all exclusions and limitations as may be set forth therein.

V. DETERMINATION OF FINAL TERMS.

After the execution of this Agreement, if the Parties agree to revise the ECMs listed in the EAR, and to make all associated revisions to this

Agreement, including, without limitation, to the numbers or dates, as the case may be, entered in the column entitled, "Engineering Estimates" appearing in lines 5 through 11 of Section I, the Parties shall, upon such agreement, make any associated revision and enter in the column entitled, "Final Installed Figures" the agreed-to numbers and/or dates. The Parties shall then re-execute Section I upon the entry of such numbers and/or dates and **ESCO** shall revise the Termination Values in Appendix D consistent with Appendix D and the Final Installed Project Total Capital Cost then appearing in line 6 of Section I above.

All such revisions shall be voluntary and the Parties shall not be required, absent mutual consent, to revise the ECMs listed in the EAR after the execution hereof; provided, however, **ESCO** shall not be required to install ECMs affected by Uncontrollable Circumstances. Such ECMs shall be deleted from the Project unless the Parties agree to all necessary changes to the Project numbers and/or dates required to adjust to such circumstances. Changes to Project numbers and/or dates in connection with Uncontrollable Circumstances shall be entered by **ESCO** in the column entitled "Final Installed Figures," in lines 5 through 11 of Section I. The Parties shall then re-execute Section I and **ESCO** shall revise the Termination Values set forth in Appendix D consistent with Appendix D and the Final Installed Project Total Capital Cost then appearing in line 6 of Section I.

VI. COMPENSATION.

From and after the ECM Commencement Date, except as provided in Sections I and VII with respect to **Customer**'s option to fix monthly payments, **Customer** shall pay **ESCO** an amount equal to **ESCO**'s Percentage of Energy Savings, as set forth in Section I above, multiplied by the applicable Current Market Value of Energy Savings, all as determined pursuant to Appendix C.

ESCO shall prepare and send **Customer**, and **Customer** agrees to pay, a monthly invoice calculated pursuant to Section VII below.

VII. BILLING.

ESCO will submit monthly invoices to **Customer** in amounts determined in accordance with Section VI. In the event **Customer** exercises its option under Section I to fix its monthly payments to **ESCO**, **Customer** shall pay to **ESCO** the **Customer** Fixed Monthly Payment defined in Section I. If **Customer** does not exercise such option, monthly payments shall be estimated as set forth in this Section VII and paid on such estimated basis, subject to reconciliation as provided hereinafter. Monthly invoices shall be paid within thirty (30) calendar days following receipt. Reconciliation payments, or refunds, as the case may be, shall be due within thirty (30) calendar days of receipt of a reconciliation invoice.

Subject to reconciliation, invoice amounts shall be estimated for each year following the ECM Commencement Date. In the first such year, monthly payments shall be 1/12 of the product of **ESCO**'s Percentage of Energy Savings and the Current Market Value of Energy Savings expected in such first year as set forth in line 8 of the column entitled "Final Installed Figures" in Section I above. Such fixed amounts shall be reconciled and adjusted as necessary every six (6) months based on the difference between the Current Market Value of Energy Savings measured pursuant to Appendix C and the estimated value then applicable. **ESCO** shall prepare and submit to **Customer** a reconciliation invoice within thirty (30) calendar days of the expiration of each such six-month period. The estimate of the value of Energy Savings then in effect at the end of any year shall continue until replaced with reconciled amounts in the succeeding year. In the second year and following years, the estimate of the Current Market Value of Energy Savings expected in such year shall equal the expected value (in nominal dollars for the billing year in question) of the Energy Savings actually delivered in the prior year.

VIII. NOTICES.

All notices to be given by either Party to the other shall be in writing and must be delivered or mailed by registered or certified mail, return receipt requested, or sent by a courier service which renders a receipt upon delivery addressed as set forth above in Section I or such other addresses as either Party may hereinafter designate by a Notice to the other. Notices are deemed delivered or given and become effective upon mailing if mailed as aforesaid and upon actual receipt if otherwise delivered to the addresses set forth in Section I.

IX. APPLICABLE LAW.

This Agreement and the construction and enforceability thereof shall be interpreted under the laws of the state of ________________.

X. FINAL AGREEMENT.

This Agreement, together with its appendices and attachments, shall constitute the full and final Agreement between the Parties, shall supersede all prior agreements, communications and understandings regarding the subject matter hereof and shall not be amended, modified or revised except in writing. **This Agreement shall bind the Parties as of the date on which it was executed.** Any inconsistency in this Agreement shall be resolved by giving priority in the following order: (a) Amendments to the Agreement, in reverse chronological order; (b) the Agreement, Sections I through XI; (c) General Terms and Conditions of the Agreement; (d) Appendices to the Agreement; and (e) Attachments to, or other documents incorporated into, the Agreement.

XI. INCORPORATION OF GENERAL TERMS AND CONDITIONS:

THIS AGREEMENT IS SUBJECT TO THE GENERAL TERMS AND CONDITIONS ATTACHED HERETO AND MADE A PART HEREOF. IN ADDITION, THE PERFORMANCE OF THIS AGREEMENT BY ESCO IS SUBJECT TO THE PROVISIONS OF THE UTILITY AGREEMENT.

IN WITNESS WHEREOF, and intending to be legally bound, the Parties hereto subscribe their names to this instrument hereinabove in Section I as of the date of execution first written above.

SUMMARY OF GENERAL TERMS AND CONDITIONS

Section 1.	Operating the Premises and Maintaining the Use of the ECMs.
Section 2.	Equipment Location and Access.
Section 3.	Construction.
Section 4.	Ownership of ECMs.
Section 5.	Condition of ECMs.
Section 6.	Transfer of ECMs.
Section 6.1.	Material Shortfall in Energy Savings.
Section 6.2.	Customer Purchase Option.
Section 6.3.	Purchase Upon Customer Default.
Section 6.4.	Expiration of the Term.
Section 6.5.	Transfer Without Encumbrance.
Section 7.	Requirements of Utility Agreement.
Section 8.	Insurance.
Section 8.1.	Insurance on the Premises and the ECMs.

Section 8.2. Risk of Loss of the ECMs.
Section 8.3. ESCO's Insurance Requirements.
Section 9. Hazardous Materials and Activities.
Section 10. Events of Default.
Section 10.1. Events of Default by Customer.
Section 10.2. Events of Default by ESCO.
Section 11. Remedies Upon Default.
Section 11.1. Remedies Upon Default by Customer.
Section 11.2. Remedies Upon Default by ESCO.
Section 11.3. Limitation of Remedies.
Section 12. Representations and Warranties of Both Parties.
Section 13. Compliance With Law and Standard Practices.
Section 14. Assignment.
Section 15. Taxes.
Section 16. Severability.
Section 17. Effect of Waiver.
Section 18. Usage of Customer's Records.
Section 19. Air Emission Rights, Credits or Allowances.

APPENDICES:

"A" Description and Address of Customer's Facility(ies).
"B" Energy Audit Report and List of ECMs.
"C" Method of Savings Calculation.
"D" Termination Values.
"E" Requirements of Utility Agreement.

15

Steam System Optimization: A Case Study

The steam system optimization (generation, distribution, use and condensate return) offers a large opportunity for action to comply with the new levels of energy efficiency standards. Superior design and improved maintenance practices are the two main sources of savings in steam systems. Increased competition no longer permits an industry to survive with energy waste that could be eliminated.

This chapter highlights the study findings of the steam system in a plant from the food industry. The steam system operates with an annual budget of $1.9 million. Normal steam demand ranges between 80,000 to 85,000 lb/hr.

The steam system analysis identified energy savings worth $270,000 per year. The optimization measures were in two categories:

- No cost/low cost optimizations that can be done through better maintenance and improved operating conditions.

- Major improvements that require a significant investment and include the modification of the process and major equipment.

INTRODUCTION

Nearly half of industrial energy is used to generate steam. Improving steam system efficiency will contribute significantly to the profitability for every plant. There are proven energy savings techniques that

---

Presented at the 22nd World Energy Engineering Congress by Nevena Iordanova, Ven V. Venkatesan, and Michael Calogero

capture significant energy saving benefits upon implementation. Some of them are industry wide, others are specific for each industry.

Steam systems consist of several components such as:

- Steam generation
- Steam distribution
- Steam usage
- Condensate collection and return

A comprehensive steam system optimization addresses these interrelated areas, identifies the problems, and recommends all necessary corrective measures to eliminate them.

OVERVIEW OF THE SITE STEAM SYSTEM

This chapter discusses the steam system of a typical food industry plant in the USA.

Steam Load and Cost

ABC Plant spends over $1,920,000 for steam generation (fuel, water, electricity and chemicals) every year. During 1997, steam generation averaged from 80,000 to 85,000 lbs/hr. At the end of 1997, with the commissioning of a new process reactor, the steam consumption will be increased by 7,000 lb/hr. The steam cost, excluding labor, as calculated by ABC Plant engineers, varies between $3.00 and $4.08 per 1000 lb depending on the condensate disposition (see Table 15-1).

Table 15-1. Steam cost.

Steam Cost Based on Condensate Disposition	*$/1000 lb*
100% Condensate Return	$3.00
Steam Injected to Process	$3.28
20% Condensate Return	$3.86
Condensate to Sewer	$4.08

Steam Generation

ABC Plant has two (2) boilers to meet its steam demand. Both boilers are natural gas and waste fuel fired, with the capacity of 150,000 lb/hr and 70,000 lbs/hr at 165 psi and 370°F. At normal operating conditions they work at average steam output respectively of 80,000 lb/hr and of 10,000 lb/hr (standby). There are meters available to measure steam and natural gas flow, flue gas stack temperature, excess O_2, pressures and other important parameters controlling the combustion and boiler operation. Table 15-2 shows the readings for June 10th, 1998, that are used for further calculations.

By the end of 1998, a new natural gas fired boiler with the capacity of 30,000 lb/hr at 600 psig will be installed to serve the needs of the new process reactor. It will operate at 7000 lb/hr and will back up the other boiler during annual shutdowns.

Table 15-2. Stack gas and sir temperatures.

Date	*Stack Gas to Atmosphere* (SGA)*	*Air to Main Burners* (AMB)*	*Stack Gas to Air Heater* (SGH)*
10-Jun	370	500	740
	320	436	627
	365	464	698
	340	457	665
	325	428	615
	341	461	704
Average	344	458	675

*Location of the measurements see in Figure 15-1.

Steam Distribution System and Utilization

Steam is generated at 165 psig and distributed at 165 psig and 35 psig. In the buildings, pressure-reducing stations and temperature-regulated valves additionally reduce and control the pressure to the required levels for different consumers.

ABC Plant utilizes steam in two ways:

- indirect use, returning the condensate after process heating, hot water generation and comfort heating.
- direct use in XXX moisturizers, XXX steamers, XXX water tanks, XXX process and deaerator.

The following differentiation of the average steam consumption by area is established:

Medium Pressure (MP) Steam (165 psig)	Building	Average Flow lb/hr
1. X	15, 16	15,000
2. XX	16, 17, 18	30,000
3. XXX	18, 19, 20, 20A, 22	20,000
4. XXXX	10	20,000
Low Pressure (LP) Steam System (35 psig)		

The 35 psig steam system is used for building air space heating throughout the plant, HVAC systems and in the deaerator.

Steam distribution lines are predominantly accessible and insulated.

There are about 360 steam traps in the whole complex. The average steam trap age is 10 years. Improper steam trapping, applications and piping practices were observed during the visit. More detailed data and comments about the steam traps and their performance were provided in a separate steam trap report.

Condensate return

Condensate recovery is about 20%, due to a large percentage of the total steam load used for direct injection. Obvious wastes of condensate were observed during the visit. The drainage of hot condensate, due to condensate return system problems, is one of the main concerns of plant personnel.

At the ABC Plant, condensate from the whole plant is collected at a Central Condensate Receiver (CCR) in the basement of Buildings 15 and 16. All the condensate return lines discharge to a common header above the CCR. It also collects the make-up water (MUW) for the boiler after the heat recovery of the XX waste water and the boiler blowdown. An 8″ vent maintains the tank at atmospheric pressure. Through electrical pumps the mix of condensate and MUW is routed to the deaerator, and then to the boiler.

The Medium Pressure Condensate (CM) and Low Pressure Condensate (CL) is either returned by direct discharge (steam traps) to the CCR or through pumps from the farthest points of the plant (Nitrogen and Mix XX areas). The MP steam users working in a modulated pressure control were identified as the most difficult part of the system to return condensate to the CCR.

SAVINGS OPPORTUNITIES

A thorough review of the system confirmed there are energy saving potentials in the boiler system, steam distribution system and in the condensate return system. The following paragraphs highlight eight energy savings opportunities identified at the site and describe the measures proposed to realize these savings.

1. Install Economizer in the Boiler Stack to Recover the Heat from Flue Gases

To recover the heat from the flue gases, ABC Plant has installed an air preheater in the boiler stack. During the study, the boiler was operating at an average load of 80,000 lbs/h, with 7.3% oxygen and 340°F stack temperature. A forced draft (FD) fan supplies the combustion air to the air preheater, and then to the boiler. The ambient air is preheated from ambient temperature to 460°F, while the flue gases are cooled by more than 360°F. Although there is heat recovery, the flue gas temperature is higher than optimum and shows a possibility for further heat recovery.

One of the most effective ways to recover energy from flue gases is to use an economizer to preheat the lower temperature water (make-up water) going into the boiler. At 165 psig, the temperature of the steam and water in the boiler is about 370°F and each pound of water needs 290 Btu/lb to reach the boiling point.

Hot flue gases contain a lot of wasted energy. It is beneficial to put this energy back into the boiler with the heat recovery equipment.

Preheating the incoming feedwater to the deaerator (after the CCR) in an economizer will improve the boiler efficiency. Since economizers are fairly expensive, they must be justified by a high level of boiler utilization. Once installed, they are usually trouble-free and require little maintenance.

It was proposed to install an economizer in the stack of the boilers to recover a portion of the heat that is carried away by the flue gases leaving the stack. For proposed modifications see Figure 15-1.

Installing an economizer will result in an annual fuel saving up to $72,900. The saving calculations are based on the average stack gas temperature (340°F) recorded by ABC Plant personnel and 85,000 lbs/h steam load.

The resulting increase of boiler efficiency will reduce the fuel consumption by 26,130 mcf per year and the total carbon emissions by 442 tons per year.

Total investment cost for materials and labor is expected to be less than $182,000 and the payback period for this project will be less than 2.5 years.

2. Install Vent Condenser to the Deaerator

Site observations identified steam loss through the deaerator vent, which is designed to release the non-condensable vapors. As informed

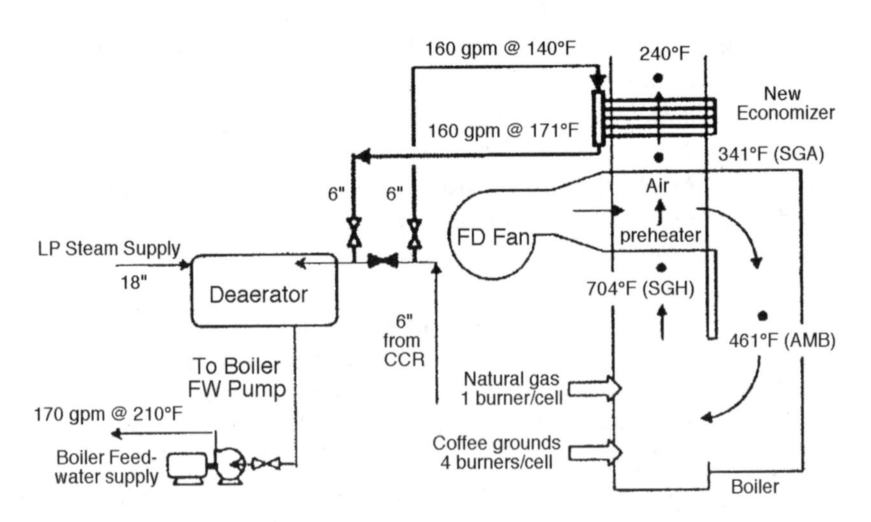

Figure 15-1. Heat recovery of the flue gases.

by the operating personnel and measured by the gages, the deaerator pressure is normally maintained at 3-4 psig. Even at normal operating conditions considerable steam will escape. Installing a vent condenser can recover this escaping steam, still allowing the non-condensable vapors to escape. There is no measurement of flash vapors and steam escaping through the vent; however, reliable estimations of the steam flow are used to size the vent condenser.

It is proposed to install a vent condenser at the top of the deaerator. The proposed scheme is shown in Figure 15-2.

The proposed vent condenser will be cooled by incoming mixed treated MUW and condensate from the Central Condensate Receiver before it enters the deaerator. The savings estimations for the vent condenser are based on 330 lbs/hr of steam, while the vent condenser may be designed to handle up to 660 lbs/hr of steam under upset conditions. Recovered condensate from the vent condenser will flow down to the deaerator by gravity.

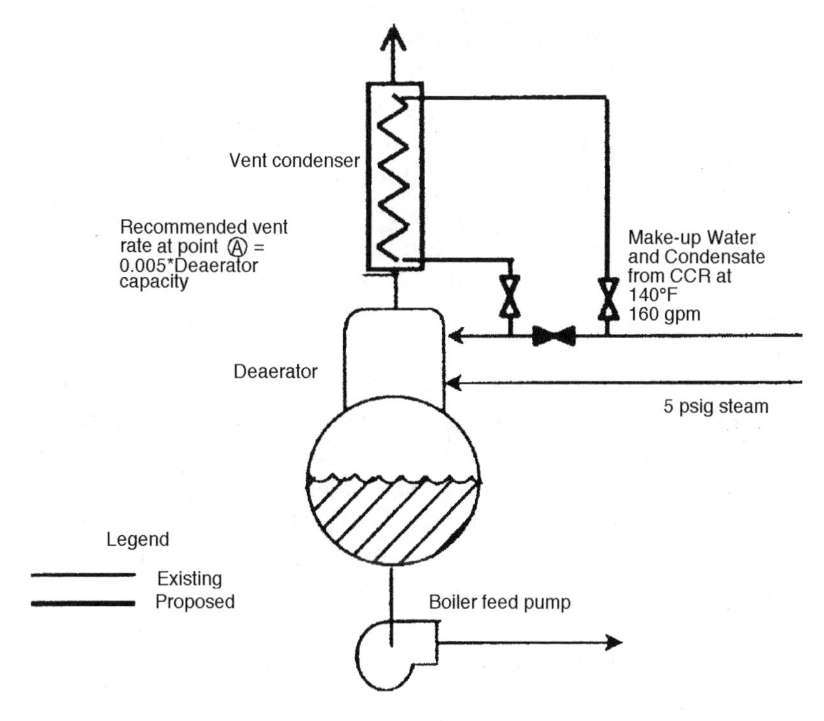

Figure 15-2. Proposed heat recovery at the deaerator.

Reduced pressure and recovery of vent steam at the deaerator will save in fuel, water and chemicals totally $10,400 annually.

Recovering the vented steam will reduce fuel consumption by 3,140 mcf/year and will reduce the total carbon emissions by 53 tons/year.

Saving water by increasing the condensate return will lead to reduction in raw water consumption by 350,000 gallons per year.

Total investment cost for materials and labor will be less than $18,000. The simple payback period is less than 2 years.

3. Quench the Flash in the Central Condensate Receiver Vent

Survey observations identified considerable steam loss through the CCR vent, which is designed to release the flash steam from the receiver and to maintain atmospheric pressure.

Discussion with operating personnel confirmed problems with the CCR and excessive pressure and steam CCR venting. Recently a new arrangement of the condensate return lines was established. A large header collects the condensate from the incoming return lines and the MUW enters the CCR next to the header inlet. This arrangement helps to absorb some of the heat from the flash steam, but a considerable amount escapes through the vent. At normal operating conditions, greater than a 5-foot plum was observed.

It is proposed to minimize losses due to vented flash steam by quenching it with make-up water. The proposed modification is shown in Figure 15-3.

Recovery of vented flash steam at the CCR will save fuel, water and chemicals equaling up to $10,900 annually. The savings estimations are based on flash steam velocity of 8 fps through the 8″ vent (375 lbs/hr of steam).

Recovering the heat of vented flash steam will reduce the fuel consumption by 3500 mcf per year and the total carbon emissions by 59 tons per year.

Saving water by increasing the condensate return will lead to a decrease in raw water consumption by 390,000 gallons per year.

Total investment cost for materials and labor will be less than $3,000. The simple payback period is less than 4 months.

4. Recover Heat from Cooling in Air Compressors

Four compressors (2 Joys and 2 Suns) supply more than 3000 cfm of compressed air at 90 psig to the plant. They are installed at

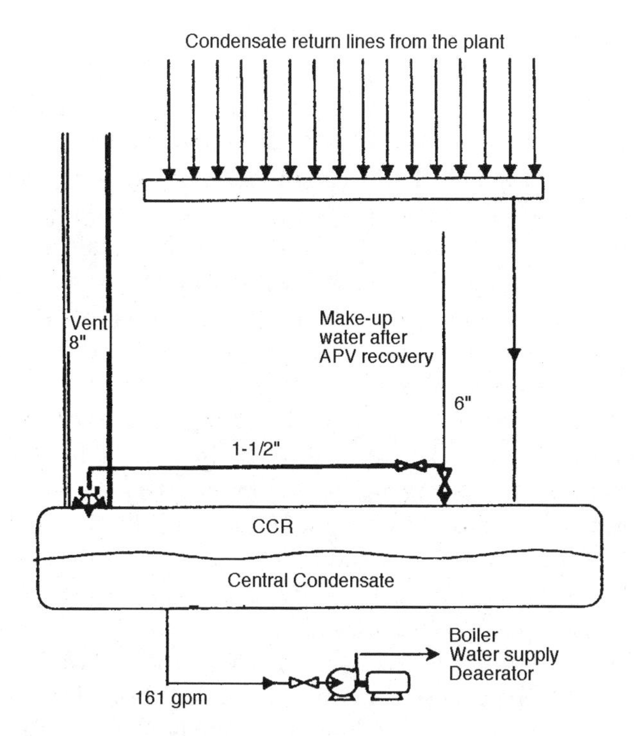

Figure 15-3. Quenching flash steam in central condensate receiver.

the ground floor of Building 23 and operate continuously throughout the year. These are two-stage compressors with an oil cooler, an intercooler and an aftercooler. At present, circulating cooling water recovers heat from the above coolers and rejects it to the atmosphere through a cooling tower. Only one of the compressors (Joy #1) consumes approximately 115 gpm of cooling water. The cooling water enters these heat exchangers at 87°F and leaves at 108-114°F. ABC Plant personnel had a big concern about the oil cooler performance at higher ambient temperatures. The set point for shutdown of the compressors is 140°F and the oil temperature reached 136-140°F during the survey.

Trouble-free compressor operation is ensured when required temperatures are maintained in the coolers. Their performance is highly affected by scaling problems at the jackets. At present, circulating cooling water that flows through the heat exchanger is chemically treated for hardness. Hence, no scaling problems are experienced. However, at present, the cooling tower rejects 11 million Btu of heat annually from this compressor alone, while make-up water is heated using steam in the deaerator. Steam system efficiency can be improved by matching this heat rejection with a suitable heating need. However, the matching need to recover the compressor's heat rejection should use chemically treated water and should be in continuous service. ABC Plant boilers and process consume cold make-up water at an average rate of 220 gpm throughout the year. Softened water is the best quality treated water and chances of scaling are negligible. Make-up water from the softening plant is available at ambient temperature. In addition to the recovery of the heat that is rejected to the atmosphere, this method of cooling will assure the unimpeded work of the compressors. Furthermore, scaling problems are minimal in once-through cooling water systems.

It is proposed to preheat the soft water from the softeners, by passing it through the oil coolers and part of intercoolers and aftercoolers of air compressors. Accordingly, a 6″ soft water supply line and an insulated 6″ return line will be provided from softeners to the compressor building to recover this heat. These two lines will be tied-in parallel to the existing circulating water line. This configuration will allow ABC Plant to revert back to the existing circulating cooling water system in case of an emergency. Since soft water is passed through the coolers of the compressor, a chance of scaling is negligible. Suitable check valves and correct changeover sequence will prevent cooling tower water contamination of the boiler feedwater system.

There is no special equipment involved and no interference with any heat recovery process is anticipated due to the proposed modifications. The suggested modifications are shown in Figure 15-4.

Recovering heat that is rejected at present to the atmosphere will save 21,000 million Btu annually. This will save fuel in the boilers and reduce the cooling tower load considerably while the compressor reliability is maintained, if not improved. At 80% boiler efficiency and 1997 fuel costs, this will save ABC Plant $57,300 annually.

Recovering the heat from compressor cooling will reduce the fuel consumption by 20,500 mcf per year and reduce the total carbon emissions by 346 tons per year.

Total estimated investment to implement the proposed modifications will be $62,000 and the investment can be recovered in 1.1 year.

5. Optimize Condensate Return System at XX Area

The predominant problem throughout the facility is the conden-

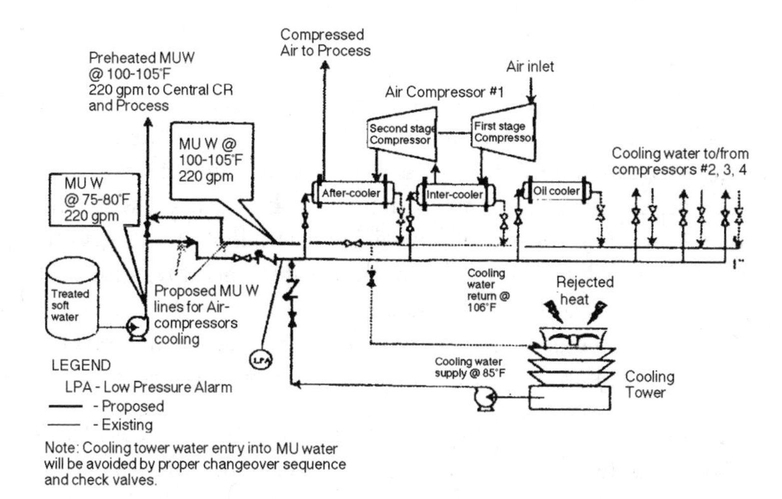

Figure 15-4. Heat recovery from compressor cooling.

sate return system. Considerable condensate draining is required, combined with venting steam at the CCR tank.

Many steam and condensate system inadequacies were observed, including incorrect and missing drainage of supply lines, misapplications, inappropriate sizing and condensate removal from heat exchange equipment. This led to drainage problems, condensate flooding and energy/performance losses as equipment became damaged (water-hammered and/or corroded), which reduce efficiency and require increased manpower for maintenance.

XX area uses MP and LP steam for the process. The LP steam is supplied to the Moisturizers (#1 to #4) for hot water heat exchangers (HE's) and for direct steam injection. All other process users are MP steam users. All heat exchangers (process heating) are on-demand systems using steam modulating pressure control valves to control steam pressure and temperature. During no-load conditions, steam flow is reduced to zero. This modulating steam pressure causes drainage problem with the heat exchangers because the systems are not trapped properly. The above systems are not efficient and are maintenance intensive.

In wintertime, comfort heating equipment uses LP steam. It is reduced through a PRV at the 5th floor of Building 20.

Problems in returning the condensate and operating the equipment forced the plant personnel to open the bypass and bleed valves, and to drain the condensate to the floor. At modulated steam pressure supply, the steam trap cannot respond immediately to the process requirements as the motive force for every steam trap is pressure differential. At start-up, low inlet pressure prevents drain traps from functioning properly. Live steam from failed steam traps in the condensate return system further restricts the steam trap function as the back pressure increases and makes the pressure differential even smaller. If drain valves are not opened, the obstructed condensate creates back-pressure problems, leaks and safety relief valves discharges (Figure 15-5).

Several problems were observed and noted about the improper operation and installation of the condensate return system in this area:

- No air vents are installed at heat exchangers and steam traps. This decreases heat transfer efficiency, lowers the process speed and creates opportunities for air locks, which prevents the condensate removal.
- No differentiation between low and medium pressure condensate return is established in the original design of the system.
- There is no separate return of modulated and constant pressure condensate. This allows the discharged higher pressure condensate to create back pressure for the traps operating at lower pressure. This keeps them from working properly

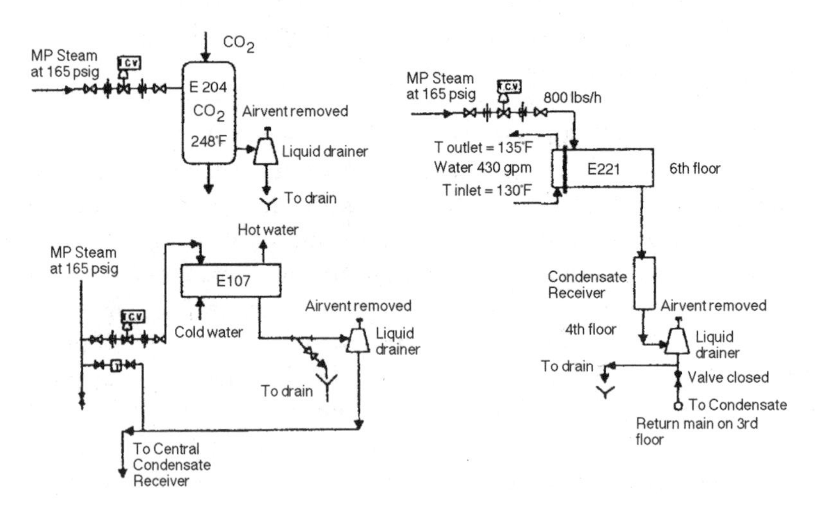

Figure 15-5. Existing piping.

- Pipes and return lines are unreasonably restricted or undersized.

Energy efficiency of the steam supply and condensate return system depends on the effective usage of steam traps, air vents and properly sized and piped condensate return lines.

Properly sized and installed condensate return lines assure reliable operation of the main process equipment and maximum amount of returned condensate.

Maximizing condensate return to the boilers reduces the quantity of make-up water, as well as the amount of chemicals for water treatment. Also condensate contains valuable energy and by returning it, it reduces the fuel burned in the boiler.

To rectify the condensate return problems, it was proposed:

1. To facilitate the condensate return where there is lower steam supply pressure, it is proposed to install steam driven pump (pumping trap) after each separate heat exchanger ahead of the existing steam trap. An equalizing line will equal the pressure in the pumping trap and the CR. If a receiver is not available a new one should be installed. This will assure the condensate return and the back-pressure will not affect the work of the heat exchanger.

2. To remove the air from the system air vents have to be installed at the steam traps and at the heat exchangers. To prevent vacuum formation vacuum breakers have to be installed. The existing steam trap will handle the condensate when the inlet steam pressure is higher than the back pressure in the condensate return lines and will stop the steam from flowing through the pump trap. Figure 15-6 is an illustration of the modifications proposed in 1 and 2.

3. To use the heat in high pressure condensate it is proposed to install a flash tank in the middle of line CL-5123, right after the tie-in of line CL-3309 (E-215 condensate return). It will collect the condensate from several heat exchangers. The flash steam has to be routed to LP steam header to steam users with continuous operation, such as the HE for the moisturizers and the direct steam injection. Other potential users of LP steam are the concentration tank, TK-112 and the hot water HE. During winter, it will be used in the comfort heating equipment. Another

option for summertime use of flash steam is to be routed to the CCR and to condense, while preheating the MUW. Figure 15-7 is an illustration of the proposed modification.

Implementing the above recommendations will save ABC Plant a minimum $74,300 annually. Separation of the high temperature condensate will prevent thermal shock in the pipes and will create a safe working environment. It will improve the system reliability and will reduce maintenance costs.

Improving the condensate return system will decrease fuel consumption by 12,270 mcf per year and will reduce the total carbon emissions by 207 tons per year.

Increasing the condensate return will decrease raw water consumption by 5,890,000 gal per year.

The required investment to implement the proposed modifications is estimated to be $137,000. The payback will be less than two years.

6. Prevent Thermal Shock in Aquachem and Nitrogen Areas

At present, in the aquachem area, condensate from two HEs (outside) is returned through pumping traps and discharged to the nitrogen

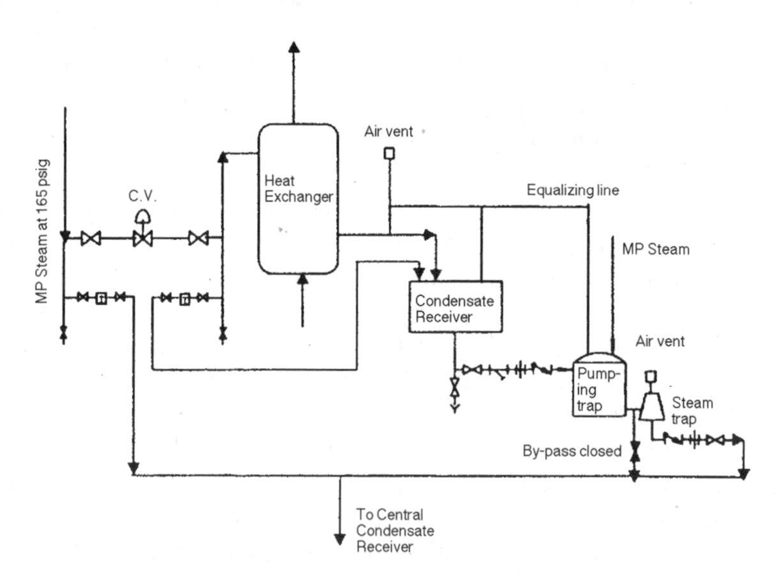

Figure 15-6. Proposed condensate removal from the HE working under modulated pressure.

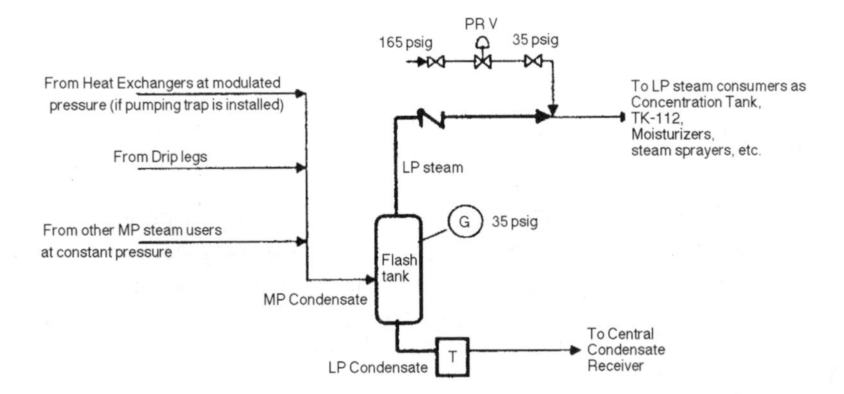

Figure 15-7. MP Condensate flash steam usage.

tank to evaporate the nitrogen. The pumping traps will discharge the condensate to the condensate return line when the inlet pressure is lower than the condensate return pressure. In the same tank steam is directly injected at low pressure. A PRV reduces the pressure from 160 to 60 psig. A drip leg from the main line also discharges in the tank. The temperature in the tank is maintained at 145°F. The condensate is collected through an overflow pipe, which is routed to the condensate collection tank and pumped with an electrical pump to the condensate return header and further to the CCR (see Figure 15-8). Water hammer occurs in the return line after the pump. The condensate is drained from the receiver to the sewer.

Thermal shock is a term that describes the interference between fluids with different pressures and temperatures. When condensate at steam supply pressure passes through the trap and enters the condensate header at lower pressure, flash steam is released. One pound of steam occupies more than 1500 times the volume of a pound of water. When the flash steam mixes with cold condensate it collapses. However, when the flash steam collapses, water is accelerated into the resulting vacuum from all directions. Water hammer results from the collapse of this trapped steam. The localized sudden reduction in pressure caused by the collapse of the steam bubbles has a tendency to chip out pipe and tube interiors. Oxide layers that otherwise would resist further corrosion are removed, resulting in accelerated corrosion. Over a period of time, this repeated stress and wear on the pipe would weaken it to a point of rupture.

It is proposed to:

1. Install steam traps after the pumping trap at the outside HE. This will prevent the steam from blowing through the pump trap when the inlet pressure is higher than the condensate return pressure.

2. Reroute the pumped condensate (140°F) in a separate 2″ line to the CCR to prevent water hammer by mixing cold and hot condensate. This will also help the discharge of the steam traps in the building because there won't be any back pressure created from the electrical pump.

Figure 15-9 reflects the above recommendations.

Implementing the above listed recommendation will save ABC Plant $4,300 yearly. It will also improve the system reliability and safety, and will reduce maintenance costs.

Improved condensate return system will reduce fuel consumption by 255 mcf per year and reduce the total carbon emissions by 4 tons per year.

Improving the condensate return system will lead to a decrease in raw water consumption by 520,000 gallons per year.

Total investment to implement the above listed recommendation will be $15,000. The payback will be less than 3.5 years.

7. Recover the Flash Steam from X's Heat Exchangers

At present, in Building 18, four (4) HEs continuously heat fresh

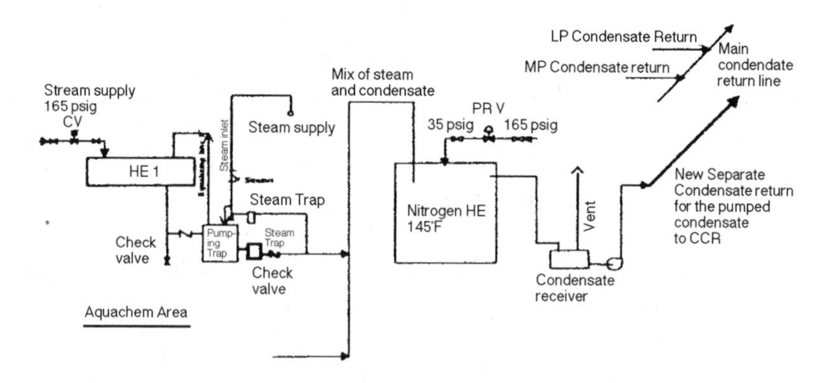

Figure 15-9. Proposed condensate return modifications.

water from 115°F to 335°F. Three of them are operating while the fourth is on standby, The fresh water flow is 30 gpm through each heat exchanger. Each heat exchanger consists of two separate shell and tube heat exchangers with common steam control valve and separate steam traps. The condensate is returned directly to the CCR. The heat content of the high-pressure condensate is not used at present and the flash steam is vented through the vent of the CCR. See existing arrangement in Figure 15-10.

It is proposed to rearrange the condensate return system in a way that the MP condensate can be used. The flash steam from it will preheat the incoming fresh water. This will decrease the steam flow through the PRVs and reduce the consumption of MP steam. To implement this proposal the existing system will not be disturbed. A new heat exchanger will be installed on the common water supply line ahead of the four existing HE and a flash tank at the condensate discharge side downstream of the existing steam traps. The HE will use 15 psig flash steam from 165 psig condensate. A new 3" pipe originating from the flash tank will supply the LP steam to the new HE. Safety drain traps will be installed on the HE's outlet to assure proper drainage at start-up

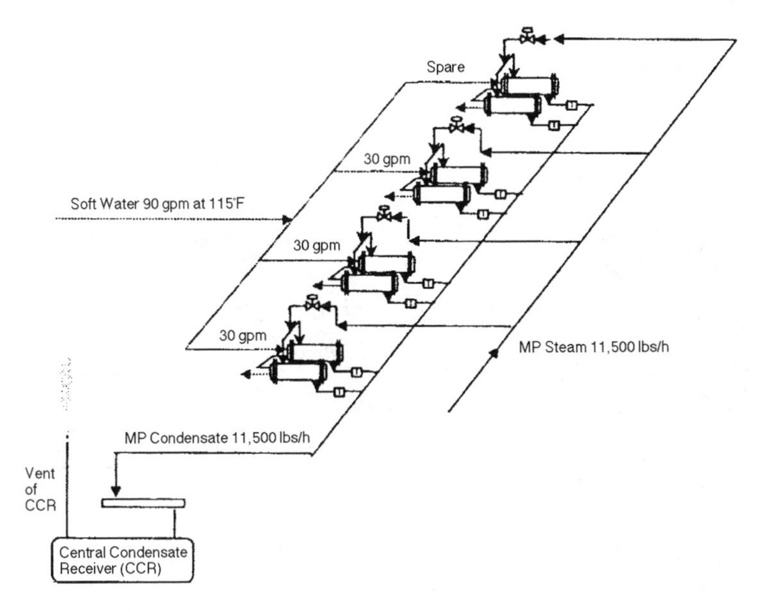

Figure 15-10. Existing arrangement at heat exchangers.

operation. See proposed modifications in Figure 15-11.

Implementing the above recommendation will benefit ABC Plant with savings of $28,700 annually.

Improving the condensate return system and using the flash steam from MP condensate will reduce the fuel consumption by 10,270 mcf per year and the total carbon emissions by 173 tons per year.

Total investment to implement the above listed recommendation will be $51,000 and the payback period is less than 1.8 years.

8. Recover Pressure (Energy) Instead of Reducing Steam Pressure through PRV

The deaerator consumes LP steam year round, at average load of 5,000 lb/hr. All the steam is reduced from 165 psig to 35 psig through a pressure-reducing valve (PRV), while an existing steam turbine driven feedwater pump is abandoned. Due to high maintenance cost and operational problems, this pump has not been in use for more than 10 years.

Reducing steam pressure through PRV wastes useful available energy. When steam flow through a PRV is small and fluctuating, recovering mechanical energy may not be economically feasible. However, the deaerator needs about 5,000 lb/hr steam constantly throughout the year.

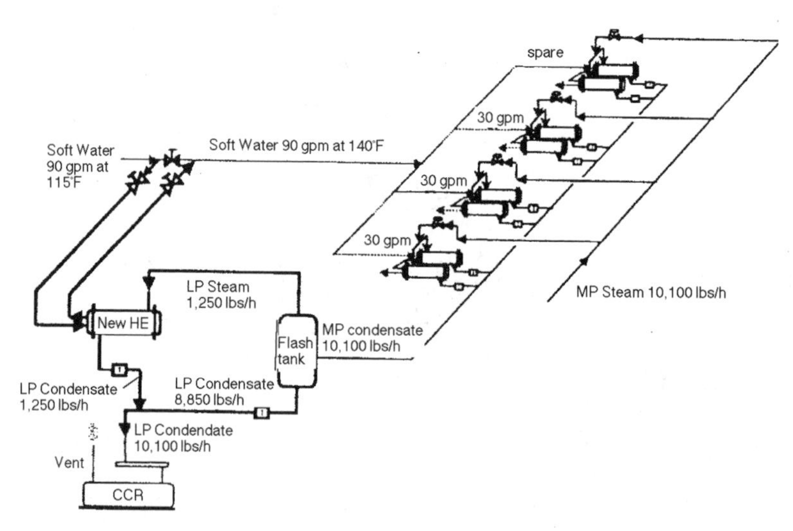

Figure 15-11. Proposed heat recovery at heat exchangers.

It is possible to utilize the pressure (energy) available in steam to drive the existing pump; that will reduce electrical consumption. An advantage of this is the simplicity of replacing one standby electrical motor, without disturbing any of the existing arrangement. Also, the present system will turn on when steam flow drops, and this will be a smooth transition. One pump can be replaced with the proposed steam turbine and the necessary steam line tie-in connections. The steam turbine will work as a parallel drive to the existing feedwater pumps. This pump is located one level below the present PRV and the 165 and 35 psig steam headers. It will be in service continuously. The configuration for the turbine drive installation is shown in Figure 15-12.

The power available by reducing an average of 5000 lb/h will be in the range of 70 hp. The recovered pressure (energy) from steam is expected to save $10,800 annually.

When the LP steam usage is reduced, due to the future improvements in the deaerator and condensate recovery systems, the benefits are expected to decrease if the deaerator is the single user of LP steam. This saving opportunity should be considered after identification of other LP steam users (additionally to the deaerator). This consideration will assure the required steam load for the pump drive and the constant head at the pump discharge even when the deaerator LP consumption changes.

Then utilizing the pressure (energy) available in this steam is technically and economically feasible, but not recommended at this stage of the study.

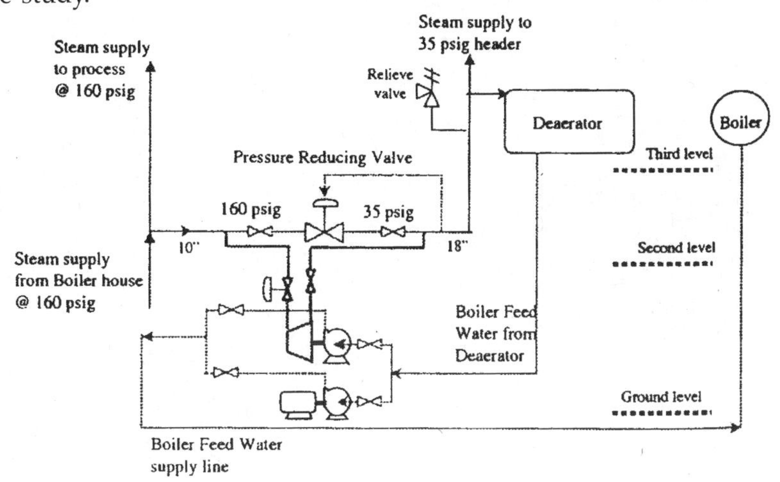

Figure 15-12. Steam turbine drive to BFW pump.

CONCLUSION

The ABC Plant has a potential to **capture $270,000** in annual energy savings by implementing the above eight recommendations.

These recommendations require no major process modification. Some of them could be implemented through a periodic maintenance program. The others requiring new equipment can be done during the plant turnaround.

Optimization of the steam system by implementing the above recommendations will also **reduce the total carbon emission** to the environment by **1,284 tons** annually.

References

1. 1997 Operating data from client's food processing facilities
2. Turner, Wayne C., *Energy Management Handbook,* 1992.
3. Armstrong International Inc., *Steam Conservation Guidelines for Condensate Drainage,* Handbook N-101, 1994.

16

Cost Containment Design for Commercial Geothermal Heat Pumps

A geothermal heat pump (GHP) system consists of four distinct subsystems: (1) the ground heat exchanger loop, (2) the in-ground and building interior piping loop that connects the heat exchangers to the heat pumps, (3) the water source heat pumps and (4) the air distribution system. When engineers, contractors and vendors attempt to contain installation costs, often they have concentrated on reducing the cost of item (1), the ground loop. Reducing the size of the ground loop will significantly lower operating efficiency and will negatively impact system life and maintenance requirements. However, two of the other subsystems that typically receive little attention are the piping loop and the air distribution system. Modifications to traditional piping and ducting methods can both reduce installation costs and improve operating efficiency. This chapter will discuss the design methods that will lead to these benefits. The chapter will also integrate these methods with a widely used ground loop sizing procedure.

The geothermal heat pump is a heating and cooling concept that still faces barriers to realizing its full potential. There are many examples of GHP installations that have provided high levels of customer satisfaction as noted by the rapidly increasing number of schools using this technology (GHPC Site List, 1999). There are even a few instances of systems where customers are satisfied not only with performance and dependability but also installation costs. However, in many areas GHPs cannot shed the reputation (and reality) of having a high installation cost. There are also examples of systems that have

---

Presented at the 23rd World Energy Engineering Congress by Stephen P. Kavanaugh

not lived up to expectations because of misguided attempts to reduce first costs.

There are pathways to affordable GHPs that do not compromise their inherent efficiency, dependability, and comfort. These pathways include simplicity of design, quality of installation, access to information, and competition in the market.

Simplicity of Design

GHP systems are inherently efficient and comfortable. The best designs are elegant in their simplicity. Systems need not contain redundant and expensive components. Complexity rarely improves performance and typically leads to customer dissatisfaction and increased service requirements.

Quality of Installation

The ground loop should be installed by a well-trained and experienced loop contractor. This quality must also be extended to the heat pump equipment and piping system in the building. Fortunately, this quality can be attained with simple components.

WHY GHPs? WHY NOW?

Geothermal heat pumps are becoming increasingly popular in the commercial sector where the added premium compared to minimum efficiency equipment is much smaller than in residential applications. The most popular application is in schools as indicated in the site list published by the Geothermal Heat Pump Consortium (GHPC, 1999). This increase in commercial application is surprising given the modest level of support provided by the major HVAC manufacturers and the federal government.

One primary reason for this increase is indicated by the graphic on the following page. Although GHPs require ground loops, which are outside the scope of the traditional HVAC industry, the components in the building can be extremely simple and dependable. In fact, the primary equipment and related components are even less complicated than those used with residential HVAC systems. Furthermore, the recommended high density polyethylene (HDPE) piping material was developed for the natural gas and oil field industry, which are much more demanding applications. The net effect is a very dependable, simple and low maintenance system that has very low energy costs.

However, the challenges to wider acceptance remain. Drill rig operators must adapt to becoming ground loop contractors and then be invited to the wedding with HVAC contractors and engineers. This unlikely union will not always be welcomed to share in the HVAC family inheritance. However, the HVAC industry would be wise to prepare as more potential customers begin to hear about the quality of offspring the sometime not-so-happy couple will produce.

DESIGN METHODS TO REALIZE ADVANTAGES

For best results the three major components of a geothermal heat pump (GHP) should be designed as a system. The experienced GHP design team will be able to adjust heat pump specifications, the piping/pump network, and the ground loop to optimize the system. For example, it would be prudent to improve the heat pump EER from 11 to 14 (COP = 3.2 to 4.1) at a cost of $150 per ton ($43/kW) rather than to increase the loop length by 50 ft. per ton (4.3 m/kW) at $7 per ft. ($23/m)

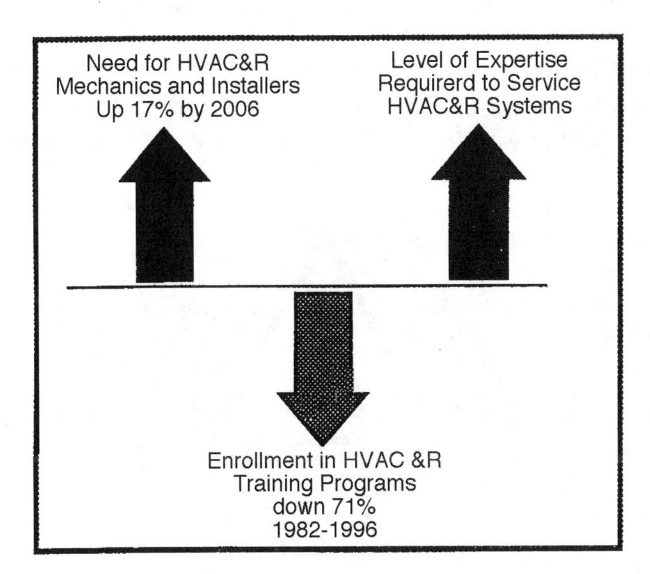

Figure 16-1. Possible reasons for increase of commercial geothermal heat pumps.

to achieve the same level of performance. Figure 16-2 illustrates the primary characteristics of an efficient and low maintenance system.

Heat Pumps

An EER of 13 (COP=3.8) at ARI 330 conditions should be a minimum. This type of equipment should not be expensive since this level of efficiency can be easily achieved with the same compressors, fans, and motors used in a 10 SEER air-source heat pump. Higher EER (14 to 18) single speed units can often be justified. However, multi-speed (or multi-stage) units are usually unnecessary in commercial applications. Although multi-capacity units currently available in the U.S. have high "rated" efficiencies, they have much lower high-speed efficiencies. Latent capacity is often compromised to obtain the high efficiency rating.

Ground Loops

Tools are available for loop design, which is performed in conjunction with heat pump selection. Closed loops must be of sufficient length, depth and bore separation to provide design capacity and efficiency during extreme conditions. Bore grouts must have good thermal properties and protect groundwater. Open loop wells must be properly designed and isolation (plate) heat exchangers are required.

Piping and Pumps

Required pump motor size should fall in the range of 5 to 7-1/2 hp per 100 tons (0.0016 to 0.0021 kW_e/kW_t). If pumps are larger,

Figure 16-2. Features of a high efficiency, low maintenance GSHP (OTL, 1998)

head loss or flow rates are excessive. Variable speed pump control is recommended for central systems, but there are other options that can minimize pump energy to below 10% of total energy use. Continuously operating fixed-speed pumps are discouraged, even in primary-secondary loops.

SOFTWARE

An evaluation of the five design programs available for vertical loops was conducted (Thornton, et al. 1997). One of the programs produced acceptable results compared to field data from a large sample. Modifications to the other programs were made and subsequent tests from other sites resulted in better agreement (Shonder, et al., 1999 and 2000). Therefore, design practices with regard to ground heat exchangers are becoming more accepted in the engineering community.

Figures 16-3, 16-4, and 16-5 are selected screens from one of the available design programs (EIS, 1997). This program begins with the completion of a heating/cooling load calculation program. The user selects one of the manufacturers shown in Figure 16-3 and the program corrects the operating characteristics to user-specified water loop temperatures and flow rates. The program then selects the smallest unit of the chosen manufacturer that will satisfy the load.

The user must then enter information about the ground loop characteristics. Figure 16-4, which is used to describe the ground loop pipe, is one of the three screens for this purpose. Another screen is necessary to describe the thermal properties of the ground and a third is needed to describe the overall layout of the piping grid.

Figure 16-5 demonstrates one of the output screens for this program. This screen shows the required ground loop length required to maintain the user-selected loop temperatures. Additional outputs provide equipment capacity and efficiency information, should the user want to see the impact of using different loop temperatures, equipment, ground coil arrangements, piping options, bore grouts, and other alternatives. The program also provides the option of designing a hybrid cooling tower-ground loop system for buildings with high cooling loads.

A critical need in the design procedure of closed-loop GSHPs, or ground-coupled heat pumps (GCHPs), is an accurate knowledge of soil/

HEAT PUMP MANUFACTURERS

( ) Addison (1.50 to 10 ton)
( ) Carrier (.50 to 10 ton)
( ) ClimateMaster E (.50 to 10 ton)
( ) ClimateMaster P (1.50 to 5 ton), E (.50 to 1.50, 6-10)
( ) Comfort-Aire (1 to 5 ton)
( ) FHP GT/SL (.75 to 20 ton)
( ) FHP SX (2-6 ton), SL (.75 to 2, 7 to 20 ton)
( ) Hydro Delta
( ) Mamouth (1 to 20 ton)
( ) McQuay (1 to 5 ton)
( ) Trane WPD (1, 1.50, 8 to 24 ton), E (2-5 ton)
( ) Trane WPD (1, 1.50, 8 to 20 ton), F (2-6 ton)
( ) WaterFurne SX (.75 to 10 ton)
( ) WaterFurn AT (.75 to 6 ton), SX (8 to 10)
( ) Other

Figure 16-3. Heat pump manufacturer selection screen.

BORE HOLE/PIPE RESISTANCE

[Main Screen] [Next Screen]

Bore Hold Diameter = [ ] inches

Grout/Backfill Property Table Conductivity [ ]

Backfill/Grout Conductivity = [ ] Btu/hr-ft-F

HDPE U-Tube Nominal Diameter = [ ] inches SDR = [ ]

Tube Flow Regime = [( ) Turbulent] [( ) Transition] [( ) Laminar]

Resulting Eqv. Diameter = [ ] Ft. Bore Resistance = [ ] hr-ft-F/Btu

Figure 16-4. Typical input screen for ground loop description.

MAIN OUTPUT SCREEN

Length for minimal ground water movement = 7170 ft.
30 bores @ 239 ft. each
(Design based on COOLING = net heat rejection to the ground)

Required lengths for high rates of groundwater movement
Cool: L = 5940 ft (198 ft/bore) Heat: L = 4210 ft (140 ft/bore)

Unit Inlet (clg.) = 85.0°F	Clg. Load/dmd = 342 MBtuh/23 kW
Unit Outlet (clg.)= 94.5°F	Ht. Load/dmd = 243 MBtuh/17 kW
Unit Inlet (htg.) = 50.0°F	EER (Ht. Pump/Sys.) = 14.9/14.1
Unit Outlet (htg.) = 43.9°F	COP (Ht. Pump/Sys.) = 4.2/3.9
Ground temp = 62°F	Liquid Flow Rate = 86 gpm

U-tube Data	Ground Properties
Equivalent Dia. = 0.25 ft.	Thermal cond. = 1.3 Btu/h-ft-°F
Bore Separation = 20 ft.	Thermal diff. = 0.75 ft^2/day

Figure 16-5. Main output screen.

rock formation thermal properties. These properties can be estimated in the field by installing a heat exchanger, as shown in Figure 16-6, of approximately the same size and depth (and with the same grout or fill) as the heat exchangers planned for the site. Heat is added in a water loop at a constant rate. Water flow, inlet and outlet temperatures are measured at regular intervals. These data are compared with a mathematical model of the heat transfer processes occurring in the borehole and surrounding soil. The model depends primarily on the thermal conductivity and diffusivity of the formation and attempts to minimize the impact of other parameters such as thermal conductivity of the grout, tube dimensions, and tube location in the bore (Kavanaugh, 2000).

Heat pump selection, interior piping, pump control, and other design details have not been evaluated in detail as the loop design programs. Kavanaugh and Rafferty (1997) include details and recommendations specific to this application. However, these recommendations have not been widely adopted. Many engineers prefer to adapt recommended practices for water loop heat pump (WLHP) and chilled water piping, which are primarily central systems. The building piping loop is

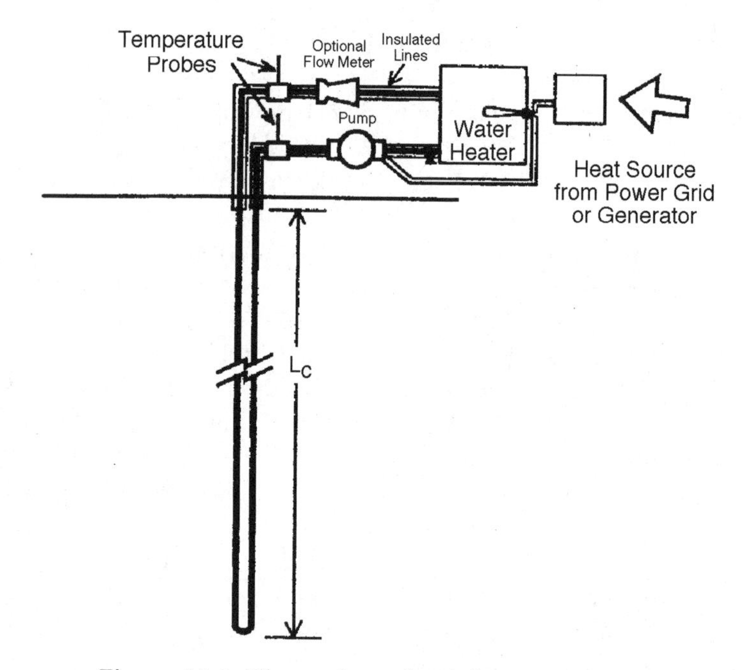

Figure 16-6. Thermal conductivity test system.

connected to a central ground loop. Interior piping is most often carbon steel. Pumps are arranged in primary-secondary loops and variable speed drives are only occasionally used. A survey indicates the pump energy is a disproportionate percentage of the total (Cane, 1995).

Although the central loop approach typically will result in the smallest total ground loop size, it will not necessarily be the lowest cost option. Larger ground loop piping and interior piping can exceed ground loops costs (OTL, 1998). Additionally, the required corrosion inhibitors for the carbon steel piping are not always acceptable for in-ground use. Finally, most of the system head loss typically occurs in the headers connecting the ground loop to the interior heat pumps. Vertical ground loop losses are typically less than 10 ft. (3 m) of water and heat pump losses are of a similar magnitude. In a central system, total losses are in the 60 to 120 ft. (18 to 36 m) range.

Two alternatives are shown in Figure 16-7. These will require 5% to 20% more ground loop but overall costs can be substantially reduced, since large diameter ground loop and interior metal piping are eliminated. If pumps with bronze, stainless steel, or epoxy-coated components are specified, the need for corrosion inhibitors is eliminated since the entire piping loop is HDPE. Pumping head is substantially reduced as shown in Table 16-1. The subcentral loop is shown to indicate one loop of several in an educational application. Some diversity is available since six zones are served by a common loop. The pumps can be low head circulators since the length of the headers is small. However, a check valve is required on every unit to prevent back-flow through units not in operation. The pump is activated only when its unit is running.

The second option in Figure 16-7 is a system with individual units with individual loops. Total ground loop length will be greatest since no diversity is available. While this option may appear to be simplistic, it is often the most superior option in many applications because of dependability, efficiency, lower cost and it can be installed and serviced

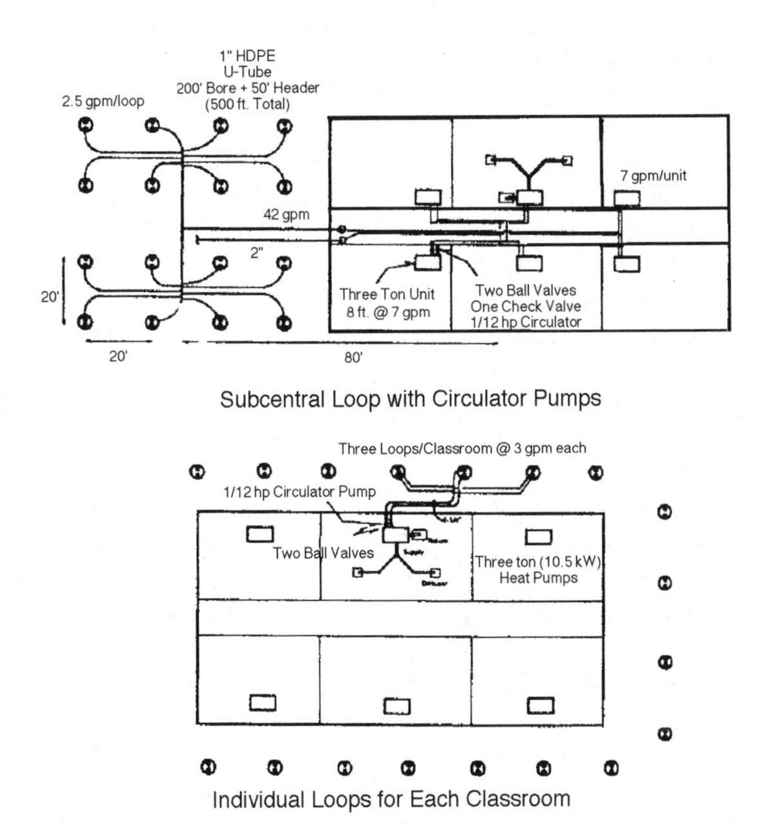

Figure 16-7. GHP loop options for low maintenance and pumping cost.

by personnel with limited HVAC experience. An experienced GHP engineer summarized his design philosophy with the statement,

I talk to the people in our office and encourage them to design systems that keep in mind what our local work force is capable of doing. The number of quality people is going down. They are still out there but there is a greater chance that the contractor that is low bid on the project is not going to have people with a high level of skill. So we need to keep it simple. Mike Green, PE (OTL, 2000)

Table 16-1. Head Loss Summary for Subcentral GHP Loop (PDI, 1999)

Water @ 60°F	Density = 62.4 lb/ft^3		Viscosity = 1. 14 cp		
No.	Component	gpm	Leq	ft/100	Δh (ft)
4	2″ HDPE T Branch	42	15.2	4.29	2.61
2	2″ HDPE Butt L	42	12.3	4.29	1.06
1	10-Ton Header End	2.5	18	.48	0.09
1	1″ HDPE SDR 11 Pipe	2.5	500	.48	2.42
1	1″ HDPE UniCoil	2.5	10.2	.48	0.05
1	1″ HDPE DR 11 Pipe	7	60	2.98	1.79
2	1″ HDPE Tee Branch	7	7.1	2.98	0.42
2	1″ Zone Valve (Ball)	7	Cv=35		0.18
1	1″ Swing Check Valve	7	Cv=21		0.26
1	Coil's Rated PSI	7			8
1	2″ PE DR 11 Pipe	42	160	4.29	6.87
Total			23.8 ft. of water (7.3 m)		

CHALLENGES IN THE U.S. MARKET

The penetration of these systems has occurred primarily in schools, offices, and other larger buildings in regions with different climates. Local popularity appears to happen by chance rather than by logical climatic or energy costs. As with any new technology, several challenges remain.

Challenge #1—Containment of High First Cost

First costs are higher than conventional low-end systems. GHPs cannot compete with this type equipment with regard to installation costs. However, first costs of GHPs in commercial applications have been reported and shown to be equal to, and in some cases lower than, higher quality systems like VAV and four-pipe chilled water that are common in the U.S. (Cane, 1995, Kavanaugh and Rafferty, 1997)

Challenge #2—Maintenance of Performance without Compromising Ground Coil and Equipment Quality

Too often individuals assume that since the system is "geothermal," it has to be better. Inexperienced individuals specify high cost equipment and expect it to operate at rated performance regardless of the ground loop design and installation quality. Another common practice, when the ground loop costs appear to be excessive, is to specify lower quality heat pumps and inferior water circulation systems.

Challenge #3—Development of Qualified Designers

HVAC designers are caught between tightening construction budgets, increasing need for compliance (indoor air quality, ASHRAE Standard 90.1, etc.), and much greater legal liability. Often the last thing they want to do is try something "new." This is especially true if they must devote a great deal of time learning how to design GHPs.

Challenge #4—Development of Qualified Loop Contractors

The investment in time and equipment required to be a proficient ground loop contractor is significant. Water well and environmental drillers can make much more per linear foot of bore. Although it is true that a contractor can drill more linear feet on a GHP project, the attractiveness of the work is not great because:

1. Jobs are few and far between in many areas and crews must travel to stay in business.

2. It is hard, dirty work, and employee turnover is great.

3. Often the loop contractor is blamed for high GHP costs.

4. The loop contractor is usually the first person asked to cut his price when bids exceed the budgeted building cost.

Challenge #5—Competing with Technologies that Generate Higher Vendor Profit

The attractiveness of GHPs to vendors, who have significant influence in the decision of what type system is used in buildings, is low. The increasing complexity of HVAC systems has caused the engineers to rely heavily upon the resources of manufacturers to design systems. However, the simplicity of GHP systems makes the engineer more independent. Thus, the total profit per job for the equipment supplier has the potential of being much lower and it may be viewed as being more profitable to sell conventional equipment.

ACKNOWLEDGMENTS

This chapter is part of the Strategic Outreach Program of the Geothermal Heat Pump Consortium in conjunction with the project sponsored by the Tennessee Valley Authority.

References

1. Geothermal Heat Pump Consortium, 1999, *GeoExchange Site List* (RP-01 1), Washington, DC.
2. *Air Conditioning, Heating, & Refrigeration News,* 2000, "How bad is the tech shortage," Business News Publishing, Troy, MI, Feb. 14.
3. *Outside the Loop Newsletter,* 1998. "Think system to get the most from GSHPs," University of Alabama. Tuscaloosa. Vol. 1, No. 2.
4. Kavanaugh, S.P. and K. Rafferty, 1997, *Ground Source Heat Pumps: Design of Geothermal Systems for Commercial and Institutional Buildings.* ASHRAE, Atlanta.
5. Thornton, J.W., T.P. McDowell, and P.J. Hughes, 1997, "Comparison of Practical Ground Heat Exchanger Sizing Methods to a Fort Polk Data/ Model Benchmark," ASHRAE Transactions, Atlanta, Vol. 103, Pt. 2.
6. Shonder, J.A., V. Baxter, J.W. Thornton, and P.J. Hughes, 1999, "A New Comparison of Vertical Design Methods for Residential Applications." ASHRAE Transactions, Atlanta, Vol. 105, Pt. 2.
7. Shonder, J.A., V. Baxter, P.J. Hughes, and J.W. Thornton, 2000. "A Comparison of Vertical Ground Heat Exchanger Design Software for Commercial Applications." ASHRAE Transactions, Atlanta, Vol. 106, Pt. 1.
8. Energy Information Services, 1997, GchpCalc, V. 3.1, Tuscaloosa, AL.
9. Kavanaugh, S.P., 2000, "Field Test for Ground Thermal Properties—

Methods and Impact on GSHP Designs." ASHRAE Transactions, Atlanta, Vol. 106, Pt. 1.

10. Cane, R.L.D, 1995, Operating Experiences with Commercial Ground-Source Heat Pumps, 863RP (Research Project Report), ASHRAE, Atlanta, 1995.
11. *Outside the Loop Newsletter,* 1998, "Monticello, Iowa, High School-First Cost," University of Alabama, Tuscaloosa. Vol. 1, No. 2.
12. *Outside the Loop Newsletter,* 2000. "The KIS Philosophy of Austin's Mike Green, P.E.," University of Alabama, Tuscaloosa, Vol. 2, No. 4.
13. PDI. 1999. "GHP pipe and fitting calculator," Phillips Driscopipe Inc., Richardson, TX.

17

Energy Audit Case Study: Log Home Manufacturer

While industry requirements may vary greatly depending on the product a plant is producing, most facilities use the same types of energy sources. Thus, the business of energy efficiency is by its nature quite cross-cutting.

This chapter presents the findings from an industrial assessment performed at a log home manufacturer. It is provided as an example to demonstrate a typical format and samples of energy efficiency calculations.

The company prepares kits for log homes that are delivered to a client's site where they are assembled. The plant is located in a cool climate with average year-round temperature of 41 degrees Fahrenheit. They operate with approximately 100 employees.

The energy consumption in the facility came in five forms. These were electricity, diesel fuel, kerosene, fuel oil and propane. The consumption amounts were determined by reviewing twelve months of energy bills. Table 17-1 summarizes energy consumption.

Table 17-1. Energy usage summary.

Energy Source	*Total MMBtu*	*Percent of Total Energy*	*Total Cost $*	*Percent of Total*
Electricity	3,623.21	47.3	$127,796.34	77.6
Diesel	1,887.00	24.6	$18,375.89	11.2
Kerosene	464.48	6.1	$5,614.14	3.4
Fuel Oil	1,200.83	15.7	$11,976.80	7.3
Propane	487.42	6.3	$846.81	.5
Total	7,662.94	100	$164,610	100

The total energy used is equivalent to approximately 7,662.94 million Btus. Total energy costs for this period were $164,610. The energy assessment recommendations contained in this report could save an estimated 800 million Btus each year, or 10.4% of the client's total energy usage. The annual cost savings related to energy would amount to approximately $21,000/yr, which represents about 12.7% of total energy costs.

Table 17-2 summarizes the report recommended actions along with potential savings and implementation costs.

Table 17-2. Summary of Savings and Costs for Recommendations

EMO. No.	*Description*	*Potential Savings ($/yr)*	*Resource Conserved*	*Impl. Cost ($)*
1	Install High Efficiency Lighting	$2,066	Electricity	$9,070
2	Repair Air Leaks	$2,282	Electricity	$500
3	Shut Off Unused Equipment	$2,630	Electricity	$0
4	Use Outside Air for Compressor Intakes	$1,248	Electricity	$200
5	Manufacturers Tax Rebate for Energy	$13,000	n/a	$0
6	Install Limit Switch on Reciprocating Compressor	$640	Electricity	$200
Totals		$21,866	xx	$9,970

GENERAL BACKGROUND

Facility Description

The company considered in this report produces log homes and secondary wood products. These products are distributed nationally. Typically, 60 employees work in the log homes business and 38 work in the secondary wood products business. Annual sales are approximately $14 million. Approximate operating schedules of the various areas considered in this report are given below:

OPERATING SCHEDULE BY AREA

Area	*Operating Schedule*	*Number of Operating Days (days/wk)*	*Total Operating Hours (hr/yr)*
Office and Production	7:00-3:30	5	2,210
Kilns	24 hours/day	8 months/year	5,376

The facility considered in this assessment consists of twenty-nine independent structures with thirteen primary occupied and heated structures, they are as follows:

1 – Cedar Ideas – Office & Mfr
2 – Cedar Ideas – Dip Room
3 – New House Line
4 – Purlin Mill
5 – Drill Mill
6 – Maintenance Shop
7 – Ten Foot Mill
8 – Picket Mill
9 – Dry Kiln
10 – Six Foot Mill
11 – Door Building
12 – Chop Mill
13 – Band Saw Mill

The buildings vary in size from 7000 ft^2 to 2500 ft^2 with the average at 3000 ft^2. May of the buildings were built in the 1980s with updates in the 1990s. The newest building is the new house line which was completed in 2004. Most building are 2 by 6 construction with 6″ of rolled fiberglass insulation in the walls and ceiling. The new house line has 9″ of insulation in the walls. The lighting is primarily 8′, 2 tube, T-12 fluorescent fixtures utilizing magnetic ballasts with some use of incandescent and HID lamps. Heat is provided by individual furnaces or wood stoves in each building.

Process Description

A simplified description of the manufacturing process performed at this facility is given in this section. It is not intended to be a complete detailed description, but rather to provide general information on the process.

Manufacturing operations required for the home plant are listed below:

Receipt of tree length logs
Cutting to length with slasher
Removal of outer bark and wood in saw mills
Planing in the planar mill for tongue and groove formation
Drying in the dry kiln
Notching of wood in the new house line
Shipped to destination

Manufacturing operations for the post and rail line is similar but differs in that after the slasher, some wood goes to the post and rail line mill and then to shipping.

Manufacturing operations required for spruce wood processing are:

Spruce wood starts in a debarker
Stored by length and air-dried
Processed in the purlin mill
Shipped to destination

Major Energy Consuming Equipment

The following list is an approximate summary of the major energy-consuming equipment at this facility.

1. Electricity
   A. Air Compressors:
      (1 of) Ingersoll-Rand 25 hp Screw Compressor in New House Line
      (1 of) Ingersoll-Rand 25 hp Piston-type Compressor in 10′ Mill
      Misc. 10 and 15 hp reciprocating compressors in each mill
         and in shop

   B. Major Process-Related Equipment: Connected Horsepower

Six Foot Mill: 250 hp miscellaneous.
Ten Foot Mill: 300 hp miscellaneous
Planar Mill: 300 hp miscellaneous
Small Planar Mill: 80 hp miscellaneous
New House Line: 100 hp miscellaneous
Post and Rail: 130 hp miscellaneous
Purlin Mill: 150 hp miscellaneous

C. Lighting
42 kW total installed Capacity

2. Facility Heat
Oil-fired hot air furnace in each building
Purlin mill has infrared propane heat
Shop has a used oil burner

Energy Forms and Use in the Plant:

Electrical energy is used for operating the majority of process-related equipment, for lighting, and for compressing air.

Fuel Oil and Kerosene is used for facility and process heating.

Diesel fuel is used for yard equipment

Propane is used for space heating

No other energy sources or fuels are consumed at this facility.

ENERGY ACCOUNTING

An essential component of any energy management program is a continuing account of energy use and its cost. This can be developed by keeping up-to-date records of energy consumption and associated costs on a monthly basis. When utility bills are received, it is recommended that energy use and costs be recorded as soon as practical. A separate record will be required for each type of energy used, i.e., gas, electric, oil, etc. A combination will be necessary, for example, when both gas and oil are used interchangeably in a boiler. A single energy unit should be used to express the heating values of the various fuel sources so that a meaningful comparison of fuel types and fuel combinations can be made. The primary energy used in this report is the Btu, (British thermal unit), or million Btus (MMBtu). The conversion factors are:

ENERGY UNIT	ENERGY EQUIVALENT
1 kWh	3,413 Btu
1 therm	100,000 Btu
1 cu ft natural gas	1,016 Btu*
1 gallon No. 1 oil (kerosene)	137,000 Btu*
1 gallon No. 2 oil (diesel)	140,000 Btu*
1 gallon No. 5 oil	148,000 Btu*
1 gallon No. 6 oil	150,000 Btu*
1 gallon gasoline	130,000 Btu*
1 gallon propane	92,000 Btu*
1 ton coal	20,000,000 Btu*
1 ton refrigeration	12,000 Btu/hr
1 hp h (electric)	2,545 Btu
1 hp h (boiler)	33,500 Btu

*Varies with supplier

The value of energy and cost records can be understood by examining the data representing your facility on the following pages. The energy usage of the facility is summarized in Table 17-1 and charted in Figure 17-1. A pie chart illustrating the percentage of energy use for twelve previous months is shown in Figure 17-2 and another pie chart illustrating the percentage of energy costs for the twelve previous months is shown in Figure 17-3. From these figures, trends and irregularities in energy usage and costs can be detected and the relative merits of energy conservation and load management can be assessed.

In addition to plotting the monthly energy consumption and cost, it may be desirable to plot the ratio of monthly energy consumption to monthly production. An appropriate measure of production should be used that is consistent with the company's record-keeping procedures. The measure of production used can be gross sales, number of units produced or processed, pounds of raw material used, etc. It is important that the same time period be used for energy consumption and production.

ENERGY BILLS FOR THIS CLIENT WERE DEEMED PROPRIETARY. THE REPORT ORIGINALLY SHOWED DETAILED SPREADSHEETS OF THE CLIENTS VARIOUS ENERGY BILLS. ONLY THE SUMMARY CHARTS ARE INCLUDED FOR REFERENCE.

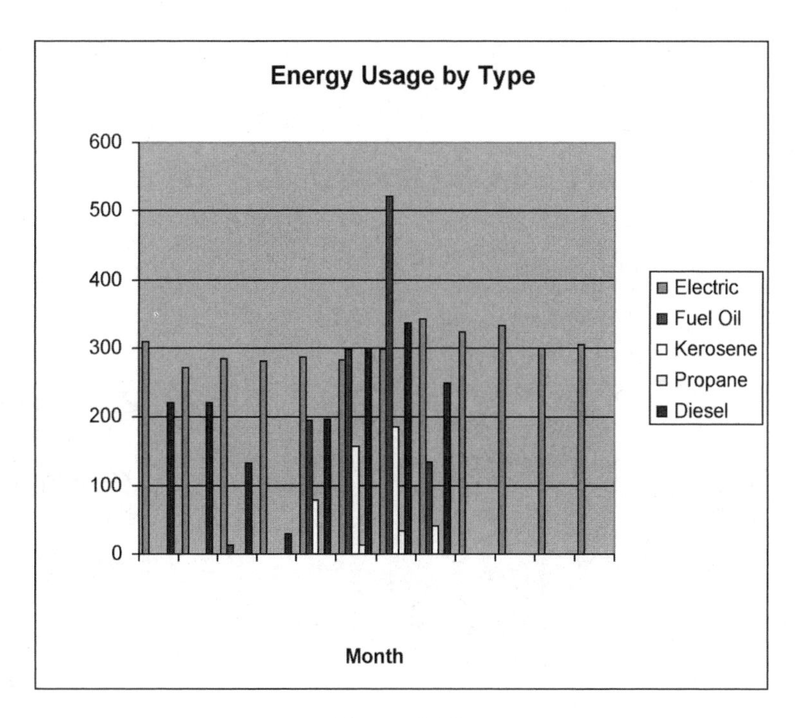

Figure 17-1.

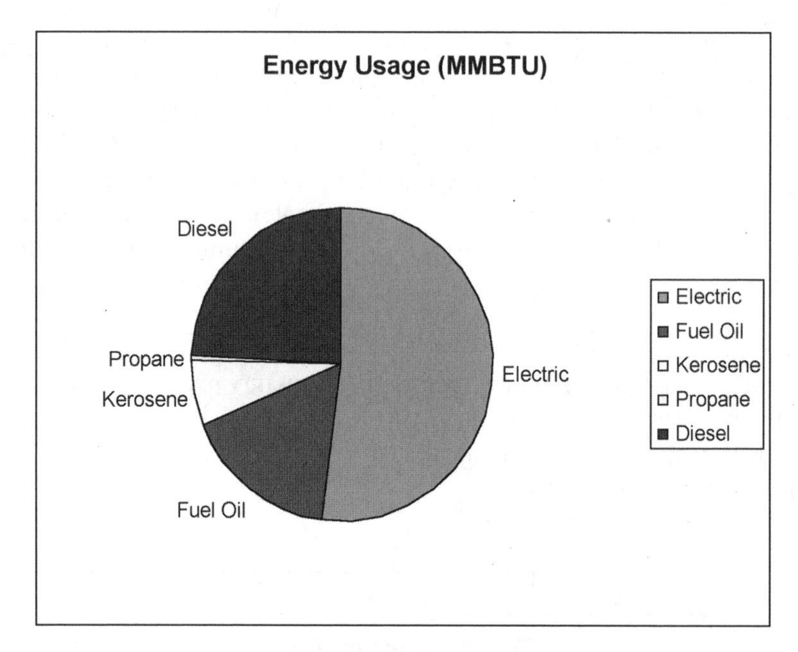

Figure 17-2.

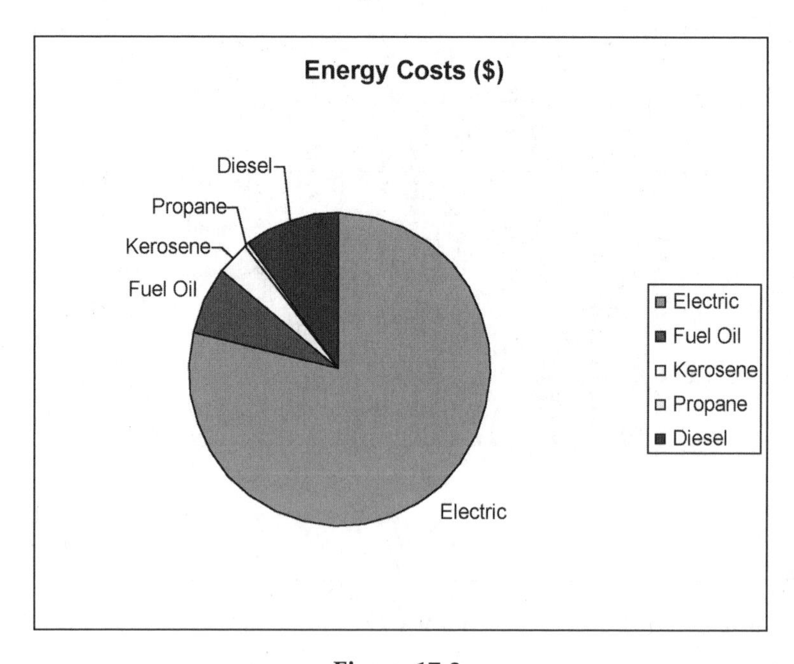

Figure 17-3.

ENERGY REDUCTION RECOMMENDATIONS

EMO No. 1—Install High Efficiency Lighting

Recommended Action

Replace the existing lighting systems with higher efficiency systems and maintain fixture cleanliness. High efficiency fixtures use less energy than standard fixtures with equivalent light output, which when combined with properly cleaned fixtures that allow full light intensity, result in energy and cost savings. In areas of critical lighting, replacement of the full fixture including reflectors is recommended.

Estimated Energy Savings (kWh) = 18,791 kWh/yr
Estimated Energy Savings (MMBtu) = 64.13 MMBtu/yr
Estimated Cost Savings = $2,066/yr
Estimated Implementation Cost = $9,070
Payback Period = 4.5 years

Lighting Summary

A detailed tabular breakdown of the lighting calculations for this facility is shown in the EXISTING LIGHTING and PROPOSED LIGHTING tables, which provide an area-by-area analysis. The values given under the EXISTING LIGHTING heading are the result of a lighting survey conducted during the facility visit. Values in the PROPOSED LIGHTING table were calculated based on replacement of all existing fixtures with high efficiency systems. A lighting code table is provided below to assist in identifying specific areas which will be recommended for conversions throughout this EMO.

Area Descriptions

	CEDAR IDEAS FABRICATION SHOP
A1	Fabrication Shop
A2	Office
A3	Wood Shop
A4	Benches
A5	Nailing Room

	CEDAR IDEAS FABRICATION SHOP
B1	Paint Room 1
B2	Paint Room 2
B3	Paint Room 3
B4	2nd Floor

	CEDAR IDEAS FINISH SHOP
C1	Break Room
C2	1st Floor Finish Shop
C3	2nd Floor Finish Shop

	MAIN PRODUCTION
D1	New House Line
D2	Purlin Line
D3	Post & Rail
D4	10' Mill
D5	Picket Mill
D6	6' Mill
D7	Door Building
D8	Chop Mill
D9	Band Saw
E1	Drill Mill Main Area
E2	Drill Mill Maintenance Shop
E3	Drill Mill Office

EXISTING LIGHTING

D6	2	75	75	150	1,220	2,500	75 Watt Incandescent Bulbs
D7	3	110	250	750	8,000	12,000	(2) 110 Watt F96T12 Magnetic Ballast
D7	1	250	250	250	3,600	4,200	250W Incandescent Flood Light
D8	1	250	250	250	3,600	4,200	250W Incandescent Flood Light
D8	7	60	138	966	5,750	12,000	(2) 60 Watt F96T12 Magnetic Ballast
D8	1	250	300	300	4,000	3,000	Halogen Light
D9	3	60	138	414	5,750	12,000	(2) 60 Watt F96T12 Magnetic Ballast
D9	6	250	250	1,500	4,000	3,000	Halogen Light
D9	2	250	250	500	3,600	4,200	250W Incandescent Flood Light
D9	2	250	300	600	4,000	3,000	250 Watt Metal Halide

Code	Quantity	Lamp Watts	System Watts	Total Watts	System Lumens	Rated Lamp Life (Hours)	Lighting Description
E1	5	110	250	1,250	8,000	12,000	(2) 110 Watt F96T12 Magnetic Ballast
E2	2	60	138	276	5,750	12,000	(2) 60 Watt F96T12 Magnetic Ballast
E3	8	60	138	1,104	5,750	12,000	(2) 60 Watt F96T12 Magnetic Ballast
E4	3	60	138	414	5,750	12,000	(2) 60 Watt F96T12 Magnetic Ballast
E4	1	75	75	75	1,220	2,500	75 Watt Incandescent Bulbs

EXISTING LIGHTING (Continued)

Code	Quantity	Lamp Watts	System Watts	Total Watts	System Lumens	Rated Lamp Life (Hours)	Lighting Description
C1	2	60	138	276	5,750	12,000	(2) 60 Watt F96T12 Magnetic Ballast
C2	36	60	138	4,968	5,750	12,000	(2) 60 Watt F96T12 Magnetic Ballast
C3	18	60	138	2,484	5,750	12,000	(2) 60 Watt F96T12 Magnetic Ballast

Code	Quantity	Lamp Watts	System Watts	Total Watts	System Lumens	Rated Lamp Life (Hours)	Lighting Description
D1	101	59	110	11,110	5,900	15,000	(2) 59 Watt F96T8 Electronic Rapid Start Ballast
D2	1	60	138	138	5,750	12,000	(2) 60 Watt F96T12 Magnetic Ballast
D3	2	75	75	150	1,220	2,500	75 Watt Incandescent Bulbs
D3	8	60	138	1,104	5,750	12,000	(2) 60 Watt F96T12 Magnetic Ballast
D4	3	60	138	414	5,750	12,000	(2) 60 Watt F96T12 Magnetic Ballast
D4	3	75	75	225	1,220	2,500	75 Watt Incandescent Bulbs
D5	3	110	250	750	8,000	12,000	(2) 110 Watt F96T12 Magnetic Ballast
D5	8	110	250	2,000	8,000	12,000	(2) 110 Watt F96T12 Magnetic Ballast
D6	7	110	250	1,750	8,000	12,000	(2) 110 Watt F96T12 Magnetic Ballast

EXISTING LIGHTING (Continued)

Code	Quantity	Lamp Watts	System Watts	Total Watts	System Lumens	Rated Lamp Life (Hours)	Lighting Description
D6	2	75	75	150	1,220	2,500	75 Watt Incandescent Bulbs
D7	3	110	250	750	8,000	12,000	(2) 110 Watt F96T12 Magnetic Ballast
D7	1	250	250	250	3,600	4,200	250W Incandescent Flood Light
D8	1	250	250	250	3,600	4,200	250W Incandescent Flood Light
D8	7	60	138	966	5,750	12,000	(2) 60 Watt F96T12 Magnetic Ballast
D8	1	250	300	300	4,000	3,000	Halogen Light
D9	3	60	138	414	5,750	12,000	(2) 60 Watt F96T12 Magnetic Ballast
D9	6	250	250	1,500	4,000	3,000	Halogen Light
D9	2	250	250	500	3,600	4,200	250W Incandescent Flood Light
D9	2	250	300	600	4,000	3,000	250 Watt Metal Halide

Code	Quantity	Lamp Watts	System Watts	Total Watts	System Lumens	Rated Lamp Life (Hours)	Lighting Description
E1	5	110	250	1,250	8,000	12,000	(2) 110 Watt F96T12 Magnetic Ballast
E2	2	60	138	276	5,750	12,000	(2) 60 Watt F96T12 Magnetic Ballast
E3	8	60	138	1,104	5,750	12,000	(2) 60 Watt F96T12 Magnetic Ballast
E4	3	60	138	414	5,750	12,000	(2) 60 Watt F96T12 Magnetic Ballast
E4	1	75	75	75	1,220	2,500	75 Watt Incandescent Bulbs

PROPOSED LIGHTING

The following tables illustrate the high efficiency lighting systems that are recommended to replace the existing lighting systems described earlier. The specifications for the new systems were gathered from manufacturer's literature.

Code	Quantity	Lamp Watts	System Watts	Total Watts	System Lumens	Rated Lamp Life (Hours)	Lighting Description
A1	9	59	110	990	5,900	15,000	(2) 59 Watt F96T8 Electronic Rapid Start Ballast
A2	1	59	110	110	5,900	15,000	(2) 59 Watt F96T8 Electronic Rapid Start Ballast
A3	***Current***	***lighting***	***is***	***energy***	***and***	***cost***	***effective***
A4	2	59	110	220	5,900	15,000	(2) 59 Watt F96T8 Electronic Rapid Start Ballast
A5	16	59	110	1,760	5,900	15,000	(2) 59 Watt F96T8 Electronic Rapid Start Ballast

Code	Quantity	Lamp Watts	System Watts	Total Watts	System Lumens	Rated Lamp Life (Hours)	Lighting Description
B1	10	59	110	1,100	5,900	15,000	(2) 59 Watt F96T8 Electronic Rapid Start Ballast
B2	5	59	110	550	5,900	15,000	(2) 59 Watt F96T8 Electronic Rapid Start Ballast
B3	3	59	110	330	5,900	15,000	(2) 59 Watt F96T8 Electronic Rapid Start Ballast
B4	3	59	110	330	5,900	15,000	(2) 59 Watt F96T8 Electronic Rapid Start Ballast

Proposed Lighting (Continued)

Code	Quantity	Lamp Watts	System Watts	Total Watts	System Lumens	Rated Lamp Life (Hours)	Lighting Description
C1	2	59	110	220	5,900	15,000	(2) 59 Watt F96T8 Electronic Rapid Start Ballast
C2	36	59	110	3,960	5,900	15,000	(2) 59 Watt F96T8 Electronic Rapid Start Ballast
C3	18	59	110	1,980	5,900	15,000	(2) 59 Watt F96T8 Electronic Rapid Start Ballast

Code	Quantity	Lamp Watts	System Watts	Total Watts	System Lumens	Rated Lamp Life (Hours)	Lighting Description
D1	***Current***	***lighting***	***is***	***energy***	***and***	***cost***	***effective***
D2	1	59	110	110	5,900	15,000	(2) 59 Watt F96T8 Electronic Rapid Start Ballast
D3	2	15	15	30	1,000	10,000	15 Watt Compact Fluorescent
D3	8	59	110	880	5,900	15,000	(2) 59 Watt F96T8 Electronic Rapid Start Ballast
D4	3	59	110	330	5,900	15,000	(2) 59 Watt F96T8 Electronic Rapid Start Ballast
D4	3	15	15	45	1,000	10,000	15 Watt Compact Fluorescent
D5	3	86	160	480	8,200	18,000	(2) 86 Watt F96T8 Electronic Rapid Start Ballast

Proposed Lighting (Continued)

D5	8	86	160	1,280	8,200	18,000	(2) 86 Watt F96T8 Electronic Rapid Start Ballast
D6	7	86	160	1,120	8,200	18,000	(2) 86 Watt F96T8 Electronic Rapid Start Ballast
D6	2	15	15	30	1,000	10,000	15 Watt Compact Fluorescent
D7	3	86	160	480	8,200	18,000	(2) 86 Watt F96T8 Electronic Rapid Start Ballast
D7	1	42	46	46	3,200	12,000	42 Watt FF42QT Compact Fluorescent
D8	1	42	46	46	3,200	12,000	42 Watt FF42QT Compact Fluorescent
D8	7	59	110	770	5,900	15,000	(2) 59 Watt F96T8 Electronic Rapid Start Ballast
D8	1	42	46	46	3,200	12,000	42 Watt FF42QT Compact Fluorescent
D9	3	59	110	330	5,900	15,000	(2) 59 Watt F96T8 Electronic Rapid Start Ballast
D9	6	42	46	276	3,200	12,000	42 Watt FF42QT Compact Fluorescent
D9	2	42	46	92	3,200	12,000	42 Watt FF42QT Compact Fluorescent
D9	*Current*	*Lighting*	*is*	*energy*	*and*	*cost*	*effective*

Proposed Lighting (Continued)

Code	Quantity	Lamp Watts	System Watts	Total Watts	System Lumens	Rated Lamp Life (Hours)	Lighting Description
E1	5	86	160	800	8,200	18,000	(2) 86 Watt F96T8 Electronic Rapid Start Ballast
E2	2	59	110	220	5,900	15,000	(2) 59 Watt F96T8 Electronic Rapid Start Ballast
E3	8	59	110	880	5,900	15,000	(2) 59 Watt F96T8 Electronic Rapid Start Ballast
E4	3	59	110	330	5,900	15,000	(2) 59 Watt F96T8 Electronic Rapid Start Ballast
E4	1	15	15	15	1,000	10,000	15 Watt Compact Fluorescent

Anticipated Savings

The energy savings from lighting upgrades, ES1a and ES1b are realized by replacing the existing conventional lighting systems with high efficiency lighting systems where applicable. Where the quantity of lights remain the same in the table below, the energy savings are calculated using the wattage reduction between the existing and proposed systems as follows:

$$ES_{1a} = (WR \times H)/K$$

Where,

WR = Estimated Total wattage reduction, Watts

H = Hours of Use, hrs.

K = Conversion factor (1,000W/kW)

kW demand reduction is equivalent to the energy savings due to retrofit.

Cost savings, CS, are determined by the following equation:

$$CS = ES \times \text{Cost of Electricity } (\$.11/\text{kWhr})$$

Fixture Upgrade

Summary of Energy Savings Resulting from Fixture Upgrades

Area Code	Quantity Existing	Quantity Proposed	System Watts Existing	System Watts Proposed	Total Wattage Reduction	Hours of Use (hrs/yr)	Energy Savings (kWhr/yr)	Cost Savings ($/yr)
A1	9	9	1,242	990	252	2,080	524	$57.66
A2	1	1	138	110	28	2,080	58	$6.41
A4	2	2	276	220	56	2,080	116	$12.81
A5	16	16	2,208	1,760	448	2,080	932	$102.50

Area Code	Quantity Existing	Quantity Proposed	System Watts Existing	System Watts Proposed	Total Wattage Reduction	Hours of Use (hrs/yr)	Energy Savings (kWhr/yr)	Cost Savings ($/yr)
B1	10	10	1,380	1,100	280	2,080	582	$64.06
B2	5	5	690	550	140	2,080	291	$32.03
B3	3	3	414	330	84	2,080	175	$19.22
B4	3	3	414	330	84	2,080	175	$19.22

Area Code	Quantity Existing	Quantity Proposed	System Watts Existing	System Watts Proposed	Total Wattage Reduction	Hours of Use (hrs/yr)	Energy Savings (kWhr/yr)	Cost Savings ($/yr)
C1	2	2	276	220	56	2,080	116	$12.81
C2	36	36	4,968	3,960	1,008	2,080	2,097	$230.63
C3	18	18	2,484	1,980	504	2,080	1,048	$115.32

The anticipated annual cost savings for the facility, CS, due to the recommended retrofit of the lighting systems can be estimated by the total energy savings (kWhr) and demand reduction (kW demand). However, since the company is not charged for demand, the cost savings are directly related to the energy savings.

CS = CS1

*From Fixture Upgrade table.

Fixture Upgrade (Continued)

Area Code	Quantity Existing	Quantity Proposed	System Watts Existing	System Watts Proposed	Total Wattage Reduction	Hours of Use (hrs/yr)	Energy Savings (kWhr/yr)	Cost Savings ($/yr)
D2	1	1	138	110	28	2,080	58	$6.41
D3	2	2	150	30	120	2,080	250	$27.46
D3	8	8	1,104	880	224	2,080	466	$51.25
D4	3	3	414	330	84	2,080	175	$19.22
D4	3	3	225	45	180	2,080	374	$41.18
D5	3	3	750	480	270	2,080	562	$61.78
D5	8	8	2,000	1,280	720	2,080	1,498	$164.74
D6	7	7	1,750	1,120	630	2,080	1,310	$144.14
D6	1	1	150	30	120	2,080	250	$27.46
D7	3	3	750	480	270	2,080	562	$61.78
D7	1	1	250	46	204	2,080	424	$46.68
D9	1	1	250	46	204	2,080	424	$46.68
D8	7	7	966	770	196	2,080	408	$44.84
D8	1	1	300	46	254	2,080	528	$58.12
D9	3	3	414	330	84	2,080	175	$19.22
D9	6	6	1,500	276	1,224	2,080	2,546	$280.05
D9	2	2	500	92	408	2,080	849	$93.35

Energy reduction equals 18,791 kWhr/yr. This is equal to 64.13 MMBtu.

Therefore,

CS = $2,066

Implementation

The cost of implementation is based on replacement of the necessary existing fixtures with high-efficiency lamp systems as previously described. The costs involved with this recommendation include the following:

- Lamp Replacement
- Ballast and/or Fixture Replacement (where necessary)
- Labor
- Disposal costs
- Occupancy Sensors (where recommended)

The implementation costs of the lighting systems described below may actually be overly conservative for several reasons. We have taken into account costs that can be neglected in certain circumstances as listed below:

- If some systems are approaching their rated life span, new systems would be required regardless of this recommendation; therefore, only the *difference* in cost between existing and proposed systems should be considered.

- If some systems are approaching their rated life span, labor would be required to change them, regardless of this recommendation; therefore labor costs would be eliminated.

Additionally, we have not accounted for the benefits of the increased rated life of certain proposed lighting systems (the increased rated life of several of the proposed systems implies that fewer purchasing and labor costs would be incurred throughout the years). Since these exceptions are difficult to identify, we can only generate a general upgrade report. The facility management can then follow through where they feel necessary.

The following table lists the costs associated with upgrades and modifications to the lighting systems.

Implementation Costs

Code	Quantity (n)	Lamp & Ballast Replacement ($/fixture) [A]	Labor Cost ($/fixture) [B]	Disposal Cost ($/fixture) [C]	Sensor Cost ($/sensor)	Cost Per Fixture ($/each) [A+B+C]	Total Cost (bulbs & ballast) (n x $)	Total Cost (Full Fixture Replacement)
A1	9	$30.52	$10.00	$1.30	NA	$41.82	$376.38	$709.20
A2	1	$30.52	$10.00	$1.30	NA	$41.82	$41.82	$78.80
A4	2	$30.52	$10.00	$1.30	NA	$41.82	$83.64	$157.60
A5	16	$30.52	$10.00	$1.30	NA	$41.82	$669.12	$1,260.80

Code	Quantity (n)	Lamp & Ballast Replacement ($/fixture) [A]	Labor Cost ($/fixture) [B]	Disposal Cost ($/fixture) [C]	Sensor Cost ($/sensor)	Cost Per Fixture ($/each) [A+B+C]	Total Cost (n x $)	Total Cost (Full Fixture Replacement)
B1	10	$30.52	$10.00	$1.30	NA	$41.82	$418.20	$788.00
B2	5	$30.52	$10.00	$1.30	NA	$41.82	$209.10	$394.00
B3	3	$30.52	$10.00	$1.30	NA	$41.82	$125.46	$236.40
B4	3	$30.52	$10.00	$1.30	NA	$41.82	$125.46	$236.40

Code	Quantity (n)	Lamp & Ballast Replacement ($/fixture) [A]	Labor Cost ($/fixture) [B]	Disposal Cost ($/fixture) [C]	Sensor Cost ($/sensor)	Cost Per Fixture ($/each) [A+B+C]	Total Cost (n x $)	Total Cost (Full Fixture Replacement)
C1	2	$30.52	$10.00	$1.30	NA	$41.82	$83.64	$157.60
C2	36	$30.52	$10.00	$1.30	NA	$41.82	$1,505.52	$2,836.80
C3	18	$30.52	$10.00	$1.30	NA	$41.82	$752.76	$1,418.40

Implementation Costs (Continued)

Code	Quantity (n)	Lamp & Ballast Replacement ($/fixture) [A]	Labor Cost ($/fixture) [B]	Disposal Cost ($/fixture) [C]	Sensor Cost ($/sensor)	Cost Per Fixture ($/each) [A+B+C]	Total Cost (n x $)	Total Cost (Full Fixture Replacement)
D2	1	$30.52	$10.00	$1.30	NA	$41.82	$41.82	$78.80
D3	2	$18.31	$1.00	$1.30	NA	$20.61	$41.22	$18.31
D3	8	$30.52	$10.00	$1.30	NA	$41.82	$334.56	$630.40
D4	3	$30.52	$10.00	$1.30	NA	$41.82	$125.46	$236.40
D4	3	$18.31	$1.00	$1.30	NA	$20.61	$61.83	$18.31
D5	6	$55.84	$10.00	$1.30	NA	$67.14	$402.84	$779.04
D5	8	$55.84	$10.00	$1.30	NA	$67.14	$537.12	$1,038.72
D6	7	$55.84	$10.00	$1.30	NA	$67.14	$469.98	$908.88
D6	2	$18.31	$1.00	$1.30	NA	$20.61	$41.22	$18.31
D7	3	$55.84	$10.00	$1.30	NA	$67.14	$201.42	$389.52
D7	1	$84.00	$15.00	$1.30	NA	$100.30	$100.30	$84.00
D8	1	$84.00	$15.00	$1.30	NA	$100.30	$100.30	$84.00
D8	7	$30.52	$10.00	$1.30	NA	$41.82	$292.74	$551.60
D8	1	$84.00	$15.00	$1.30	NA	$100.30	$100.30	$84.00
D9	3	$30.52	$10.00	$1.30	NA	$41.82	$125.46	$236.40
D9	6	$84.00	$15.00	$1.30	NA	$100.30	$601.80	$84.00
D9	2	$84.00	$15.00	$1.30	NA	$100.30	$200.60	$84.00

The overall annual cost savings of $2,066 would pay for the total implementation costs for the facility ($9,070) within approximately 4.5 years.

EMO No. 2 - Repair Compressed Air Leaks

Recommended Action

Repair leaks in compressed air lines on a regular basis.

Estimated Energy Savings in kWh = 20,746 kWh/yr

Implementation Costs (Continued)

Code	Quantity (n)	Lamp & Ballast Replacement ($/fixture) [A]	Labor Cost ($/fixture) [B]	Disposal Cost ($/fixture) [C]	Sensor Cost ($/sensor)	Cost Per Fixture ($/each) [A+B+C]	Total Cost (n x $)	Total Cost (Full Fixture Replacement)
E1	5	$55.84	$10.00	$1.30	NA	$67.14	$335.70	$649.20
E2	2	$30.52	$10.00	$1.30	NA	$41.82	$83.64	$157.60
E3	8	$30.52	$10.00	$1.30	NA	$41.82	$334.56	$630.40
E4	3	$30.52	$10.00	$1.30	NA	$41.82	$125.46	$236.40
E4	1	$18.31	$1.00	$1.30	NA	$20.61	$20.61	$18.31
Total							$9,070.04	$15,290.60

Estimated Energy Savings in MMBtu	=	70.81 MMBtu/yr
Estimated Cost Savings	=	$2,282/yr
Estimated Implementation Cost	=	$500
Simple Payback	=	2 months

Background

As compressed air delivery systems age, it is not uncommon for 20% of the compressor work to be associated with air leaks. The best method of determining the amount of energy lost to leaks is to observe the compressor during times when there is no use of air, such as when plant personnel are on break. If the compressor is cycling off and on, or loading and unloading, and there is no perceived load, then the compressor is working to make up for air leaks in the system. By timing the cycles one can obtain an estimate of the lost work. If it is not feasible to employ this method then one can listen for air leaks and, on locating a leak, measure the size of the opening. Then, with the size of the opening and the operating pressure, the volume of air flow through the leak can be calculated.

While performing the energy audit, the latter of these two methods was used to determine the volume of air flow through leaks. During our observations we found and noted a total of 5 air leaks, with openings ranging from 1/16″ to 1/8″ of an inch in diameter. These air leaks originated from various sources within the facility. Air leaks were identified in Cedar Ideas buildings 1 and 2 (small), Ten foot mill (medium), Picket mill (medium), and the Chop mill (medium). From this information it was possible to calculate the total annual energy costs based on a total of 2,080 hrs/yr production time [(8 hours/day) (5 days/week) (52 weeks/year)].

Anticipated Savings

Compressed air is currently generated at approximately 106 psig (120.7 psia). Five air leaks were noted at this pressure. The annual energy savings for these leaks, ES, are estimated as follows[1]:

$$ES = L \times H \times CO$$

Where,

L	=	power loss due to leaks, Hp
H	=	annual time during which leak occurs, hr/yr
CO	=	conversion factor, 0.746 kW/Hp

The power loss from leaks, L, is estimated as the power required to compress the volume of air lost, V_f, from atmospheric pressure, P_i, to the compressor discharge pressure, Po, as follows:

$$L = \frac{P_i \times C_2 \times V_f \times \left(\frac{k}{k-1}\right) \times N \times C_3 \times \left[\left(\frac{P_o}{P_i}\right)^{\frac{k-1}{k \times N}} - 1\right]}{E_v \times E_m}$$

Where:

Pi	=	inlet (atmospheric) pressure, psia
Po	=	compressor operating pressure, psia
C2	=	conversion constant, 144 in^2/ft^2

[1]*Compressed Air and Gas Handbook, Third Edition*, Compressed Air and Gas Institute, New York, 1961.

C3 = conversion constant, 3.03×10^{-5} hp·min/ft·lb
Ev = air compressor volumetric efficiency, 0.875 no units
Em = compressor motor efficiency, 0.8 no units
Vf = volumetric flow of free air, cfm
k = specific heat ratio of air, 1.4, no units
N = factor based on type of compressor considered, no units
N = 1 for single stage reciprocating compressors
N = 2 for two-stage reciprocating compressors
N = 1.25 for screw compressors (polytropic efficiency of 80%)

The volumetric flow rate of free air exiting the hole is dependent upon whether the flow is choked. When the ratio of atmospheric pressure to line pressure, Pi/Po, is less than 0.5283, the flow is said to be choked (i.e. it is traveling at the speed of sound). The ratio of 14.7 psia atmospheric pressure to 120.7 psia line pressure is 14.7/120.7 = 0.1218, thus the flow is choked. The volumetric flow rate of free air exiting the leak under these conditions is calculated as follows.

The Volumetric flow rate, V_f, is:

$$V_f = m/\rho$$

For choked flow, the mass flow-rate, m, is:

ρ = density of air, 0.075 lbm/ft^3

$$m = C4 \times C5 \times C_d \times A \times P_l \times (T_l + 460)^{-0.5}$$

Where,

C4 = isentropic flow constant, 0.53
C5 = conversion constant, 60 sec/min.
Cd = coefficient of discharge for square edged orifice, 0.6, no units
A = total surface area of leaks, square inches
Pl = line pressure, psia
Tl = average line temperature, °F

Hole Location	*Pressure (psia)*	*Hole Diameter (inches)*	*Hole Area (inches^2)*
Cedar Ideas Building 1	106	0.0625	0.0031
Cedar Ideas Building 2	106	0.0625	0.0031
Ten Foot Mill	106	0.1250	0.0123
Picket Mill	106	0.1250	0.0123
Chop Mill	106	0.1250	0.0123
		Total Area	**0.0431**

Therefore,

$$V_f = ((0.53)(60)(0.6)(.0431)(106)(70+460)^{-0.5})/0.075$$
$$V_f = 50.49 \text{ cfm}$$

The power loss due to air leaks is then:

$$L = 14.7 \times 144 \times 50.49 \times \left(\frac{1.4}{1.4-1}\right) \times 1 \times 3.03 \times 10^{-5} \times \frac{\left[\left(\frac{120.7}{14.7}\right)^{\frac{1.4-1}{1.4 \times 1.0}} - 1\right]}{.875 \times .8}$$

$$L = 13.37 \text{ Hp}$$

Therefore, the total annual energy savings, ES, is:

$$ES = (13.37 \text{ hp})(2{,}080 \text{ hr/yr})(0.746 \text{ kW/hp})$$
$$ES = 20{,}746 \text{ kWh/yr}$$

The annual cost savings, CS, can be estimated as follows:

$$CS = (ES) (\text{unit cost of electricity, \$/kWh})$$
$$CS = (20{,}746 \text{ kWh/yr})(\$0.11/\text{kWh})$$
$$CS = \$2{,}282/\text{yr}$$

Implementation Cost

Implementation of this AR involves:

1) replacement of couplings and/or hoses
2) replacement of seals around filters
3) shutting off air flow during lunch or break periods
4) repairing breaks in lines, etc.

Assuming that the repair work and procedural changes can be done by in-house facility maintenance and management personnel, it is estimated that these leaks can be eliminated for approximately $500. Thus, the estimated cost savings of $2,282/yr. would pay for the implementation cost in approximately 2 months. Additionally, air leak inspections should be incorporated into the routine preventive maintenance procedure and employees should be informed as to the importance of notifying maintenance personnel when leaks are discovered.

EMO No. 3—Turn Off Hog Motors During Breaks

Recommended Action

Turn off the hog motors in the six foot and ten foot mills during the morning and afternoon breaks as well as during lunch.

Estimated Energy Savings (kW-hr)	=	5,818.8 kW-hr
Estimated Energy Savings (MMBtu)	=	19.86 MMBtu
Estimated Annual Savings	=	$640/yr
Estimated Implementation Cost	=	$200
Simple Payback	=	4 Months

Background

During the site visit it was observed that the Trask-Decrow, 20 hp compressor in the ten foot mill was on during production hours but was idling approximately 75% of the time. This unit can be operated more efficiently by installing a pressure switch that will turn off the unit once it has reached a set point. Then, when the air pressure drops below a low point setting, the unit will turn back on. In discussions with the Trask-Decrow area representative, it was confirmed that if the unit is called upon for air less than six times per hour, it should be operated on a pressure switch. If it is cycled more often than that, one needs to consider the strain on components due to high cycle operation.

Anticipated Savings

The savings for the period of 6/04-5/05 are determined as follows:

ES = IH x .746 kW/1 hp × OH

Where,

ES = total energy savings, kW-hr/yr
IH = idling horsepower (energy consumption while idling)
OH = operating hours (while at idle)

Therefore,

ES = 5 hp x .746 × (2080 × 0.75)
ES = 5,818.80 kW-hr

The cost savings are determined by multiplying the energy savings (ES) by the average electricity cost of $0.11/kW-hr.

CS = 5,818.80 kW-hr × $0.11/kW-hr = $640

Implementation Cost

The costs for installation of a pressure limit switch assuming the existing starter will still be used were estimated to be approximately $200.

EMO No. 4—Use Outside Air Compressor Intake

Recommended Action

Duct outside air directly to the intake of the 75 hp and 30 hp Ingersol Rand air compressors to improve operating efficiency (see figure No. 9 below).

Estimated Energy Savings (kWh) = 18,905 kWh/yr
Estimated Energy Savings (MMBtu) = 64.52 MMBtu/yr
Estimated Cost Savings = $1,248/yr
Estimated Implementation Cost = $150
Simple Payback = 1.5 months

Background

The facility uses the 75 hp air compressor for production support and the 30 hp compressor for backup. By utilizing an outside air supply, it is possible to reduce energy requirements for compression. Outside air is (on average) cooler, therefore more dense, and requires less energy to be compressed.

Anticipated Savings

The compressor work for the usual operating conditions in manufacturing plants is proportional to the absolute temperature of the intake air. Thus, the fractional reduction in compressor work, WR, resulting from lowering the intake air temperature, is estimated as:

$$WR = (WI - WO)/WI$$

Or

$$WR = (TI - TO)/(TI + 460)$$

Where,

WI = work of compressor with inside air, hp

WO = work of compressor with outside air, hp

TI = average temperature of inside air, °F

TO = annual average outside air temperature, °F

The compressor intake temperature was measured as 88°F. The yearly average outdoor temperature for the client's location was 45°F. The fractional compressor work reduction, WR, due to lowering the compressor intake air temperature is estimated as:

$$WR = (88 - 45)/(88 + 460)$$
$$WR = 0.0785$$

It is estimated that the 75 hp compressor operates for 3,500 hours annually, which was estimated as the following;

$$(14 \text{ hrs/day}) \times (5 \text{ days/wk}) \times (50 \text{ wks/yr}) = 3{,}500 \text{ hrs/yr}$$

The average mean power for the 75 hp compressor was estimated at approximately 75% rated load.

The annual energy savings, ES, for the 75 hp compressor is estimated as:

$$ES = (kW) \times (H) \times (WR)$$

Where,

kW = compressor average mean power, 68.81 kW
H = annual operating hours, 3,500 hrs/yr
WR = compressor work reduction factor, 0.0785 (no units)

Thus,

ES = (68.81 kW) × (3,500 hrs/yr) × (0.0785)
ES = 18,905 kWh/yr (64.52 MMBtu/yr)

An energy savings table is shown below:

kWh	KW Demand	Other
18,905	0	None

The annual cost savings for the compressor, CS, is estimated as:

CS = (ES) × (unit cost of electricity)

Therefore,

CS = (18,905 kWh/yr) × ($0.066/kWh)
CS = $1,248/yr*

Implementation Cost

The most common material used for ducting outside air to the compressor intakes is plastic PVC pipe. One end of the pipe is attached to the air cleaner intake or other appropriate intake port, and the other end is routed through a wall or ceiling to the outside. Care must be taken to include a drip leg and/or hood on the intake line to prevent moisture from entering the compressor. Properly insulating and sealing the wall penetration, to prevent air infiltration, should not be overlooked.

It should be a fairly simple procedure to adapt PVC pipe to the air compressor intake. The total cost for materials and labor to make these modifications to the compressor is estimated as $150. The cost savings of $1,248/yr* would pay for the implementation cost of $150 within approximately 1.5 months.

EMO No. 5 - Eliminate Sales Tax on Fuel Oil

Recommended Action

Obtain a rebate (and possible refund) for taxes paid on fuel consumption at the production facility for fuel oil.

Estimated Energy Savings	=	N/A
Estimated First Year Savings	=	$471.27
Estimated Annual Savings	=	$1,413.80/yr
Estimated Implementation Cost	=	None
Simple Payback	=	Immediately

Background

As of July 1, 1993, ninety-five percent of the sale price of fuel and electricity purchased for use at a manufacturing facility is tax free in this state. This exemption applies if the fuel or electricity is used exclusively for the production facility.

The manufacturer's exemption law applies according to the following:

July 1, 1993 and thereafter, the noted 95% of cost will be taxed at 0%.

Anticipated Savings

According to the fuel bills for the facility, the total fuel oil taxed for the twelve-month period between 6/04 and 5/05 was 2,015 gallons. Over this period, a sales tax of $496.07 was accrued. The tax paid on fuel oil for 6/04 – 5/05 is shown in the Summary Table.

Summary Table

Billing Date (Mon-Yr)	Energy Usage (gal)	Sales Tax ($)
Sept	12.24	$612.39
Oct	0	$280.26
Nov	195.84	$541.09
Dec	298.22	$248.27
Jan	521.21	$261.99
Feb	135.19	$227.47
TOTAL	1162.7	$496.07

The savings for the period of 6/04-5/05 are determined as follows:

$$CS = T \times .95$$

Where,

CS = Total savings, $/yr

T = Total taxes on fuel oil ($496.07/yr)

Therefore,

$$CS = (\$496.07/yr \times .95)$$

$$CS = \$471.27$$

Since this exemption was implemented in 1993 the facility is eligible for a refund that is retroactive 36 months, which is shown below.

$$FCS = CS \times 3yr$$

Where,

$$FCS = \text{First years cost savings}$$

Therefore,

$$FCS = \$471.27 \times 3yr$$
$$FCS = \$1{,}413.80$$

Implementation Cost

There are no costs associated with this recommendation. The following steps are recommended to obtain a rebate (and possible refund) for taxes paid on fuel consumption at the production facility for fuel oil.

1. Contact the State Sales Tax Division to request the *Blanket Certificate* form. Once completed, this form should be submitted to the local utility that provides the electricity as proof of tax exempt status.

2. A written request to the fuel oil provider to obtain a refund on this tax overpayment. This refund is retroactive 36 months. If the fuel oil provider refuses to provide reimbursement, the manufacturer should apply to the State Sales Tax Division for repayment (include copies of related receipts and bills).

18

Economic Evaluations for Power Quality Solutions

SUMMARY

Power quality issues are continuing to become a greater concern, particularly to sensitive customers with critical operations. Measures can be taken to improve the power quality of a given circuit, however the economic feasibility of the applications should be considered. As a result, the *system compatibility* must be examined to determine the behavior of a customer's equipment within the electrical environment. A power quality investigation and an economic analysis can then determine what mitigation tactics are most effective and whether they should be applied.

With the increased interest and awareness in power quality, there is a vast increase in the number of mitigation devices now available. Many of these devices are expensive; therefore, making a decision to apply mitigating devices is not an easy one. A system and process study must be done to ascertain the level of susceptibility a plant's equipment has. Once the operation has been characterized, the most appropriate mitigation device(s) can be determined. A cost analysis must then be done to determine the cost of interruptions associated with power quality problems, and the number of interruptions that will be prevented. These savings are compared to the cost of the mitigation device.

The intent of this chapter is to illustrate a variety of techniques used to evaluate power conditioning devices. An analysis of the cost due to power quality interruptions is presented. The effectiveness of the mitigation device is examined in determining an appropriate cost. Mitigation strategies to alleviate the most problems at the lowest cost are discussed. Several techniques are presented for computing payoff time.

Presented at the 20th World Energy Engineering Congress by Stephen Middlekauff

INTRODUCTION

Power quality is a greater concern now than it has been in the past. More industrial customers are contacting their utilities about process interruptions and plant downtime. It would appear that the quality of the utility electrical service has deteriorated, however this may not be the case. In fact, with clever protection schemes and more aggressive preventive measures, the quality of the electrical service is probably better than ever. The dawn of the power electronic era has created a higher level of sensitivity to power fluctuations. Modern industrial equipment such as motor drives, programmable logic controllers (PLCs), and other automation products are all much more susceptible to power disturbances than ever before.

The result of this greater sensitivity is a greater awareness of power quality. The market is now flooded with power conditioning devices, with a very wide range of sizes, applications, and capabilities. All of them, however, have one thing in common: they claim to prevent power disturbances from reaching your equipment. Some are very inexpensive, but may not do an adequate job in protection of sensitive equipment. Other solutions may carry a hefty price tag, and may not be worth the expense. Many parameters must be considered in evaluating the cost savings of a power conditioning device.

THE PRINCIPLE INVESTIGATION

Voltage sags, surges, momentary interruptions, and other power anomalies may all look the same to an industrial customer. His process has been interrupted. A one cycle voltage sag may have the same effect as a one minute outage. In determining the specifications for power conditioning equipment, the nature of the disturbances must first be determined. The most accurate way for determining the behavior of the electrical environment is to monitor the power using a high speed, power quality monitor.

If the disturbance is not steady-state, an extended monitoring period may be necessary to capture the culprit. If the disturbance is seasonal in nature, such as lightning from frequent summer storms, monitoring may need to be done over an entire season to best characterize the types of activity seen on the system. Typically, if there are problems occurring with electronic equipment, there may be clues to help pinpoint the cause.

Fuses blown or breakers tripped may be a clue to overcurrents caused by voltage spikes. Conversely, this may be the result of inrush current caused by equipment recovery after a voltage sag. Equipment simply shutting down may be from temporary loss of voltage. Actual damage to electronic components, such as shown in Figure 18-1, may indicate an overvoltage or voltage transient. Electronic motor drives often have some fault code to indicate why they shut down. If it offers information about its DC bus voltage, this could help indicate what is happening on the system.

All of these offer important clues as to what is occurring. This also assists in determining what type of monitoring equipment should be used. Typically, customer inquiries refer to specific pieces of equipment, so the monitoring should be done as close to the affected equipment as possible.

DETERMINING THE PHENOMENON

A careful log of equipment behavior must be taken during monitoring to determine the types of disturbances that are actually causing the problem. The monitor may record numerous events, however only a portion of them may actually be causing problems. For example, a piece of equipment may "ride-through" most voltage sags recorded on the monitor, but is being damaged or interrupted by overvoltages caused by capacitor switching transients. Event logs help the investigator match

Figure 18-1. Results of an impulsive transient[1].

equipment misoperation or damage with specific events that occur on the system.

CHOOSING THE RIGHT EQUIPMENT

Using the information gathered in the investigation will determine the power conditioning equipment most effective for the problem. Data from the monitoring will also give an idea as to how often the problem events are occurring. This will be useful in deciding how much to spend on a particular device.

Monitoring should determine what type of phenomenon (e.g., sags, transients) are causing the problem. Power conditioning equipment should be selected to best combat that particular type of event. Some equipment will mitigate several types of disturbances, so it might be a greater investment to purchase one of these devices. For example, the rotary UPS shown in Figure 18-2 will not only uphold the voltage during sags and momentary interruptions, but can also isolate its load from the system, helping to deter noise and transients.

Another factor to consider in choosing mitigation equipment is the quality of the power conditioning device itself. If you believe you have found the right device, try and get some test data on it. Ask a sales representative to give a demonstration of the unit. The best option is

Figure 18-2. A rotary UPS system.

to borrow a device and try it for a period of time, while monitoring, to ensure its performance.

ECONOMIC ANALYSIS

As mentioned earlier, these power conditioning devices can come with a high price tag. Therefore it is important to justify the purchase of one. The easiest way is to use the following equation:

$$\text{Payoff Time} = \frac{P}{C \times E} \tag{18-1}$$

where P is the price of the device, C is the cost associated with each power disturbance event, and E is the number of events seen over a time period (i.e., month, year, etc.). This equation will compute a payoff time for the device in the time unit chosen in the E parameter.

For instance, suppose you are having 12 loss-producing events per year, with each event causing $4000 in lost production and wasted material. A UPS system costing $100,000 would alleviate all the interruptions your facility sees. The payoff time would be just over two years.

This is a simple formula, however the difficulty lies in determining values for the parameters. Each one must be closely examined.

Cost of Power Disturbances

This would initially seem like a straightforward value. Most industrial customers would take into account lost production time and wasted raw material. For example, the value of the lost material shown in Figure 18-3 would be fairly easy to quantify. In some cases there may be damage to the equipment itself, so there is a cost associated with repair or replacement. These dollar amounts should be fairly easy to obtain.

The amount of lost production time can be influenced by a number of things. If the affected equipment completely shuts down, it must be restarted. If the customer is lucky, it will be as simple as cycling the power, however it is usually not that simple. Many PLCs or electronic motor drives may need to be reprogrammed. Software alarms may need to be reset. Intricate, complex processes such as those in the textiles and plastics industries may require some time to run the material back

Figure 18-3. Waste material resulting from an interruption[1].

through the automated process to come back to full production.

Another contributor to lost production time is cleanup. In a plastics operation, the cleanup after a process shutdown can be quite extensive. Figure 18-4 shows a typical plastic pipe manufacturing process. Each section of the process is dependent on the other sections for steady and smooth material flow. The extruder heats and melts the

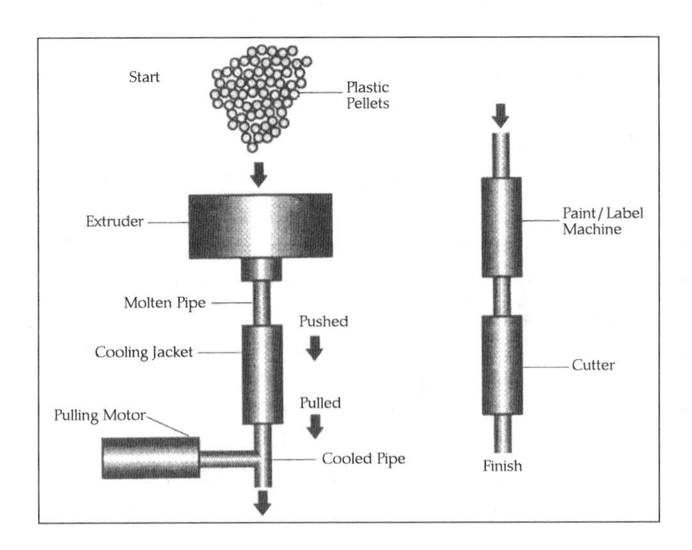

Figure 18-4. A Plastic pipe manufacturing process.

pellets to form the molten pipe as shown. If the extruder gets shut down unexpectedly, molten plastic is left in the machine and spills on the floor. It then hardens and creates a difficult mess to clean up. The time it takes to scrape the plastic out of the extruder is extensive, and there is a potential for damage to the extruder.

If there is damage to any of the equipment, it will need to be repaired or replaced. This also contributes to downtime, and the length may be unforeseen. Parts may need to be ordered, and the equipment may be inoperable until they arrive.

Wasted material is another contributor to the cost of power disturbances. In an example like the plastic process, the material wasted is obvious. The material lost in the extruder will most likely be ruined. Any material in the rest of the process will probably be damaged in some way as well, and must be discarded. The product near its final stage in the labeling and cutting process may be ruined if the label is not aligned, or if it is cut at an incorrect length due to speed variations or shutdowns due to power disturbances.

More difficult is calculating a cost for wasted material when there is not a tangible product. What if the equipment is a computer database, where work and information are lost due to a disturbance? Associating costs with this may be difficult. One would have to account for the time for the work lost, as well as the time it would take to redo the work.

Some other factors contributing to the cost of power disturbances may not have dollar amounts directly associated with them. For many automated processes, power disturbances may cause slight speed variations in electric motors, causing subtle undetectable quality variations in the final product.

Overvoltages and overcurrents may cause insulation damage to transformers, cables, or any other equipment. This will affect the life of the equipment. Electronics may slowly break down over time from repeated surges, then fail at a later date for no apparent reason.

Annoyance is another factor that should be accountable, but has no direct cost. Even if lost production costs are not high, the aggravation factor can run high when frequent restarting is necessary. Unmanned operations may need to be manually restarted, perhaps at inconvenient times.

Last but not least, safety should be considered. Equipment shutting down abruptly, or going unstable, could create dangerous situations.

Computing a cost for the sake of justifying power conditioning equipment can be difficult, however for most large processes it can be obvious. Some manufacturers have quoted losses of up to $1 million per disturbance. Even for large, complex mitigation devices, the equipment could pay for itself after just one disturbance.

Frequency of Events

This number may also be difficult to ascertain. The difficulty lies in obtaining an accurate number within a short period of time, which is important since a solution is needed quickly. If the problem disturbance is occurring regularly, an estimate will be easier to determine. However, many disturbances are seasonal, happening more frequently during certain times of the year.

For example, capacitor switching transients may occur: more during the summer as voltage support is needed more because of the added load from chillers and air conditioning systems. This can make the decision for power conditioning equipment difficult. A user may realize he's having problems in August and spend a good deal of money on some mitigating equipment. He may then not have any more problems until the following May and he finds his device is not suitable for the application.

In addition, the data may even change from year to year. Statisticians will tell you that the larger your data sample, the more accurate your study will be. Unfortunately, the end users do not have time to monitor for several years; the problem needs to be fixed today.

The Price for Protection

The simple equation given previously can be solved in several ways. Once the system has been characterized and costs per event have been estimated, the user can calculate how much he would like to spend given a desired payoff time. Alternatively, he could calculate how long it will take for several options to pay for themselves.

A mitigation strategy will help determine the expenditure for the device. There are many "cure-all" devices available designed for installation at the plant electrical service entrance. These devices claim to combat a variety of disturbances, and can cost in the thousands of dollars per kVA. Thus for a typical plant load, these solutions can cost in the millions of dollars.

For many industrial processes, it is only several components that are causing a production interruption. In a long, automated process line, each stage of the process is dependent on the other to function properly. Often there will be one master electronic controller for the entire line. If it detects that one component has failed or shut down, it will shut the entire process down. Therefore, it could be more cost-effective to identify the weak points within a process line, and apply mitigation devices to them.

There is some difficulty in identifying which devices are most sensitive. When a process fails, all the equipment may shut down, and it's impossible to tell which equipment failed first. Monitoring at different points throughout the process can help identify the susceptible devices during system events. Another option is to simulate power disturbances on the systems for quicker collection of sensitive data. Some utilities have devices which can generate disturbances onto the line for controlled testing of equipment to see which devices fail first.

For large facilities with many sensitive devices such as electronic motor drives, it may not be cost-effective to mitigate every single device. In this case a power conditioning device which protects the entire facility may be more cost-effective. A strategy must be chosen to accurately estimate the cost of the mitigation equipment.

GRAPHICAL ANALYSIS

A method to further analyze Equation 18-1 is by graphically plotting several curves. After the monitoring phase has gathered some data, a *characterization* curve can be plotted. This curve, shown in Figure 18-5, illustrates the numbers and types of events the plant equipment is seeing over a time period. It summarizes the expected number of events for a given magnitude and duration.

An example point is shown on the graph in Figure 18-5, indicating a sag of approximately 2.5 cycles and 0.57 p.u. magnitude. From the curve, the user could predict 10-15 events per year of similar magnitude and duration.

The activity log is an important step for the next curve. Correlating equipment shutdown or failure with the disturbances monitored can illustrate more clearly the *susceptibility* of the equipment. The example susceptibility curve shown in Figure 18-6 illustrates what combinations of magnitude and durations will cause equipment to fail. This will

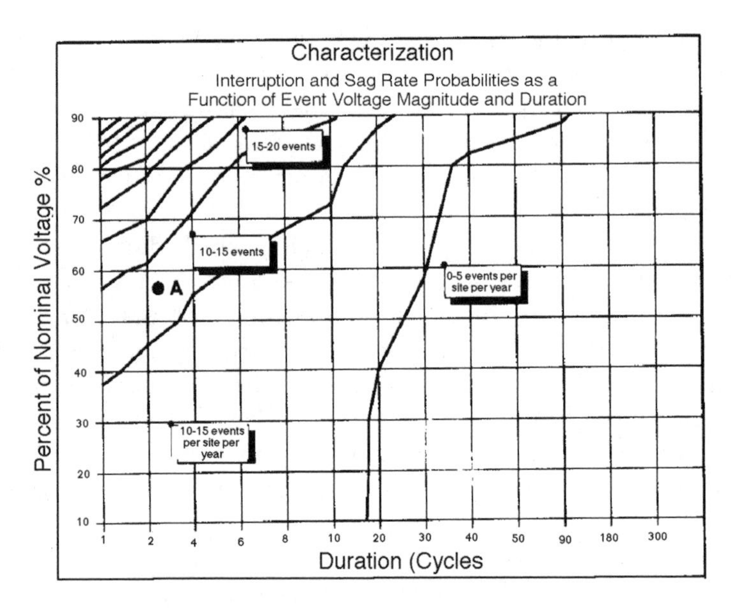

Figure 18-5. An example characterization curve[2].

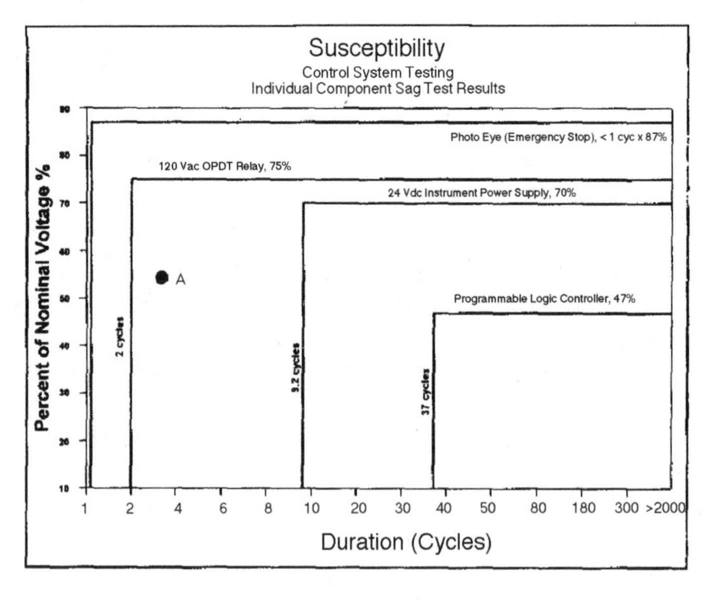

Figure 18-6. An example susceptibility curve[2].

pinpoint the most sensitive equipment, allowing the user to focus the mitigation strategy.

The example sag shown in point A on Figure 18-6 would cause the photo eye and the 120V relay to trip, however the power supply and PLC would ride through the event.

This now introduces the concept of *system compatibility.* System compatibility analyzes the behavior of equipment in its given electrical environment. Overlaying the characterization and susceptibility curves results in a *compatibility* curve. This will help predict the number of equipment shutdowns over a given time period[2]. For the sag shown in point A in Figure 18-7, this example could expect the photo eye and 120V relay to trip the process roughly 10 times per year.

Suppose these curves apply to the example scenario previously used to illustrate the equation. The photo eye and relay have been identified as causing 10 trips per year. It is found that a $300 UPS will eliminate these trips. If there are 10 process lines, a total of $3000 will be spent on mitigation equipment. Using the previous cost of $4000 for lost production and wasted material, the UPSs will pay for themselves after just one event. This is also a big savings from the $100,000 the user could have spent to protect the entire plant.

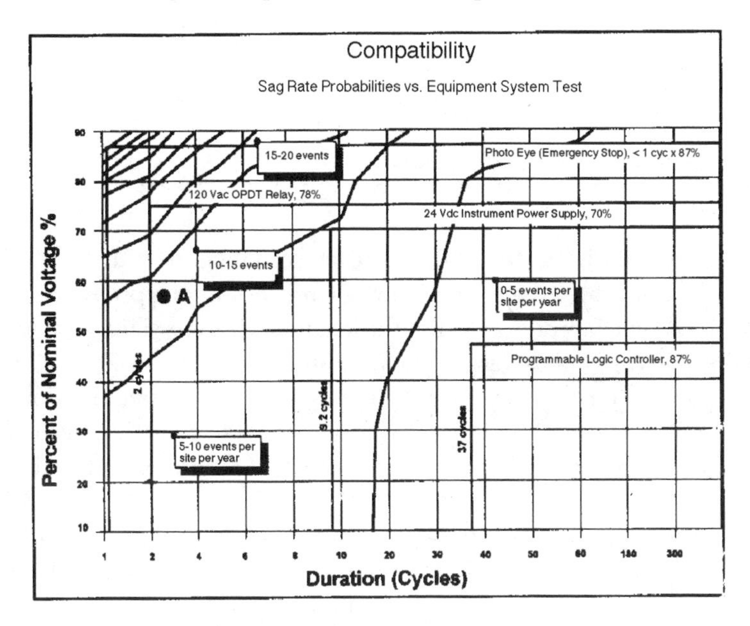

Figure 18-7. An example compatibility curve[2].

This will help in the mitigation process, giving the user a better idea for how many process interruptions he may see over a time period. Associating this with the cost information will help determine the user's payoff time for a chosen power conditioning device.

A MORE DIRECT APPROACH

Taking the system compatibility approach one step further will paint an even more accurate picture of the effectiveness of a power conditioning device. Given the characterization and susceptibility data, the user can weight different categories of disturbances on the basis of whether or not they will cause an interruption. Table 18-1 is an example table of weighting for voltage sags.

Table 18-1. An example table for different magnitude voltage sags[3]

Category of Event	*Weighting Value*
Interruption	1.0
Sag below 50%	0.8
Sag between 50% and 70%	0.4
Sag between 70% and 90%	0.1

Using this weighting technique in conjunction with the characterization data can give a prediction of process interruptions per year. Table 18-2 is a sample table for the characterization data, taking into account the weights assigned to each category.

Multiplying the events per year within each category times its weight will result in the number of probable loss-producing events. For the example shown, the prediction will be 5.4 loss producing events per year.

Given a specific power conditioning device, the user can make a quick estimate for the saved shutdowns per year. For example, suppose the user chooses a device that will mitigate all sags where the nominal voltage stays above 50%. Then he can generate new data for the ex-

Table 18-2. A Table of characterization data[3]

Category	*Weighting*	*Events Near*
Interruption	1.0	0.5
Sag (<50%)	0.8	1.5
Sag (50-70%)	0.4	4.0
Sag (70-90%)	0.1	21.0

pected events per year, and multiply each category by the weights. For this example, he will now expect zero events for sags 50% or above, so his total loss-producing events will now be 1.7 per year.

The user will achieve much greater accuracy if the categories are broken down even further. Sags could be broken down on the basis of number of phases affected, or whether they occurred on the same or alternate feeder.

The above analysis could be done for a number of power conditioning devices. A comparison could then be made of the resulting number of loss-producing events versus the price of each device. From this analysis a decision could be made on which device would be most cost-effective[3].

CONCLUSION

As power quality becomes a greater concern for industrial customers, solution providers are actively flooding the market with mitigation devices. Justifying the purchase of these devices is difficult, as many come with a high price tag. An even bigger obstacle is determining just what you are getting for your money. Careful investigation and analysis techniques can simplify the search for the right power conditioning device.

It is important to determine exactly what type of disturbance is causing problems for the end-users process. Furthermore, the user should attempt to determine the effects of different severity of specific types of events. Using engineering judgment, weights or probabilities can be given to the various categories of disturbances seen at the customer's facility. This can be used to calculate a more accurate prediction of loss-producing events seen over a time period.

Using gathered information, analyzed numerically or graphically, can help determine the best mitigation device for the application. It is usually more cost-effective to mitigate individual susceptible equipment. However, in the case of large facilities with a large number of sensitive devices, it may be more beneficial to protect the entire facility.

References

1. Jeff G. Dougherty and Wayne L. Stebbins, *Power Quality: A Utility and Industry Perspective.* IEEE IAS Textile Fiber and Film Technical Conference, Greenville, SC, May 1997.
2. IEEE P1346, *Recommended Practice for Evaluating Electric Power System Compatibility with Electronic Process Equipment.*
3. Robert J. Gilleskie, *Utility Expertise in Providing Power Quality and Other Engineering Services.* EPRI PQA '97, Stockholm, Sweden, June 1997.

19
Purchasing Strategies for Electricity

SUMMARY

A new age is dawning for lower-cost energy use and supply. The deregulation of the electric industry is creating new pricing structures that will change how we calculate the payback of alternatives for cutting the cost of energy in our facilities.

Energy users can help themselves navigate these choices. By understanding the concepts inherent in deregulation, learning to use new energy tools, influencing the deregulation process, applying new technologies, and (most important of all) by being as creative as possible, smart users will grasp new options for energy cost reductions.

AT&T VS. MCI: A PARADIGM

To better understand how electric utilities will be transformed, think about how long-distance phone service has changed. When AT&T was forced by a federal judge to divest itself of many of its divisions, long-distance was separated from local service, and new providers such as MCI and Sprint became household names. While, after 12 years, deregulation of that industry is not yet complete (local service is still generally a monopoly), during that time long-distance use has nearly quadrupled, while the average price of a long-distance call fell by more than 50%. To satisfy consumer demand for communication services, a vast new array of technologies was also born. How many of us anticipated the proliferation of fax machines, cellular phones, pagers, and on-line services that would result from a single court order? While we can also expect the cost of power to fall, the future of electricity similarly holds much more than price reductions.

Presented at the 20th World Energy Engineering Congress by Lindsay Audin

FACTORS IMPACTING POWER PRICES

Electricity is generated by utilities and independent power producers (IPPs), both regulated to some degree by state public utility commissions (PUCs), and then transmitted through high-tension lines criss-crossing North America in a giant network. These lines are owned by utilities and regulated by the Federal Energy Regulatory Commission (FERC). Once voltage is stepped down at substations, power is distributed through local utility-owned lines and meters regulated by PUCs.

With the exception of municipal utilities (which are controlled by local governments), PUCs determine how to distribute these costs to end user classes. Most of our bills break out only charges for electric consumption and demand (and perhaps a fuel adjustment), but the true cost of power includes many other components, including transmission, distribution, and a variety of ancillary services (such as voltage support, spinning reserve, and load following). Bills may also include taxes, social programs, and other charges that are not apparent to end users.

To develop the prices we pay, the PUCs apply a standard based on the utility's costs for providing a service, plus a guaranteed rate-of-return to ensure a ready supply of investment capital. All of these costs and profits are "bundled" together to create tariff pricing. While theory dictates that charges should be based on the true cost of service, politics and other pressures often result in cross subsidies in which one rate class (e.g., industrial) is charged more to contain prices charged to another (e.g., residential).

While the electric rates we pay are controlled by PUC tariffs, utilities and IPPs buy and sell electricity among themselves, and such wholesale prices vary with time, climate, power plant outages, fuel prices, and other factors. The base cost of power seen by a utility is therefore a mix of its own generating costs and the price it pays for electricity delivered to it through the transmission system from other power providers. This base cost is subject to commodity market conditions usually not visible to end users. It is increasingly being influenced by factors such as commodity trading techniques (e.g., futures and financing plans), ways to adjust user load profiles (such as real-time pricing), and transmission system constraints (that can drive prices up to the highest local generating cost).

THREE GENERAL RELATIONSHIPS

There are three general relationships that clarify how new techniques both interact and can be applied to control energy pricing. They are:

- time - load - price
- generation - transmission - natural gas options
- load shaping - financing methods - user technologies.

Time, Load, and Price

As deregulated retail power prices begin to vary like those at the wholesale level, they will become more time-sensitive. Since one's demand for power generally changes during the day and the week (and by season) we can expect the average price to also change with time and use (unless controlled by other factors, discussed below). As a result, load profile shapes (i.e., a graph of power versus time) will influence pricing, with flat profiles generally having the lowest average cost. Utilities generally typify such patterns via load factor, defined as average demand divided by peak demand. A high load factor would indicate a flattened profile while low load factors would occur where a peak demand is relatively brief, and is surrounded by much lower demand during the rest of the day. While most non-industrial building power demand varies with time, it usually does so in predictable patterns. Knowing the shape of your typical load profile can often reveal ways to cut the present and future cost—and price—of power, while also helping your power supplier offer the best and most secure pricing.

Like Sprint's "dime-a-minute" long distance rate, marketers will likely offer highly simplified rates that smooth out such time-based price variations, but subscribing to such options will not yield the lowest *average* power costs. Rates that vary widely over time may provide the lowest *average* price, and techniques that cut, level, or shift peak demand will help reduce those prices.

Transmission, Generation, and Natural Gas Options

In some cities and states, peak loads exceed transmission capacity many hours each year. When low-cost power can't be brought in, prices could be bid up to the highest local generating costs. A good example of such constraints appeared during the early hot spell of June 1997 in the PJM (Pennsylvania-New Jersey-Maryland) power pool. While daily bulk wholesale generation prices (which make up 40% to 65% of most bills) generally don't vary from one end of the pool to another by more than $.01/kWh, June saw *variations exceeding $.13/kWh* when transmission constraints blocked cheap power from reaching high-cost areas.[1]

In some urban areas with older power systems (such as New York and San Diego), transmission constraints could yield similar results. Such areas with constraints are sometimes called "load pockets" during the period of constraint (which may exceed 1000 hours a year). In the United Kingdom, which uses a national power pool supplied by deregulated generators, power suppliers have also found ways to "game" the system to purposely congest transmission, thereby driving up the price of their product.[2]

To address such possibilities, energy marketers have begun examining and/or promoting new local generation (or cogeneration) facilities, either at customer-owned sites or through repowering of obsolete utility plants inside the load pockets. Natural gas generators with very low emission levels have become quite cost-competitive for both peak shaving and as base load power, opening the door to competition during transmission constraints. Similarly, a variety of technologies (discussed later in this chapter) exist to reinforce existing transmission systems. A recent study[3] found that a small investment toward improving transmission capacity in California could have a major impact on limiting power prices.

One way to shift peak demand is to substitute natural gas for electricity during peak pricing periods. As will be discussed below, a variety of technologies exists for using gas to directly provide horsepower, cooling, air compression and other power-intensive needs. Such convertibility will create truly interruptible energy rates, allowing clever end users to contract for both interruptible gas and power, attaining the lowest possible energy prices. Under these circumstances, transmission, generation, and natural gas options will compete with each other, driving all prices down over time.

Load Shaping, Financing, and User Technologies

A variety of choices is emerging to configure loads in advantageous ways. While most have existed in one form or another, deregulation will allow marketers to help end users gather—and segregate—their loads more readily through metering and contractual means.

Coincident metering has been used to cut the average cost for power at facilities with many meters on differed accounts held by one customer. At Columbia University in New York, for example, several dozen accounts existed on one property, the result of gradual expansion without attention to energy costs. Each account peaked at a different time, but (due to tariff construction) the sum of the bills was the same as though all buildings had peaked at the same time. By combining the accounts under one master demand meter, the average cost of power was cut by over 10%. Such combination will become easier under deregulation as usage (for the power commodity, as versus transmission and distribution) can be contracted under one account.

Load isolation, while not favored by utilities, can allow an end user to segregate loads that would be cheaper under utility tariff-based power than under market-based power, thereby reducing the average cost for all loads.

Demand cooperatives allow end users to work together (typically through an organizing vendor) to obtain lower utility rates by pooling their interruptible loads and agreeing to curtail them when requested by the utility. Creating such a cooperative eases the difficulty of ensuring that any one load must be interrupted by allowing the demand to apply to a group of loads, only a few of which would be interrupted at any one time.

District-wide systems serve multiple customers with (for example) chilled water from a central facility that, in effect, transfers many individual electric chiller loads (which lower load factor) to a high load factor central facility that uses both electric and gas-driven units to minimize total cooling costs. Existing chillers may remain, but are bypassed by a connection to a common chilled water loop serving numerous facilities.

Bill Consolidation allows many accounts held by one customer to be gathered for both coincident metering and attainment of cheaper energy block load rates previously beyond any one account.

Aggregation involves the gathering of different customer accounts through a third party for purposes of bulk power purchasing, coincident metering, bill consolidation, transmission capacity reservation, and expert load analysis. Aggregators work along the same lines as MCI, buying co-ops, credit card handlers, and other organizations that compete for the privilege of bringing many end users together. All try to provide lower prices through bulk purchasing and handling. Present-day utility customers are, in effect, already "aggregated" into rate classes, but only to the point of developing a rate based on an assumed typical load profile. Each of these options allows for re-shaping of loads (as portrayed to utilities and regulators) without major alterations to end user facilities.

In similar ways, financial tools exist that can cut or level costs more easily than with engineering. Both marketers and financial firms will (or already do) provide access to futures, options, and payment plans that ensure prices do not exceed a predetermined level. While a full discussion of financial instruments is beyond the scope of this chapter, end users need to understand the following basic concepts:

- Electric futures (which are available on the West Coast, and soon in the East) are traceable contracts for monthly blocks of firm power, purchased in advance of need, that will later be provided during normal business hours at a predetermined price. They are, in effect, negotiable promises for supply of power, though there is not necessarily a guarantee of delivery unless accompanied by a secure transmission arrangement.

- Options (typically in the form of "calls" and "puts") are contracts between suppliers and marketers (and thus end users) that allow a user of power to know that he can "call" for power and be ensured supply, or that a seller of power can be ensured a buyer when he "puts" out an offer to sell, both at predetermined prices.

- Payment plans have been offered to low-income customers by utilities for years. Now, however, marketers are ensuring level (or predetermined) monthly bills (not just pricing) through risk management techniques that involve financial and load analysis tools. While some monthly bills may be higher than in prior years, other months will be lower, and the annual total will be confined to a narrow range. Such plans are often complicated, requiring careful analysis.

- Tolling allows an end user (or his designee) to provide a generator with boiler fuel (typically natural gas) in trade for electricity at negotiated, non-tariff, rates. This process is common among power marketers also trading natural gas, and is used at times that utilities have excess generating capacity that can provide power into another utility's territory.

While all of these options provide end users with choices for controlling energy pricing, the impact of many of them can be maximized when used in cooperation with new technologies. Just as we have seen an explosion of choice in communications, it is likely that the future will see a variety of ways to create, store, and manage power. Many are already being offered, or are in the prototype stages of development. Following are several being supplied or considered by both energy marketers and equipment vendors.

Metering. Marketers are using more sophisticated power metering (e.g., ENet, MC^3) as a sales tool to ensure that their clients' power costs are minimized through a variety of methods. The new standard in metering involves wireless communications, non-invasive wiring, and hourly, quarter-hourly, or real-time monitoring. Human meter readers, obsolete for many years, will likely disappear as more cost-effective systems are installed. Such systems will allow better load control, more accurate power nominations, better pricing, and a reduction in theft and tampering (a major problem at the residential and small commercial levels).

Software. Computerized building simulations and load analyses have taken on new prominence as tools for predicting and flattening load profiles. Analyzing short intervals (1/4 hour) has become essential to maximize savings through tighter load control. Names and acronyms such as PEDA, RBOSS, and PowerManager will be heard quite often as marketers offer new services to cut electric bills and power pricing.

Energy Management Systems (EMS). As responding to real-time or market-based pricing signals becomes a common way to attain savings, an EMS takes on new importance for controlling variable loads, such as fans, pumps, DHW heaters and chillers. When tied into the metering and software tools mentioned above, the load managing power of an EMS can be greatly enhanced.

Power Storage Devices. While many power practitioners and regulators continue to assume that power cannot be cost-effectively stored, flywheel power storage units are now in use as uninterruptible power systems (UPS) to supply "clean" power, and small (2 kWh) units act as backup power for cable TV systems. Larger units (over 12 kWh) are being prototyped as peak shifters/shavers for buildings. Chemical battery technology has advanced considerably, and may play a part, as well. Depending on how stranded costs are collected, a storage system that gets "filled" with cheap off-peak power at night (when there are no transmission constraints) and "empties" to flatten peak loads during the day could be an instant moneymaker for a smart marketer and/or end user.

Distributed Generation. The ability to generate power in a pinch has always been useful. New small (<100 kilowatt) modular gas-fired turbine generators[5] can also cut billing for peak demand by operating in parallel with utility power, or (to avoid backup charges) by feeding dedicated loads, while acting as backup power when needed by the utility's transmission and distribution system, thereby increasing overall system reliability. Using ceramics and limiting movement to a single part, the Capstone Turbine is an intriguing option, while the ONSI 200 kW fuel cell (and its soon-to-be seen competitors) has already racked up an impressive operating record. Once prices on these devices come down, watch marketers and end users grab them up as a way to minimize stranded cost payments.

Transmission and Distribution (T&D) Networks. Since power is cheaper only when transmission is available to move it, pressure for new or more robust transmission will increase once large price differentials occur between adjacent areas. Thyristor-based switching of high-voltage loads on T&D systems can raise the effective capacity of existing transmission lines. Such options are among a family of Flexible AC Transmission System (FACTS) improvements under development or deployment. Even new types of underground high-voltage cables,[6] designed for use in transmission-constrained urban areas, are being rolled out to meet the expected demand for beefed up transmission.

Gas-Powered Motor Drives. Natural gas-driven devices, such as gas engine-powered air compressors and chillers, have replaced electric motor-driven units in industrial and commercial facilities, cutting their peak demand.

Advanced HVAC Systems. Chemical desiccants dehumidify outside air using natural gas, thereby cutting peak electric chiller loads. This process is already common in new buildings with large outside air loads (e.g., hospitals) and industrial processes (such as air compression). For smaller facilities configured around rooftop units, Entergy has been offering a super-high-efficiency replacement unit which takes advantage of several refrigeration engineering innovations.[7]

Many other options are either in the queue or already being sold. Try to imagine what the "fax machine" or "cell phone" of the power industry will look like. As we have seen in other industries, the combination of several new technologies often results in devices few of us could have imagined only a few years ago.

WHO OFFERS THESE OPTIONS?

Accompanying the profusion of technical choices will be a more bewildering expansion of vendor choices. Even as the merger and acquisition mania gripping the industry creates new firms out of old ones, one sees a coalescence of energy services providers (ESPs) into a few distinct groups.

- Unregulated utility subsidiaries
- Independent power providers
- Mega-wholesalers
- Equipment vendors
- Gas marketers
- Existing ESCOs
- Aggregators
- Financial firms

Of course, local utility distribution companies (UDCs) are also trying to stay afloat and retain their load.

THE COLLEGE OF POWER KNOWLEDGE

How does one cope with this continuously changing panorama? Fortunately, both the advancing energy industry and other innovations are providing some of the means to do so.

Getting Up to Speed

Start by learning the "lingo" and concepts. There's no need to enroll in college (and none of them teach this stuff). Most PUCs provide readable summaries of their decisions (both on paper and on their Web pages), and a variety of newsletters and free magazines is available to keep abreast of the latest changes. Computer-savvy managers can "surf" informative Internet sites for even quicker access.

Attending a technical conference focused on power supply issues can be very helpful to get your questions answered. Such events are also a good way to make useful future contacts. Local trade associations often sponsor panel discussions on such issues, or are open to holding them if interest is expressed by their members.

Speeding Up the Process

You may already belong to a trade or professional organization that is (or could be) taking action toward deregulation. Several local BOMA (Building Owner's and Manager's Association) chapters, for example, are already actively pursuing power issues. To properly represent your company's interests, membership in a customer group—or working through an energy "partner"—(i.e., a consultant or marketer) involved in rate proceedings can also be of great value. Your PUC can provide lists of past intervenors. On the national level, ELCON (Electricity Consumers Resource Council, in Washington, D.C.) represents many large industrial firms, and is a good resource for user-friendly information.

But all the preparation in the world does no good unless your PUC or state legislature takes the right action on this issue. Experience has shown that only intervention in the process can make that happen. While the better marketers are already involved in this process (and those that aren't don't deserve your business), customer input is essential to ensure that the results are acceptable.

Waiting on the sidelines for "the other guy," or the PUC, to release you from your utility's grip will only prolong the present situation. Trusting your utility to do the right thing (by reducing its profit margins, selling off its assets, cutting its staff and perks) is naive: no industry has ever done so without the push of competition. Watching others bear the cost of interventions, while you reap the benefits, might give you a free ride on others' success, but experience has shown that utilities take advantage of such apathy by dragging out proceedings long enough to exhaust opponents' financial resources.

On the other hand, like many customers around the country, you can help make the right changes happen by supporting intervention into the regulatory process. When a group of energy users financially sustains such actions (either directly or through an energy partner), the contribution of each is small compared to the value of hastening competition. The payback period of such efforts is typically measured in *weeks,* not years.

Become part of the changes already under way, because those who understand the opportunities will reap the benefits of that knowledge, for both their facilities and their careers!

References

1. *MegaWatt Daily,* June 27, 1997, Pasha Publications, Houston, TX.
2. "Moving to Competitive Utility Markets: Parallels with the British Experience," by Dr. George Backus and Susan Kleeman, in March/April 1997 *PowerValue* magazine, published by Intertec International Inc., Ventura, CA,
3. "The Competitive Effects of Transmission Capacity in a Deregulated Electricity Industry," by Severin Borenstein, James Bushnell, and Steven Stoft, published by the University of California Energy Institute, Berkeley, CA, April 1997
4. Planergy, Inc. Web Page, http://www.planergy.com, Austin, TX, February 1997
5. Capstone Turbine Corporation
   Web page, http://www.capstonctwbine.com, April 1997
6. Product Announcement by Southwire Corp., Power Daily, McGraw-Hill Publishing Co., New York, NY, April 8, 1997
7. Entegrity Packaged Rooftop Air Conditioner product brochure, Entergy Inc., Memphis, TN, November 1996

20

Power Quality Case Studies

CASE STUDY 1

Customer Background

The first case study involves a machine shop. This shop utilizes different machines (i.e., grinders, honers, presses, injection molding devices and a precision cutter) to produce various aircraft components. The facility was provided overhead 120/240-volt, three-phase, four-wire open-delta service (see Figure 20-1). The distribution feeder circuit serving the facility is described in Figure 20-2.

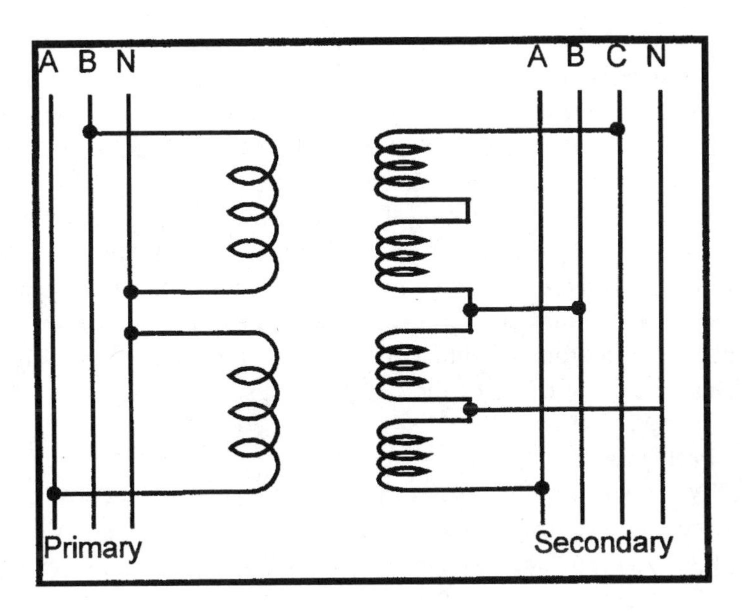

Figure 20-1. Open-wye open-delta transformer configuration.

Presented at Globalcon 2000 by Jon A. Bickel, P.E.

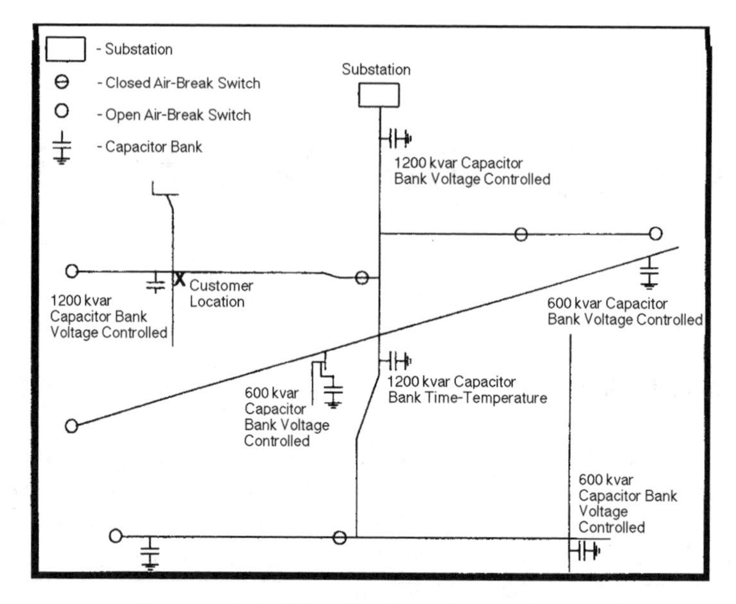

Figure 20-2. Distribution feeder layout.

The customer was experiencing four to five shutdowns of their electronically controlled precision cutter per day. The cutter, which incorporated a microprocessor-controlled adjustable speed drive (ASD), was used to create counter-balance weights for aircraft. The internal protective circuitry was shutting the cutter down, and not allowing it to operate due to an unspecified power problem. The ASD manufacturer was unable to determine the source of the problem, but the controls representative stated that the utility voltage was the source of the problem. The customer contacted the utility and requested assistance in determining the cause of the machine trips.

Tests Performed

TXU Power Quality Services personnel met with the customer and inspected the electrical facilities. A BMI 4800 Powerscope, BMI 3030A Power Profiler, and Hewlett-Packard Dynamic Signal Analyzer (DSA) Model# 3561A were installed at the main electrical panel serving the precision cutter. In order to determine the characteristics of the utility service, five of the seven power factor correction capacitor banks on the distribution feeder were sequentially switched on. The capacitor switching transients were measured with the Powerscope; the Power Profiler and DSA were used to measure voltage distortion at the customer's main electrical panel as each capacitor bank was energized and de-energized.

Test Results and Conclusions

The average voltage during the tests ranged from 238 volts to 253 volts which is within acceptable limits. Inspection of the customer's equipment indicated that the machine had internal circuitry that tripped the machine off if the high or low voltage, phase loss, or phase unbalance trip limits were exceeded. Because the voltage was not exceeding these trip limits, it was determined that a distorted voltage waveform was causing the machine's internal protection circuitry to erroneously sense phase unbalance or loss, resulting in the shutdown of the machine. The voltage distortion could be attributed to several factors including the number of capacitor banks energized on the distribution circuit, the load on the circuit, and the open-wye/open-delta winding configuration of the customer's service transformers.

The seven capacitor banks on the distribution circuit serving the customer total 6000 kvar. Six of the capacitor banks were operated by voltage-controlled switches, and one capacitor bank was operated by a time/temperature-controlled switch. The lowest level of total harmonic distortion (THD) was recorded when all capacitor banks were de-energized (THD of V_{ab} = 2.9%, V_{bc} = 3.1%, and V_{ca} = 5.3%), and the greatest level of THD occurred when five of the capacitor banks were energized (THD of V_{ab} = 4.0, V_{bc} = 4.6%, and V_{ca}= 7.7%). The distortion was comprised almost entirely of the 180-Hz component. All capacitor banks were not energized at the same time to prevent the excessive primary circuit voltage levels which were predicted.

Disturbances due to switching transients were recorded when the two 1200-kvar capacitor banks located closest to the customer's facility were energized. Although most of the capacitor banks have the potential to operate several times a day, none of the recorded switching transients caused the equipment to shut down.

The recommended maximum voltage THD level according to IEEE Standard 519 is 5 percent. Open-Phase Data no single harmonic component exceeding 3 percent. Maximum distortion levels of 7.7 percent were observed on V_{ca} during the tests. Higher distortion levels probably would have occurred if all the capacitors banks on the distribution feeder would have been energized. Load and voltage magnitude on the distribution circuit can also affect voltage distortion levels. It is important to note that the THD on the "open" phase is approximately twice that of the other two phases (actually the sum of the distortion on the other two phases). This is a characteristic inherent to the open wye-open delta transformation.

RECOMMENDATIONS

Prior to conducting these tests, a 1200-kvar capacitor bank one span from the customer's facility was removed from service. The frequency of trips on the customer's equipment was reduced from 4 to 5 occurrences per day to 1 occurrence per day. It was recommended that the size, placement and controls of the remaining capacitor banks be reviewed to determine if modifications were necessary. As illustrated by the tests, it is often possible to reduce voltage distortion on a circuit by downsizing or removing capacitor banks. Voltage distortion may also be reduced by moving capacitor banks closer to large load centers. For this particular customer, converting the customer's service from "open" transformation to a "closed" transformation was determined to be the most straightforward method for reducing voltage distortion to acceptable levels.

CASE STUDY 2

Customer Background

This study involves a plastic foam manufacturing facility. The customer manufactures several lines of polystyrene food containers including plates, take-out containers, etc. They are provided two 12.47 kV medium voltage points-of-delivery (POD). The customer had been in operation at this facility for approximately 10 years. The Plant Engineer reported experiencing various equipment failures over a period of approximately two months. The failed equipment included motors, electronic motor drives, and programmable logic control (PLC) cards.

Two 30-hp induction motors failed simultaneously during a thunderstorm. Both motors were rewound, but no failure report was generated by the motor repair company. The motors were served from the same POD; however, the customer had experienced equipment losses on both PODs. TXU Power Quality Services group was requested to assist in determining the cause of the failures.

Tests Performed

Power Quality Services personnel met with the plant engineer and the plant manager to discuss the equipment failures. A BMI 8800 Powerscope was installed on one of the main 480-volt electrical buses to monitor and record voltage and current. The customer provided a tour of the facility to point out specific equipment which had failed. The customer's main electrical room and the utility facilities serving the customer were visually inspected at the time of the plant tour. The BMI 8800 was removed after a week of surveying the customer's voltage and current at the main.

Test Results

Review of the voltage data showed that the instantaneous 6-cycle average voltage ranged from a minimum of 456 volts to a maximum of 492 volts. This voltage range is within normal equipment tolerances. Review of the current data showed that the instantaneous 6-cycle average current ranged from a minimum of 1098 amperes to a maximum of 2081 amperes. The customer did not experience any equipment failures or problems during the test period.

Visual inspection of the facility revealed that the failed motors (two 30-hp induction motors and an additional 10-hp drive motor) were located in close proximity to each other outside the facility on the East side. Failed remote I/O cards and control fuses were located throughout the facility. The customer had 480-volt transient voltage surge suppressors (TVSSs) installed at the main switchgear for each POD. The TVSSs did not appear to be U.L. listed and a let-through voltage of 1200 volts was written in ink on the units. Each of these units also had approximately 10 feet of lead length connecting it to the main 480-volt conductors.

CONCLUSIONS

Due to the number of failures within a brief time period and the fact that lightning was present in the area at the time of the failures, it is likely that the failures were the result of high voltage transients from lightning strikes. The effects of high voltage transients (such as those caused by lightning strikes) include not only immediate equipment failures, but residual component failures due to the degradation of components caused by the high voltage stress. These failure modes were consistent with the equipment failures as experienced by the customer.

A lightning strike in close proximity to the large metal tanks on the East side of the building probably caused the immediate failure of the motors and drive located in that area. It was apparent that common electrical grounding paths existed between the customer's two PODs. With this configuration, a voltage potential rise (i.e., due to lightning) can cause damaging circulating currents to flow from one source to the other. Sensitive electronic equipment (such as the remote I/O PLC cards) that are not isolated from ground currents may be damaged or stressed under these circumstances.

Index

A
absorption system 231
advanced HVAC systems 447
advanced turbines 333
after-tax analysis 56
aggregation 443
airflow hoods 73
air conditioning system design 239
air velocity 72
ammeter and voltmeter 64
anemometers 73
aquachem areas 372
ASHRAE 90.1-1989 23
ASHRAE 90.2 23
ASHRAE Standard 90-80 23

B
bill consolidation 443
blower door 72
boiler 116
    test kit 71
Bourdon tube gauges 75
building design 205

C
"cool windows" 224
capital recovery 47
Carnot cycle 111
coefficient of performance 225
coil run-around cycle 244
coincident metering 443
combustion air temperature 121
combustion analyzer 72
combustion tester 71
commercial loan 342
condensate return 368
    system 362
conduction 151, 152
convection 151, 157
cool storage 248

D
deaerator vent 364
demand cooperatives 443
depreciation 53
deregulated electric power 26
deregulated retail power 441
desiccant systems 235, 237, 238
dimensional change 75
distillation columns 148
distributed generation (DG) 325, 446
district-wide systems 443
draft gauge 74

E
economizer cycle 243
electrical conductivity 75
electrical system performance 64
electric futures 444
electrolytic 75
energy accounting 7
energy conservation 7, 115, 225, 227, 239, 243, 291, 294
Energy Efficiency 21
Energy Independence and Security Act of 2007 14
energy management 1
Energy Management Systems (EMS) 445

energy manager 12
Energy Policy Act of 1992 21
energy services providers (ESPs) 447
energy utilization 3, 150
exchangers 245

F
facility survey 61
Federal Energy Regulatory Commission (FERC) 440
Federal Power Act 24
financing alternatives 341
finned-tube heat exchangers 158
flashing condensate 117
flash steam 366, 374
flue gas 122, 127
food industry 359
foot-candle meters 67
fuel cell power systems 335
fuel inflation 58
furnace efficiency 119

G
gas-powered motor drives 446
gas analyzers 72
General Obligation Bond 342
generation 442
geothermal heat pump 379
glass considerations 205
glass thermometers 74
gradient present worth 48, 49, 56
gravimeter 75

H
heat exchangers 375
heat gain 173
heat loss 161, 162, 163, 166, 167
heat pump 226, 382
heat recovery 124, 244, 366
heat transfer 129, 130, 131
heat wheels 244
high load factor 441
humidity measurement 75
HVAC 70, 72, 75, 225

I
independent power producers (IPPs) 440
industrial energy audit 2
infiltration 222
infrared equipment 61
insulation 214
investment decision-making 31

L
leakage rates 72
life cycle costing 29, 30, 58
lighting 247
liquid chiller 228
load isolation 443
low pressure condensate 363

M
manometers 74
measuring combustion systems 71
measuring system efficiency 225
medium pressure condensate 363
metering 445
micromanometer 74
microturbines 333
mitigation 428, 432, 433, 436
Mollier Diagram 131
municipal lease 342

N
natural gas-fired DG systems 327
natural gas engine system 236

natural gas options 442
Natural Gas Policy Act (NGPA) 25
nitrogen areas 372
non-condensable vapors 364

O
optical pyrometers 74
optical technologies 223
options 444

P
partnerships 327
payback period method 29
payment plans 444
peak demand 441, 442
piping systems 142
pollution control impact 10
portable electronic thermometer 70
power conditioning 426, 428, 429, 433, 436
power disturbances 429
power quality 425, 426, 451
  monitor 426
power storage devices 446
pressure-reducing valve 376
pressure bulb thermometers 74
pressure measurement 74
pressurization 73
procurement specifications 150
project financing 342
psychrometer 69, 75
public utility commissions (PUCs) 440, 448
Public Utility Holding Company Act of 1935 25
Public Utility Regulatory Policies Act (PURPA) 24
pumps 142
purchasing strategies 439

R
radiation 151, 156
  pyrometers 74
reciprocating (recip) engines 336
refrigeration equipment 228
refrigeration systems-double bundle condenser 247
resistance thermometers 74
Resource Conservation and Recovery Act of 1976 (RC 25

S
"smart windows" 223
sample energy services agreement 346
single payment compound amount 31
single payment present worth 32
sinking fund payment 47
smoke generators 73
smoke tester 72
software 383, 445
solar control 218
specific heat concept 113
state codes 22
steam distribution system 361
steam generation 135, 361
steam leaks 294
steam load and cost 360
steam loss 364
steam pressure 113
steam system optimization 359
steam tracing 123
steam traps 294, 298, 374
steam turbine 115
suction pyrometer 71
sunlight 203, 205, 219
superwindows 224
surface pyrometers 68

surface temperatures 204
swing vane gauges 75
switchable glazing 223
system compatibility 435

T

taxable lease 342
temperature 67, 113
test and balance considerations 258
thermal barriers 220
thermal shock 372
thermal storage 236
thermocouple 74
    probe 70
thermodynamics 109
thermographs 74
thermometer 68
time value of money 31
tolling 445
transmission 441, 442
transmission and distribution (T&D) networks 446
tube heat exchanger 159, 160
two-stage absorption 234

U

uniform series compound amount 32
uniform series present worth 46

V

vapor compression 236
variable air volume 240
ventilation audit 255
volume/mix impact 10

W

waste heat 124, 125, 126, 127, 128
    recovery 131, 135
weather 10
window treatments 218